Refurbishment and Upgrading of Buildings

Second edition

Refurbishment and Upgrading of Buildings

Second edition

Christopher Gorse and
David Highfield

First edition published 2000 by E & FN Spon

This edition first published 2009 by Spon Press
2 Park Square, Milton Park, Abingdon, OX14 4RN
Simultaneously published in the USA and Canada

by Spon Press
270 Madison Avenue, New York, NY10016

Spon Press is an imprint of the Taylor & Francis Group, an informa business

Typeset in Gill Sans by
Florence Production Ltd, Stoodleigh, Devon
Printed and bound in Great Britain by
MPG Books Group, UK

British Library Cataloguing in Publication Data
A catalogue record for this book is available from the British Library

Library of Congress Cataloging-in-Publication Data
Gorse, Christopher A.
Refurbishment and upgrading of buildings / Christopher Gorse and David Highfield. — 2nd ed.
p. cm.
"First edition published 2000 by E & FN Spon."
Includes bibliographical references and index.
1. Buildings—Repair and reconstruction. I. Highfield, David. II. Title.
TH3401.H53 2009
690′.24—dc22 2008044495

ISBN10: 0–415–44123–4 (hbk)
ISBN10: 0–415–44124–2 (pbk)
ISBN10: 0–203–87916–3 (ebk)

ISBN13: 978–0–415–44123–0 (hbk)
ISBN13: 978–0–415–44124–7 (pbk)
ISBN13: 978–0–203–87916–0 (ebk)

Christine and Ruth

Women bring life into the world
Thank you

Contents

Acknowledgements

We are indebted to the many people who have helped and contributed to this work. Particular thanks is made to A. J. Peters and Nicholas Donnithorne – Rentokil Property Care, Shaun Long – Scott Wilson Group plc, Mike Armstrong – MACE Group, Paul Smith – RoK Developments, Tony Takle – Quarmby Construction Company Ltd, Ed Jagger – Brewster Bye Architects, Gordon Throup – Big Sky Contracting, Neil Gorse and Jeffrey Hobday – JORDAY Group, Adam Bowles – Dobson Construction Ltd, David Johnston, Joseph Kangwa and Ian Dickinson, Claire Walker, John Bradley and Martin Smith – Construction Knowledge Exchange, Leeds Metropolitan University.

Introduction

Finance and sustainability

Building refurbishment and upgrading (including maintenance, repair, restoration and extension) – all broadly categorised as 'repair and maintenance' by the government for statistical purposes – are major components of construction activity, having consistently accounted for just under half of the construction industry's total output for the last two decades. The government Department of the Environment, Transport and the Regions categorises construction output as 'new work' or 'repair and maintenance', and between 1994 and the first quarter of 1999 the value of repair and maintenance work has averaged, annually, some 49 per cent of the total of all work carried out by the construction industry (DETR 1999: *Information Bulletin 599*). During the period between 2002 and 2007 this trend has continued, with repair and maintenance representing just under half of the work undertaken in the UK. A full breakdown of statistics is available from the UK statistics authority (www.statistics.gov.uk). Further statistics on construction can be found at the Department for Business Enterprise and Regulatory Reform (www.berr.gov.uk).

Building owners and developers have, in recent years, come to realise the potential value of our vast stock of old, redundant and obsolete buildings as a means of providing, through their refurbishment and re-use, high-quality 'modern' accommodation more quickly, and at a lower cost, than the alternative of new construction. Refurbishment can provide 'new' accommodation in only half to three-quarters of the time needed for the alternative of demolition and new construction, and at only 50–80 per cent of the cost, resulting in considerable financial benefits to the developer.

There has also been a significant change in attitudes favouring the conservation and recycling of resources, especially in recent years with efforts to reduce the use of fuel and carbon dioxide emissions and develop a more sustainable way of existing. The wholesale demolition and redevelopment policies of the 1960s and 1970s resulted in the replacement of large numbers of basically sound buildings, often with buildings of much poorer quality, particularly in the housing sector. It is now widely recognised that it makes far greater sense to retain and refurbish buildings in preference to demolishing and replacing them. Much of this book focuses on efforts that can be considered sustainable; however, façade retention rarely offers a form of construction that could be considered environmentally friendly. Many of the buildings that have their façades retained could also have much of the original structure re-used. However, in most building projects that involve large-scale demolition, there are now greater efforts to conserve and re-use the materials obtained.

While definitions of sustainability vary, anybody involved in building has been confronted by the term (Sassi 2006). The increasing significance of man-made and natural environmental change has brought sustainability to the door of all building designers and contractors involved in maintenance and refurbishment. Sustainability continues to gain importance (Douglas 2006). The latter years of the twentieth century have seen an increasing worldwide focus on sustainability and the protection of our environment. It has been recognised that the recycling of materials and products is, in the majority of cases, more environmentally friendly than simply disposing of them when they reach the end of, what is perceived to be, their useful life. There is no better example of the environmental benefits of effective sustainability in practice than the recycling of buildings. For every building that is recycled through refurbishment, the extraction of raw materials and the manufacturing processes and energy involved in converting these into a completed replacement building are avoided, to the undoubted benefit of the environment.

Current government environmental policy on housing is also encouraging the refurbishment and re-use of buildings by prescribing that 60 per cent of the 4.4 million new homes needed by 2016 should be provided on previously developed brownfield sites in order to reduce the loss of countryside. As many of these brownfield sites – the majority of which are within towns and cities – already have redundant or obsolete buildings standing on them, an increase in refurbishment activity will inevitably result. A further means by which the government has sought to increase building refurbishment is the exemption from Value Added Tax (VAT) for schemes involving the conversion of commercial premises to housing, one result of which has been a significant increase in the refurbishment of city-centre office buildings to provide 'new' homes.

The Urban Task Force, set up by the government to establish a new vision for urban regeneration, made recommendations that did encourage building refurbishment: one of the key measures proposed by the Task Force was the implementation of a strategy for the recycling of land and buildings that included making the best use of derelict, vacant and under-used land and buildings before carrying out new development on greenfield sites. The Urban Task Force also recommended the harmonisation of taxation laws on new-build and residential conversions, which discriminated against refurbishment by imposing VAT at a punitive rate of 17.5 per cent on the renovation of empty dwellings, while new house-building was exempt (DETR 1999: *Final Report*). The lobby to reduce VAT on all repair and renovation continues, although some other incentives have been introduced (Urban Task Force 2005).

Since May 2001 a reduced rate of 5 per cent VAT applies to building work that falls into the following circumstances:

- renovation of a dwelling that has been empty for at least three years;
- converting premises into a single household dwelling;
- conversion of a dwelling that results in a change in the number of dwellings;
- converting commercial or other non-residential property into a single or multiple dwellings;
- conversion of a dwelling or multiple occupation building into a property intended for residential purposes.

Conversions that qualify for the lower rate include:

- a property that has never been lived in, such as an office block or a barn;
- living accommodation that is not self-contained, such as a pub containing staff accommodation which is not self-contained; and
- any dwelling that has been adapted in its entirety to another use, such as to offices or a dental practice.

(www.customs.hmrc.gov.uk)

Where there is 'substantial reconstruction' or 'approved alteration' in cases where the finished building is to be used as a dwelling, for relevant residential purpose, relevant housing association or for relevant charitable purpose, VAT may be zero-rated customs (www. hmrc.gov.uk). In some situations, the building should have not been lived in for at least 10 years, although this varies depending on the conversion and the change of use. VAT should not be charged on materials and services used during the alteration of buildings that qualify; however, professional fees are not zero-rated (Wood 2003). Further information by Wood on refurbishment, listed buildings and tax can be found in the articles on the Building Conservation Directory website (www.buildingconservation.com).

The legislation is complex and much of it is untested by the courts, so it is very important to seek advice when considering whether the 5 per cent rate can be applied; however, guidance on the Inland Revenue site is relatively comprehensive (www.customs.hmrc.gov.uk).

Since the Urban Task Force recommendations, the government has introduced tax incentives to encourage development in deprived areas, including remediation allowances and Stamp Duty Land Tax reductions. Via various government vehicles, local authorities now have the power to advance the social, economic and environmental well-being of their communities. However, the confusing sponsoring and funding arrangements that exist, mostly funded by the ODPM (Office of the Deputy Prime Minister, www.communities.gov.uk/corporate), have not necessarily led to the high-quality, well-designed sustainable urban development that Rogers and his team anticipated (Urban Task Force 2005). The incentives and tax laws are not simple, properly coordinated or clear, and so many developers are put off inner-city redevelopment work. Even when VAT and the new local authority incentives are taken into account, the lack of public investment in infrastructure still means that greenfield sites are favoured over sustainable brownfield development (Urban Task Force 2005).

In the near future it is to be expected that tax and other fiscal advantages will be gained by upgrading buildings so that they are energy-efficient or zero-carbon buildings. The impact on our environment caused by emissions from our existing building stock has become of increasing concern. The vast majority of existing buildings are poorly insulated and demand considerable energy to run; there is pressure to upgrade buildings to create properties that are much more energy efficient. Even though the methodology of the Evan's *et al.* (2004) ratio of 1:5:200 (cost of construction: maintenance: occupancy) has been criticised (Hughes *et al.* 2004), the ratio does give some indication of the cost of running buildings. With a significant increase in running costs, as the price of oil, coal and bio-fuels rises, moves to improve thermal performance and make use of natural ventilation and renewable energy sources mean that more properties will be upgraded. Changes to our existing building stock over the next ten years will be significant, as will refurbishment, with the aim of upgrading building performance.

Even though the tax laws are not always easy to follow, and fiscal advantages are unclear, incentives do exist. The many advantages to be gained from conversions will ensure that the refurbishment and re-use of buildings will continue to represent a major, and increasing, component of construction activity well into the twenty-first century and beyond. It is therefore essential that all of those associated with building refurbishment and re-use remain fully conversant with the key issues involved and, in particular, the complex technological aspects peculiar to this type of work. In addressing a wide range of technological problems, and the solutions used to resolve them, this book should prove to be an invaluable aid to building owners, developers, architects, surveyors, and main and specialist contractors involved with the refurbishment, maintenance, repair, restoration and upgrading of buildings. It should also be of value to students of architecture, building and construction management, and surveying in their study of this increasingly important area.

The first chapter poses the question 'Why refurbishment?' and explains the reasons for, and advantages to be gained from, opting to refurbish and re-use existing buildings, rather than demolishing and replacing them with new buildings. This should enable building owners and developers to undertake a more comprehensive analysis when appraising potential refurbishment schemes.

Chapter 2 examines the problems of ensuring that elements within existing buildings have the appropriate fire resistance. The chapter first outlines the key statutory requirements and goes on to give detailed explanations of the techniques that can be used to upgrade the fire resistance of existing elements so that they comply with relevant regulations.

Chapter 3 deals with upgrading of the internal surfaces of walls and floors, explaining the methods used to restore surface finishes that have deteriorated beyond repair, and the provision of new finishes in buildings where none originally existed, for example in agricultural barns and older factories.

Chapter 4 looks at upgrading the thermal performance of existing walls and roofs to meet the statutory requirements relating to energy conservation, and as a means of improving comfort and reducing the 'running costs' of buildings. A wide range of internally and externally applied thermal upgrading methods are explained in detail, together with the actual reductions in heat loss that are possible.

Chapter 5 addresses the problem of upgrading the acoustic performance of existing walls and floors, and the techniques that can be employed to improve their sound-insulating capabilities to comply with legislation aimed at preventing noise nuisance between different occupancies.

Chapter 6 deals with the prevention of intrusive moisture and dampness within buildings, including damp penetration through walls, ground floors and roofs. The section also explains the causes and effects of condensation, and the ways in which it can be eliminated.

Chapter 7 looks at the introduction of new floors and issues that affect movement between floors. The chapter identifies factors that need to be taken into account when introducing stairs into attic conversions.

Chapter 8 examines the causes and main forms of timber decay, including fungal and insect attack, the treatments that can be used to eradicate them, and the techniques available for preventing further attacks. A wide range of methods used for the physical repair and reinstatement of decayed structural timber elements, including beams, joists and roof members, is explained.

In some refurbishment schemes it is necessary to strengthen existing timber floors so that they can support greater applied loadings. Chapter 9 describes, and gives a comparative evaluation of, the techniques used to achieve this.

Chapter 10 looks at heavy-lifting technology, which is gradually becoming more common in the refurbishment of buildings. A brief account of its capabilities in moving whole buildings is given, followed by a detailed description of its use in raising the height of a complete roof structure.

The underpinning of existing foundations is often found to be necessary in refurbishment and alteration work, especially where older buildings are involved, or where a change of use will impose greater loads on the substructure. The underpinning of buildings is dealt with in Chapter 11. Chapter 12 explains the principal reasons for underpinning and describes a wide range of techniques used, from traditional to more complex systems.

Chapter 13 deals with façade retention, an extreme, and often controversial, form of building re-use. The key technical problems associated with erecting an entirely new structure behind a retained façade are described, along with the solutions used to solve them.

As the various topics are introduced a number of renovation, rebuild and conservation case studies are discussed. These studies provide background information about the buildings that helps to contextualise the technical and management problems discussed. The diverse nature of tasks undertaken during the renovation demonstrates how, more often than not, problems associated with refurbishment can be addressed and meet current standards. Further photographs and drawings relating to these and other renovation projects can be found on the Virtual Site (www.leedsmet.ac.uk/teaching/vsite).

Throughout the text, numerous references are made to specific proprietary systems and products and, for those wishing to obtain further information on these, the details of their manufacturers are given in Chapter 14.

A key feature of this book is the large number of detailed line drawings that support the text and enhance the reader's understanding of the techniques used in the refurbishment and upgrading of buildings. The line drawings have been produced by Adrian Riley and Christopher Gorse.

References

Department of the Environment, Transport & the Regions (1999) *Final Report of the Urban Taskforce*, London: DETR.

Department of Environment, Transport & the Regions (1999) *Information Bulletin 599: Output and Employment in the Construction Industry*, London: DETR.

Douglas, J. (2006) *Building Adaptation*, 2nd edn, Butterworth-Heinemann.

Evans, R., Haryott, R., Haste, N. and Jones, A. (2004) 'The long-term costs of owning and using buildings', in Sebastian Macmillan, (ed.), *Designing Better Buildings: Quality and Value in the Built Environment*, Taylor & Francis, pp. 42–50.

Forsyth, M. (2007) *Understanding Historic Building Conservation,* Oxford: Blackwell Publishing.

Hughes, W., Ancell, D., Gruneberg, S. and Hirst, L., (2004) 'Exposing the myth of the 1:5:200 ratio relating initial cost, maintenance and staffing costs of office buildings', in F. Khoswowshahi, *Proceedings 20th Annual ARCOM Conference, 1–3 September 2004, Edinburgh, UK*, I: 373–382, Reading: ARCOM.

Sassi, P. (2006) *Strategies for Sustainable Architecture*, Oxford: Taylor & Francis.

Urban Task Force (2005) 'Towards a strong urban renaissance: an independent report by the Urban Regeneration Task Force chaired by Lord Rogers of Riverside'. Urban Task Force (www.urbantaskforce.org, accessed 7 July 2007).

Wood, R. (2003) 'Value added tax: implications for historic buildings', in J. Taylor (ed.), *The Building Conservation Directory 2003,* Tisbury: Cathedral Communications (also available at www.buildingconservation.com).

Why refurbishment?

CHAPTER 1

1.1 General

The provision of modern accommodation through the refurbishment and upgrading of existing old, redundant or obsolete buildings, in preference to constructing new buildings, has increased considerably in recent years, and there are many reasons for this. Most of the reasons can be attributed to the specific advantages that can be gained by opting for refurbishment rather than new build, although in some cases there may be legislative constraints, such as those concerned with listed buildings, which leave developers with no choice but to retain and refurbish certain buildings.

Where a developer wishes to provide modern accommodation, and a suitable existing building is available in the right location, all of the following points should be carefully considered, since it is likely that refurbishment and re-use of the building may well be a more viable means of providing the accommodation than opting for new construction.

1.2 The availability of buildings suitable for refurbishment

Advances in industry and commerce, together with society's constant demand for improved interior environments for both work and leisure, have led to large numbers of buildings becoming outdated, redundant or obsolete, and this, in turn, has provided an abundant supply of buildings suitable for refurbishment and conversion to new uses. Examples include large numbers of old factory and warehouse buildings in industrial centres, and outdated institutional buildings such as schools and hospitals – the latter as a result of government policies in the late 1980s and 1990s, which led to the closure of large numbers of asylums throughout the country. Changes in transportation systems during the nineteenth and twentieth centuries, from river and canal transport to railways and, finally, to the motorway network, have led to large numbers of riverside and canal-side factories and warehouses, and railway buildings becoming redundant as industry and commerce have had to relocate to be near the new transportation arteries. This has led to a considerable amount of refurbishment and re-use of redundant riverside, canal-side and dockside buildings in recent years as developers have come to realise the value and desirability of waterside locations for housing, offices and bars and restaurants. As recently introduced planning policies have become more favourable to housing development in cities, there has been an increase in the refurbishment and conversion of city-centre office buildings into flats to enable people to live close to their work and reduce demands on transportation systems. Another major source of buildings suitable for refurbishment and re-use are the many churches that have become redundant since the 1970s. Changing population patterns, and the declining position of the church in people's lives, have resulted in thousands of churches becoming available for redevelopment; from large city churches to small village chapels that have been successfully refurbished and converted into residential, office, recreational and manufacturing accommodation. In addition to the above examples, the existing housing stock represents a major focus for refurbishment activity. The government estimated, from information derived from the 1996 English House Condition Survey, that in England there is a £10 billion backlog of renovation needed in local authority housing stock alone (Housing and Regeneration Policy: A Statement by the Deputy Prime Minister, John Prescott, 22 July 1998).

In the absence of open sites available for new development, particularly in the prime commercial and residential areas of most of our towns and cities, developers seeking to provide modern accommodation have no choice but to focus on existing buildings. Having located a suitable building, the developer must then decide whether to demolish it and construct a new building or to opt for a refurbishment scheme, and the remaining sections explain why the latter course is often chosen.

1.3 The quality of buildings suitable for refurbishment

A further major factor in favour of the refurbishment of old, redundant or obsolete buildings, in addition to their widespread availability, is that many of these buildings are well built and structurally sound. Many may be run-down, neglected and unfit for modern usage as they stand, but the tried and tested methods of construction used to build them have left potential developers with an abundant legacy of sound, durable structures that provide an ideal basis for refurbishment and re-use. However, it should not be assumed that such buildings are always of high structural quality, and it is essential that any building being considered for refurbishment, even though it may appear to be sound, is subjected to a detailed survey in order to confirm its quality and condition, and to ascertain the likely cost of any repairs deemed necessary and their effect on the feasibility of going ahead with a refurbishment scheme.

1.4 The shorter development period

One of the principal advantages of opting for refurbishment and re-use of an existing building, rather than demolition and new construction, is that, in the majority of cases, the 'new' accommodation will be available in a much shorter time.

The work required to refurbish an existing building will normally take considerably less time than the alternative of demolition, site clearance and the construction of a new building, unless the refurbishment involves, for example, extensive structural alterations or remodelling. In addition to the time saved during the building works phase, time is also saved during the pre-contract design and planning permission phases, which normally take much longer for new development than for refurbishment, even where a change of use is proposed for the existing building. These time savings, during the pre-contract design, planning permission and building works phases of development often mean that opting for refurbishment can provide the new accommodation in only half to three-quarters of the time needed for demolition and new construction, giving the following financial benefits:

- The shorter contract duration reduces the effects of inflation on building costs.
- The shorter overall development period reduces the cost of financing the scheme.
- The client obtains the building sooner, and therefore begins to earn revenue from it (for example, rentals, retail sales or manufacturing profits) at an earlier date.

1.5 The economic advantages

The cost of refurbishing and re-using an existing building is generally considerably lower than the cost of demolition and new construction, since many of the building elements are already constructed. However, the existing construction and its condition will have a considerable bearing on the costs of refurbishment. For example, if the existing floor-to-ceiling heights are either too low or too high for the proposed new use, the necessary adjustments may be very costly, as illustrated by the Granary Building scheme in Leeds (see Chapter 10) where the entire roof structure had to be raised by 300 mm in order to allow re-use of the uppermost storey. Also, many older industrial buildings have exposed timber floors, supported by timber or cast-iron columns and beams that require upgrading to comply with current fire-protection legislation (see Chapter 2). In addition, new fire-escape stairs and enclosures will almost certainly be required, all of which will add to the costs of the refurbishment scheme. If the building is in a poor physical condition because of neglect, deterioration or vandalism, the refurbishment will also involve the expense of repair and restoration work, which may have a significant effect on overall costs.

Against costs of this nature, the developer must weigh the potential savings achieved by re-using most of the existing elements of the building, and the shorter development period with its associated financial benefits outlined in the introduction above.

There would be little point in refurbishing and re-using existing buildings if the costs were to be greater than those of demolition and new construction, unless overriding reasons exist, as in the case of buildings that have been listed because of their architectural or historic interest. Refurbishment and re-use will be substantially cheaper than demolition and new construction only where a suitable building is selected that is in a reasonable physical condition and that does not require excessive structural alterations in order to adapt it to its proposed new use.

In the majority of cases, the decision regarding whether or not to refurbish and re-use an existing building will revolve around the potential economic advantages. It is essential, therefore, that, in the first instance, a detailed cost-appraisal of alternative refurbishment schemes versus demolition and new build is carried out, since this, above all else, will normally determine whether or not refurbishment is viable.

The most important factors that determine whether or not refurbishment is viable are:

- the expected rental income (in developments for letting);
- the expected capital value (in developments to be sold after completion);
- the estimated cost of development;
- the cost of acquiring the site;
- the cost of financing the scheme.

Expected rental income

The expected rental income from a refurbished building will depend on several factors:

- the proposed new use(s) for the building;
- the location of the building;
- the relative attractiveness of the area in which the building stands, including the amenity of the surrounding area and its accessibility;
- the quality of the accommodation and services after refurbishment, which, in turn, will depend upon the standard of refurbishment carried out;
- the level of demand for such accommodation from new firms setting up or moving into the area, and existing firms wishing to expand or improve their accommodation;
- the availability of other, similar, accommodation in the area.

Details of prevailing rentals, and the extent of demand for different types of accommodation in an area are best obtained from local property/letting agents.

Expected capital value

The expected capital value on completion of the refurbishment scheme will depend upon those factors, listed above, that affect expected rental income.

Estimated cost of development

The development cost for a refurbishment scheme will depend on several factors, the most important of which are as follows:

- proposed new use
- standard of refurbishment
- age of building
- construction of the building.

The proposed new use

The proposed new use for the refurbished building can have a significant effect upon the development costs. For example, the cost of refurbishing a late nineteenth-century warehouse to provide utilitarian manufacturing accommodation would be considerably less than converting it to high-specification, prestige office accommodation.

The standard of refurbishment envisaged

The required standard or quality of the proposed accommodation will have a significant effect on the final cost of refurbishment. For example, the existing windows may be in sufficiently good condition to warrant only minor repair and repainting, but the developer may wish to replace them with new metal or uPVC windows with sealed double glazing in order to improve their appearance and thermal properties and to reduce maintenance. Opting for the latter will clearly significantly increase the cost of refurbishment.

Decisions regarding building services can also significantly affect the final cost of a refurbishment scheme. For example, where a building is being refurbished to provide modern office accommodation, the choice of a simple hot-water radiator heating system, with openable windows to provide natural ventilation, will cost considerably less than a sophisticated mechanical heating, ventilation and air-conditioning system. Costs can also be reduced in taller buildings by installing fewer passenger lifts, or even no lifts at all in buildings of only two or three storeys.

Virtually every design decision relating to the quality, standard or amenity of the refurbished building will have a direct effect on the final cost of the scheme – the higher the specification, the higher the overall cost of the completed building. However, as stated above, in the long term, the greater costs incurred by providing a higher standard of refurbishment can be recouped by the higher rental income (or resale value) that such standards can demand.

The age of the building

Generally, the older the building and the longer it has been empty, the greater the costs of repairing and restoring the existing structure and fabric. In many cases, where the building has remained empty and neglected for a prolonged period, dampness, which is the principal agent in most forms of building deterioration, may well have penetrated, leading to timber decay and damage to finishes. Empty, neglected buildings often fall victim to vandalism, the results of which may be very expensive to rectify. These factors apart, the basic fact that a building is old will mean that certain items will need attention because of their natural deterioration over a long period of time. In addition, items that may still be in good condition will often need replacing merely because they are obsolete or outdated in their design, common examples being sanitary fixtures and fittings and lift installations.

In many refurbishment schemes it will be found that overall costs are directly proportional to the age of the building and the degree of neglect it has suffered, and the proposed refurbishment of any old, neglected building should therefore be given very careful consideration before proceeding. For this reason, it is vital that the first stage of any feasibility study should comprise a detailed survey of the building in order to establish the precise extent and costs of any works required to repair and restore the structure and fabric to a reasonable standard.

The construction of the building

The construction of the existing building can have a significant effect on the cost of refurbishment, fire protection often being a key factor, particularly where older buildings are concerned. Many older buildings have timber stairs and floors that may be supported by unprotected cast-iron or steel beams and columns, and these will have to be upgraded or replaced to comply with current fire regulations. The upgrading, or replacement, of existing timber and cast-iron or steel elements to comply with current fire regulations, including means of escape, often proves to be one of the most costly areas of building refurbishment. Chapter 2 explains the current statutory requirements with regard to fire resistance and explains the techniques used to upgrade the fire resistance of existing elements of construction.

Another common example of potentially high costs in building refurbishment is the strengthening or replacement of existing floor structures where their load-carrying capacities are inadequate to meet the requirements of the proposed new use. In such cases it will be necessary either to carry out strengthening operations or to replace the existing floors completely with new floors, usually of *in situ* or precast concrete. Chapter 9 describes the techniques that can be used to strengthen existing timber floor structures.

The cost of acquiring the site

The cost of acquiring the freehold or leasehold of the site should never exceed the difference between the capital value of the completed development and the development costs, since, if it does, a financial loss will be incurred. It is therefore very important to assess accurately the proper value of the site, which will largely be determined by the following factors:

- the location of the site, which will decide the potential users of the refurbished building;
- the uses for which planning permission can be obtained;
- the expected rental income from, or capital value of, the refurbished building;
- the total development costs.

The cost of financing the refurbishment scheme

The cost of financing the scheme will depend principally on the following factors:

- the cost of the refurbishment works;
- the duration of the scheme;
- the level of interest rates prevailing at the time of the scheme.

In the majority of cases the total interest payable on money borrowed to finance a refurbishment scheme will be significantly less than that for new construction, owing to the lower overall costs and shorter development periods generally associated with refurbishment schemes. In addition, when interest rates are higher, the refurbishment option will become even more attractive, since this will result in a greater differential between the costs of financing refurbishment and the higher costs of financing new construction schemes.

Detailed consideration of the factors discussed above will be essential if the correct decision is to be made on the type and level of refurbishment or, indeed, whether the refurbishment option is viable at all. Of equal importance to examining all of the salient factors is ensuring that designers, building surveyors and building economists with a high level of expertise in refurbishment work are commissioned to prepare alternative schemes, carry out surveys and complete cost-feasibility studies. Construction firms specialising in refurbishment work should also be consulted at the feasibility stage to advise the design team on aspects such as 'buildability'. Only if this is done can the developer be sure of opting for the scheme that will give the best value for money, while remaining within a realistic budget.

1.6 The availability of financial aid

In the majority of cases, the refurbishment option is chosen for economic reasons – because it costs the developer less than new build, and a further major incentive, therefore (and one that can make the refurbishment option even more attractive), is the availability of various forms of financial aid. Financial aid, in the form of grants and 'soft' loans, is not available for all refurbishment schemes, but in many cases, for example where substandard housing or historic buildings are concerned, where jobs are being created, or urban regeneration is taking place, it may be possible to obtain substantial grants towards the cost of the work. Financial aid for the refurbishment, repair, conservation, restoration and maintenance of buildings is available from a large number of different bodies, and it is beyond the scope of this book to identify them all. Some examples of grant-awarding sources include English Heritage, which gives grants towards the costs of repair, conservation and restoration of historic (usually listed) buildings; local authorities, which administer housing improvement grant schemes; and the government Department of the Environment, Transport & the Regions Single Regeneration Budget, which provides funding for a wide range of projects focusing on sustainable development, including bringing redundant buildings back into use and housing refurbishment.

Sources of finance range from public funds, provided by central government and local authorities, to private funds from banks, insurance companies, building societies, and so on.

The main categories of financial aid are:

- **Loans**: A loan is a sum of money lent to the recipient, which must be repaid to the lender over an agreed period, at an agreed rate of interest. The rate of interest charged for a normal loan will be at the current commercial rate.
- **Soft loans**: A soft loan is similar in principle to a normal loan, with the important exception that the rate of interest charged will be below the current commercial rate.
- **Grants**: A grant is a sum of money awarded towards the cost of a scheme, which the recipient is not required to repay.

Many sources of financial aid are subject to specific conditions and restrictions with which the recipient must comply. These vary considerably, and details should be obtained from the various awarding bodies. However, the following points apply to many of the loans and grants that are available:

- The award of a loan or grant will normally only be made if the scheme is viable.
- Awards are often restricted to schemes that would be unable to proceed without the receipt of financial aid.
- Awards for a specific aspect of the work may not be duplicated where more than one source of funding is available.
- Awards are normally made on completion of the work, unless it is a large scheme, in which case the award may be made in instalments.
- Generally, awards must be applied for, and approved, before the scheme commences, although it will normally also be necessary to have obtained planning permission.
- The majority of awards are discretionary.
- Loans and grants normally provide only a part of the total funding necessary for refurbishment schemes, and only rarely are awards of 100 per cent made. Thus, for the majority of schemes, the loan or grant will represent a maximum of half the total cost of the scheme, and often it is less than this.
- The funds available for financial aid are strictly limited in the majority of cases, with demand exceeding supply.

Generally, awards are made on a 'first come, first served' basis or by the use of specific priority criteria. It is clear, therefore, that if the refurbishment option being considered for an existing building is eligible for grant, or even soft loan, aid, and a grant or loan can be obtained, the refurbishment option will become even more attractive when compared with the alternative of demolition and new construction.

1.7 Planning permission may not be required

Under section 57 of the Town and Country Planning Act 1990, planning permission is required for 'development'. However, section 55(2)(a) of the Act states that 'the carrying out of works for the maintenance, improvement or other alteration of any building *which affect only the interior of the building, or do not materially affect the external appearance of the building*' does not constitute development. Such works, therefore, do *not* require planning permission. Thus, if the refurbishment scheme does not affect the exterior appearance of the building, there may be no need for the developer to obtain planning permission, resulting in a further shortening of the development period, and a corresponding saving in costs.

However, it should be noted that a key exception to the above, as detailed in section 55(2)(f) of the Act, is where the 'use class' of the building changes, in which case planning permission will still be required. The Town and Country Planning (Use Classes) Order 1987 (as amended in 1991) designates sixteen use classes, and any proposed change from one of these use classes to another will require planning permission, even if the external appearance of the building does not change.

An example of this might be where it is proposed to refurbish and convert a riverside warehouse (use class B8, 'Storage or distribution') into a hotel (use class C1, 'Hotels and hostels'). There are, though, numerous examples of refurbishment schemes that do not require planning permission, some of the most common examples being the interior upgrading and alteration of old, outdated office buildings to provide modern office accommodation, and the refurbishment and modernisation of unfit housing: the refurbished building remains in the same use class, and it is therefore permissible to carry out extensive interior alterations without the need for planning permission. The extent of the works might go as far as completely gutting the building and providing an entirely new internal structure, as in façade retention (see Chapter 13), provided the exterior appearance remains the same.

1.8 The effects of plot ratio control

Plot ratio control, which was introduced by the Ministry of Town and Country Planning in 1948, is a device used by planning authorities to restrict the amount of floor space provided in new buildings in relation to their site areas. For example, a plot ratio of 3:1 will restrict the floor area of a new building to three times the area of its site (see Figure 1.1). One of the principal reasons for the introduction of plot ratio control was to restrict the heights, scale and mass of buildings in towns and cities so as not to impair the amenity and development possibilities of surrounding sites and buildings. Without this form of control, buildings could completely fill their sites, with the likelihood that taller buildings would cut out daylight from their

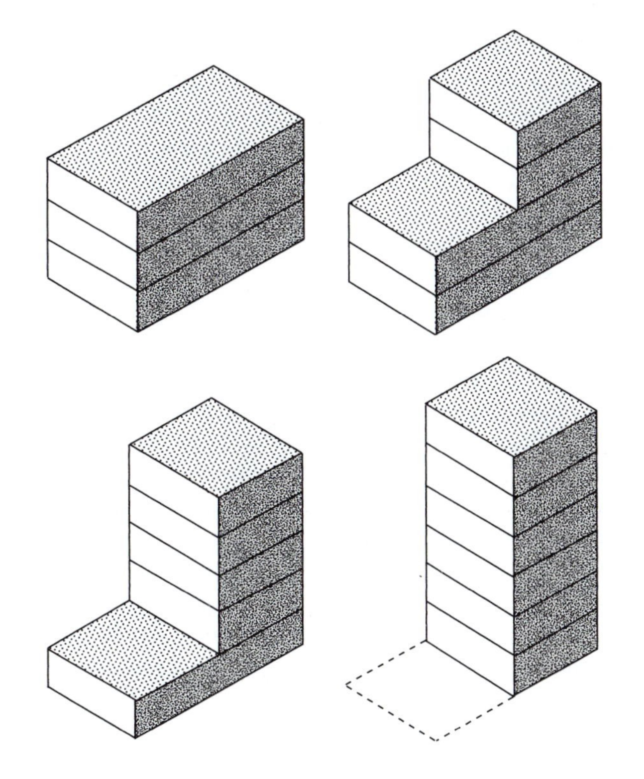

Figure 1.1
Alternative designs for a site with a plot ratio of 3:1

neighbours, turning streets into 'canyons' of building façades – a feature commonly seen in cities in the USA.

Plot ratio control is used by different planning authorities on an ad hoc basis to suit their own requirements and is most likely to be applied in the central areas of large towns and cities, where plot ratios are often restricted to between 3:1 and 5:1.

The application of plot ratio control in restricting the size of new developments often makes it advantageous to refurbish existing buildings, rather than to demolish and replace them. For example, most Victorian buildings were built to much higher plot ratios than is currently permitted by planners, some having a plot ratio of 7:1 in areas where plot ratios for new developments may well be restricted to 3:1 or 4:1. It is therefore clear that the refurbishment of such a building, retaining its existing, relatively high plot ratio, could result in the provision of around twice as much 'new' floor space as would be permitted if it were to be demolished and replaced with a new building. In such cases, where current plot ratios would prove restrictive, and where the existing building is suitable for adaptation to provide the accommodation the developer requires, it is usually well worth while giving serious consideration to opting for refurbishment rather than demolition and new construction.

1.9 Listed building legislation

Section 1 of the Planning (Listed Buildings and Conservation Areas) Act 1990 requires the Secretary of State for the Environment to compile lists of buildings of special architectural or historic interest in order that they can be protected from demolition or insensitive alteration and, therefore, preserved for the enjoyment of present and future generations. When a building has been included in the list of buildings of special architectural or historic interest it is an offence, under the provisions of the Act, to carry out works of complete or partial demolition, alteration or extension in any manner that would affect its character without first having obtained listed building consent from the local planning authority. In considering whether or not to grant listed building consent, the local planning authority must consult specified national conservation bodies, and take into account any representations made by other interested parties including local conservation groups, together with its own internal professional conservation officers.

The detailed requirements with regard to listed buildings are contained in the Town and Country Planning (Listed Buildings and Buildings in Conservation Areas) Regulations 1990 (HMSO 1990), and comprehensive guidance is given in the Department of the Environment *Planning and the Historic Environment* (PPG 15) 1994.

The number of listed buildings in Great Britain is in the region of half a million, and it is quite likely, therefore, especially in older towns and cities, that a 'developable' building might be listed, and this will almost certainly rule out the possibility of total demolition, which is only very rarely permitted. However, it is possible, in the majority of cases, to carry out 'sensitive' refurbishment and alteration (including partial demolition and/or extension) of listed buildings, provided that those features for which the building was listed are retained.

Generally, the extent of alteration permitted will depend on the grade of listing; for example, the majority (96 per cent) of listed buildings are Grade II, many of which possess only external features worthy of retention, such as the principal street façade, or possibly the entire external envelope of an isolated building. In such cases it may be possible to go as far as replacing the entire interior with a new structure, with only the external elevation(s), which led to the listing of the building, being retained. On the other hand, if the building has a Grade I listing (2 per cent of all listed buildings), any possible alterations will be severely limited since Grade I listed buildings are of exceptional interest and usually possess both interior and exterior features that must be retained.

It is clear, therefore, where a listed building is the subject of a proposed development, that the developer will be restricted to the refurbishment option within the constraints imposed by its listing, although, as has been stated, in many cases it may be permissible to carry out a considerable amount of reconstruction behind the retained external elevations. On the other hand, if the design constraints imposed by the building's listing are too great, the developer will be left with no choice but to find a different site or another building.

1.10 Conservation area legislation

Section 69 of the Planning (Listed Buildings and Conservation Areas) Act 1990 requires local authorities to determine whether there are any areas within their jurisdiction that are of special architectural or historic interest, the character or appearance of which it is desirable to preserve or enhance, and to designate those areas as conservation areas. Conservation areas may be large or small, from whole town centres to squares, terraces and smaller groups of buildings. Local authorities must then pay special attention to the desirability of preserving or enhancing the character and appearance of conservation areas within their jurisdiction. The Planning (Listed Buildings and Conservation Areas) Act 1990 brings the demolition (deemed also to include partial demolition) of buildings within conservation areas, whether listed or not, under control by applying, with modifications, the listed building control provisions of the Act. Any developer wishing to demolish (or partially demolish) an unlisted building within a conservation area must first apply to the local authority for conservation area consent, following which similar procedures to those involving listed buildings are put into effect.

It is clear, therefore, that in executing its responsibilities of preserving or enhancing the character or appearance of conservation areas, the local planning authority may impose strict constraints on the proposed development of all buildings within conservation areas, whether or not they have been listed. In most cases this will rule out the possibility of total demolition and redevelopment, and it is also unlikely that any alterations to a building will be permitted if they are likely to detract from the appearance or character of the conservation area.

Many 'developable' buildings, while they may not be listed, stand in conservation areas and, as has been stated, if these buildings form an essential part of the area's character and appearance, the development possibilities will be limited. It is therefore essential that the developer establishes what design constraints are likely to be imposed on any proposed refurbishment scheme as early as possible, since these may well be a key factor in determining whether or not it is worth while going any further in examining the feasibility of the scheme. Generally, however, it will usually be possible to go further in partially demolishing, altering or extending an unlisted building in a conservation area than would be possible with a listed building, where the design constraints are likely to be much more stringent.

1.11 The architectural advantages

There are often architectural advantages, which can be translated into financial advantages, in keeping attractive, usually older, buildings and refurbishing them to provide modern accommodation. Many older buildings possess far greater character than their modern counterparts, incorporating skilled craftsmanship and high-quality natural materials in their design and construction. Such buildings are often more attractive to certain users such as banks, insurance companies and other financial institutions as well as commercial organisations, many of which like to project a prestigious image to their customers, which is often associated with older, architecturally attractive buildings. In addition, many such buildings stand in areas where they are in close proximity to other architecturally attractive old buildings, and this adds further to their appeal and potential value, provided that their refurbishment maintains their architectural character and integrity.

1.12 Availability of the existing infrastructure

Certain refurbishment schemes can benefit from the retention and re-use of existing infrastructure, giving, in turn, further financial savings that would not be gained if the demolition and new-build option were chosen. A common example of this is the refurbishment of large, run-down housing estates in preference to their demolition and replacement with new housing. Substantial financial savings are achieved since, not only are the houses themselves re-used, but also the existing 'housing infrastructure' including roads, drainage, gas, electricity and water supplies, telecommunications, cable networks and other utility services, all, or most, of which would have to be completely renewed if the demolition and new-build option were chosen. There are many other examples where infrastructure costs can be saved, and although the greatest savings will be achieved on larger-scale projects involving more than one building, infrastructure savings are also possible when individual buildings are refurbished, since they will normally already have road access and be connected to most of the services mentioned above.

As well as achieving these direct financial savings, the avoidance of having to provide a new infrastructure will also reduce the development period, resulting in further indirect financial savings as discussed in the introduction and earlier in this chapter.

1.13 The social advantages

The refurbishment of large housing estates has important sociological advantages when compared with demolition and new build. One of the most disruptive aspects of the comprehensive, national policies of demolition and replacement of substandard housing during the 1960s and early 1970s was that established communities, many of which had existed for several generations, were broken up permanently. The 'creation' of new communities has since been recognised as a complex process, and the refurbishment of existing housing, by preserving established, stable communities, is therefore considered preferable to the alternative of wholesale clearance and new development.

1.14 The environmental advantages

A major focus of concern during the past 30 years, and one that has continued to increase in importance, is the massive worldwide consumption of energy, and its related adverse implications including global warming. One of the many ways in which worldwide energy consumption can be reduced is to recycle and re-use existing resources as much as possible, in preference to consuming even more energy by replacing them. Whenever a building is recycled by opting for refurbishment rather than new build, a considerable amount of energy is saved by avoiding the need to extract raw materials and convert them into a replacement building. 'Low-key' refurbishment, where most of the existing structure and fabric are retained, will clearly yield the greatest energy savings, but even the more drastic forms, where major alterations are made, will generally use less energy than demolition and new build.

References

Building Research Establishment (1991) *Structural Appraisal of Existing Buildings for Change of Use* (Digest 366), Watford: BRE.

Cantacuzino, S. and Brandt, S. (1980) *Saving Old Buildings*, London: Architectural Press.

Construction Industry Research and Information Association (1994) *A Guide to the Management of Building Refurbishment* (Report 133), London: CIRIA.

Corbett-Winder, K. (1995) *The Barn Book*, London: Random Century.

Davis, J., Goldsworthy, J. and Moncrieff, D. (1997) *The Directory of Grant Making Trusts*, 15th edn, Vols 1 and 2, West Malling: Charities Aid Foundation.

Department of the Environment, Transport & the Regions (1994) *Planning and the Historic Environment* (Planning Policy Guidance 15), London: HMSO.

Department of the Environment (1996) *English House Condition Survey 1996*, London: HMSO.

HMSO (1987) *Town and Country Planning (Use Classes) Order 1987*, London: HMSO.

HMSO (1990) *Planning (Listed Buildings and Conservation Areas) Regulations 1990* (Statutory Instruments, no. 1519), London: HMSO.

HMSO (1990) *Planning (Listed Buildings and Conservation Areas) Act 1990*, Ch. 9, London: HMSO.

HMSO (1990) *Town and Country Planning Act 1990*, Ch. 8, London: HMSO.

HMSO (1991) *Town and Country Planning (Use Classes) Amendment Order 1991*, London: HMSO.

Office of the Deputy Prime Minister (1998) Housing and Regeneration Policy: A Statement by the Deputy Prime Minister, John Prescott, 22 July 1998, London: ODPM.

Pickard, R.D. (1996) *Conservation in the Built Environment*, London: Longman.

Pollard, R. (1998) 'Redundant government buildings', in J. Taylor (ed.), *The Building Conservation Directory 1998*, Tisbury: Cathedral Communications, pp. 25–27.

CHAPTER 2 Upgrading the fire resistance of existing elements

2.1 General

Any refurbishment scheme, especially if it involves a building containing exposed structural elements of timber, steel or iron, is likely to need some upgrading of fire resistance if it is to comply with current regulations. Generally, the older the building, the more likely it will be to require fire resistance upgrading owing to the nature of its construction, a typical example being a nineteenth-century docklands warehouse being converted into offices. Many buildings of this type and period have open joisted timber floors supported by exposed wrought iron beams and cast-iron columns, none of which would come near to complying with current fire regulations. The timber roof structures of such buildings, often left exposed from below, and their timber staircases would also require upgrading. Another example, on a smaller scale, might be the conversion of a large Victorian house into flats. Here, the fire resistance of the existing timber stairs and the floors that would separate the newly created flats would, as existing, not comply with current fire regulations for the proposed new use and would therefore need upgrading. Further information on fire protection can be found in Emmitt and Gorse (2006), *Advanced Construction of Buildings*.

2.2 Statutory requirements

Most refurbished buildings, including those involving alterations, extensions or a change of use, must comply with the Building Regulations, and one of the most important sections affecting refurbishment schemes is Part B3: Internal Fire Spread (Structure). Compliance with Part B3 will almost certainly involve upgrading the fire resistance of some existing elements of structure or replacing them with new construction, and, in older buildings with many exposed timber, iron and steel elements, the upgrading work and associated costs can be considerable.

To comply with Regulation B3, the designer of the building can use the non-mandatory guidance contained in Approved Document B: Fire Safety, or use alternative ways of demonstrating compliance, provided the chosen solution is adequate to meet the requirements of the Regulations. Regulation B3 requires that load-bearing elements of a structure, such as columns, beams, floors and walls, must have at least the fire resistance given in the tables contained in Approved Document B to the Regulations. The tables give required minimum periods of fire resistance from half an hour to four hours, depending on the purpose group of the building and its height and size. The purpose group is a means of classifying a building according to the hazard to life that a fire would present, which is determined partly by the building's use. Buildings containing sleeping accommodation for aged or infirm people, for example, are regarded as particularly hazardous and, as a result, have more stringent requirements with regard to fire protection than, say, buildings used for the storage of goods.

2.3 Fire resistance of elements

Approved Document B to the Building Regulations specifies the required fire resistance of elements in terms of the results they achieve when subjected to standardised fire tests. The procedures for the tests, which must be carried out on properly made-up specimens of the constructions being evaluated under strictly controlled conditions, are laid down in the various parts of British Standard 476. The fire resistance tests for structural (load-bearing) elements of construction, including columns, beams, floors and walls, are contained in BS 476: Part 21:1987 *(Methods for Determination of the Fire-resistance of Load-bearing Elements of Construction)*. It should be noted that fire resistance in the contexts of both new construction and refurbishment is not a characteristic of a material, but the performance of a complete element of construction, which will normally comprise a number of different materials and components. Fire resistance is therefore determined in a test that subjects a representative specimen to heating, which simulates its anticipated exposure in a real building fire, for example floors exposed from below, walls from one side and columns on all sides. Load-bearing elements are subjected to their design loadings during the tests. The fire resistance of the specimen is the time in minutes for which it continues to meet whichever of the following criteria are relevant:

- **Load-bearing capacity**: applicable only to load-bearing elements. Failure occurs when the test specimen can no longer support its design loading. For horizontal elements (e.g., floors and beams), limits are specified for the allowable extent of vertical deflection.
- **Integrity**: applicable only to elements that separate spaces (e.g., walls and floors). Failure occurs when the specimen collapses, exhibits sustained flaming on its unexposed face, or when cracks or other openings form through which flame or hot gases can pass.
- **Insulation**: applicable only to elements that separate spaces. Failure occurs when the temperature of the unexposed (to fire) face of the element increases by more than 140 degrees C above the initial temperature, or by more than 180 degrees C, regardless of the initial temperature.

The two most common fire-resistance upgrading requirements in refurbishment work involve timber floors and unprotected beams and columns of steel or, in older buildings, cast iron.

2.4 Upgrading the fire resistance of timber floors

One of the most common examples where the upgrading of existing timber floors proves necessary is in the refurbishment and conversion to modern use of the numerous old, often redundant, factory and warehouse buildings constructed between 1850 and 1920. Such buildings are abundant in towns and cities and therefore form a large proportion of those that lend themselves to refurbishment and adaptation to modern uses, especially as many are located in prime development areas such as river- and canalsides and docklands. Typical new uses include flats, maisonettes, offices, restaurants, shops and museums, and in all cases their original timber floors will almost certainly require their fire resistance to be upgraded to comply with current regulations. In the majority of these utilitarian industrial buildings, the undersides of floors did not receive a ceiling finish and merely comprised floor-boarding on timber joists left exposed on the underside. In addition, many such floors have plain-edge boarding, which is often found to have distorted over the years, leaving gaps which, in the absence of a ceiling beneath, render the floor almost totally ineffective as a fire barrier since the floor would have minimal or zero integrity. Timber floors of this type come nowhere near to meeting the fire resistance standards required by modern regulations and therefore require extensive upgrading, in most cases to either half an hour or 1 hour depending on the purpose group, height and size of the building.

Another common example where upgrading of existing timber floors proves necessary is the conversion of large, single dwellings into self-contained flats or maisonettes. Here, the focus of attention is the floors that will separate the newly created occupancies, and these will need to be upgraded to half-hour fire resistance in two-storey buildings and one-hour fire resistance in buildings of three or more storeys. In all cases, any floor over a basement storey must have 1-hour fire resistance. Where an existing dwelling is being converted, it is clear that the floors requiring upgrading will already have some form of ceiling finish providing some degree of fire resistance, and generally the upgrading treatment will not need to be as extensive as industrial-type floors with exposed joists. However, it should be noted that a typical floor construction of plain-edge boards and a lath and plaster ceiling, often found in older buildings, will still need an upgrading treatment, even if it has to achieve only a half-hour fire resistance after conversion.

The most common technique used to upgrade the fire resistance of existing timber floors is to add a new fire-resisting layer beneath the existing joists or ceiling. This may also involve the insertion of a layer of fire-resisting material within the void, resting on the new ceiling. In certain cases, however, it may not be acceptable to cover the existing ceiling with a new layer, and in such cases an alternative technique may be used that involves filling the void between the existing floor surface and ceiling with a fire-resisting material. Both techniques are described in detail below.

Addition of a new fire-resisting layer beneath the existing joists or ceiling

A wide range of techniques and materials are available for use in upgrading the fire resistance of existing timber floors. Most, as previously stated, involve providing either a completely new ceiling where originally the joists were exposed, or an extra layer to the underside of the existing ceiling. In some cases it may also be necessary to provide an additional layer of fire-resisting material on top of the existing floorboards or between the floor joists. Typical practical upgrading techniques for a variety of existing floor constructions are given in Tables 2.1–2.4 and illustrated in Figures 2.1–2.4, indicating the fire resistances that can be achieved. Several of the techniques involve the use of proprietary materials, descriptions of which are given below.

Supalux and Masterboard

Supalux and Masterboard are rigid board materials, developed to replace asbestos-based products (www.promat.co.uk). They consist of a hydrated calcium-silicate matrix reinforced with special cellulose fibres and inorganic additives. The material is cured in high-pressure steam autoclaves. The boards are 1,220–3,050 mm long × 610 or 1,220 mm wide × 6, 9, 12, 15 or 20 mm thick.

Each board can be tightly butt-jointed, or their edges can be left slightly apart for filling and sanding. The boards do not require any surface treatment to attain their stated fire resistance, and painting does not alter fire resistance. Suitable finishes include paint, paper, tiles or plaster.

It should be noted that the upgrading techniques and specifications are provided for guidance only, and that the product manufacturers should be consulted for detailed specifications and instructions.

Table 2.1 Treatments for upgrading the fire resistance of an existing floor construction of plain-edge floorboards, 22 mm thick, on timber joists not less than 38 mm thick; no ceiling

	Upgrading treatment	*Resulting fire resistance*
1	Expanded metal lathing nailed to joists with gypsum plaster finish 16 mm thick	½ hour
2*	Expanded metal lathing nailed to joists with vermiculite–gypsum plaster finish 12.5 mm thick	½ hour
3*	Plasterboard 12.5 mm thick nailed to joists with gypsum plaster finish 12.5 mm thick	½ hour
4	2 layers of plasterboard nailed to joists with joints staggered, total thickness 25 mm	½ hour
5*	Hardboard sheet 3 mm thick nailed to floorboards. Masterboard 6 mm thick nailed to joists	½ hour
6*	Hardboard sheet 3 mm thick nailed to floorboards. Supalux boards 9 mm thick screwed to joists through 80 × 9 mm Supalux fillets and overlaid with 60 mm mineral wool mat (density 23 kg/m^3)	1 hour
7*	Hardboard sheet 3 mm thick nailed to floorboards. New Tacfire boards 9 mm thick nailed or screwed to joists through 75 × 9 mm New Tacfire cover fillets, and overlaid with 40 mm rock wool mat (density 60 kg/m^3)	1 hour

* See Figure 2.1

Table 2.2 Treatments for upgrading the fire resistance of an existing floor construction of tongued and grooved floorboards not less than 19 mm thick on timber joists not less than 38 mm thick; no ceiling

	Upgrading treatment	*Resulting fire resistance*
1	Expanded metal lathing nailed to joists with gypsum plaster finish 16 mm thick	½ hour
2	Plasterboard 9.5 mm thick nailed to joists with gypsum plaster finish 12.5 mm thick	½ hour
3*	Plasterboard 12.5 mm thick nailed to joists with gypsum plaster finish 5 mm thick	½ hour
4*	2 layers of plasterboard nailed to joists with joints staggered, total thickness 22 mm	½ hour
5*	Masterboard 6 mm thick nailed to joists	1 hour
6	Expanded metal lathing nailed to joists with gypsum plaster finish 22 mm thick	1 hour
7*	Expanded metal lathing nailed to joists with vermiculite-gypsum plaster finish 12.5 mm thick	1 hour
8	Plasterboard 9.5 mm thick nailed to joists with vermiculite-gypsum plaster finish 12.5 mm thick	1 hour
9*	Supalux board 9 mm thick screwed to joists through 80 × 9 mm Supalux fillets and overlaid with 60 mm mineral wool mat (density 23 kg/m^3)	1 hour

* See Figure 2.2

There are many other hydrated calcium-silicate fireboard materials available; the association for specialist fire protection provides useful information and contacts on its website (www.asfp.org.uk).

Autoclaved calcium-silicate boards

Autoclaved calcium-silicate boards are non-combustible, under British Standard 476; they have good performance towards moisture and humidity. Normally the sheets are processed with a smooth upper surface that is off-white in colour. They are fixed to timber by nailing or screwing and come in standard sizes from 2,134 to 3,000 mm long × 914 to 1,250 mm wide and thicknesses from 6 to 25 mm. Autoclaved calcium-silicate boards can be painted, papered, tiled (on minimum 9 mm thick boards), or finished with a proprietary decorative coating such as Artex.

Sprayed Limpet Mineral Wool – GP grade

Sprayed Limpet Mineral Wool (www.thermica.co.uk) is a blend of mineral wool and selected inorganic fillers with a binder applied. The binder is applied by spraying to form a homogeneous, jointless coating. The mineral wool can be left either as sprayed or with a tamped, stippled finish that can be coated with Limpet LD3 – a white decorative finish. This can, if required, be overpainted.

Table 2.3 Treatments for upgrading the fire resistance of an existing floor construction of tongued and grooved floorboards not less than 22 mm thick on timber joists not less than 175 × 50 mm; no ceiling

	Upgrading treatment	*Resulting fire resistance*
1	Expanded metal lathing nailed to joists with gypsum plaster finish 16 mm thick	½ hour
2*	Expanded metal lathing nailed to joists with sprayed Limpet Mineral Wool – GP grade finish 13 mm thick	½ hour
3*	Plasterboard 9.5 mm thick nailed to joists with gypsum plaster finish 12.5 mm thick	½ hour
4	Plasterboard 12.5 mm thick nailed to joists with gypsum plaster finish 5 mm thick	½ hour
5	2 layers of plasterboard nailed to joists with joints staggered, total thickness 19 mm	½ hour
6*	Expanded metal lathing nailed to joists with sprayed Limpet Mineral Wool – GP grade finish 22 mm thick	1 hour
7*	Plasterboard 9.5 mm thick nailed to joists with vermiculite-gypsum plaster finish 12.5 mm thick	1 hour
8*	Supalux board 9 mm thick screwed to joists through 80 × 9 mm Supalux fillets and overlaid with 60 mm mineral wool mat (density 23 kg/m^3)	1 hour
9	New Tacfire boards 9 mm thick nailed or screwed to joists through 75 × 9 mm New Tacfire cover fillets, and overlaid with 40 mm rock wool mat (density 60 kg/m^3)	1 hour

* See Figure 2.3

Table 2.4 Treatments for upgrading the fire resistance of an existing floor construction of plain-edge floorboards, 22 mm thick on timber joists not less than 50 mm thick; wood lath and plaster ceiling 16 mm thick

	Upgrading treatment	*Resulting fire resistance*
1*	Plasterboard 12.5 mm thick nailed to joists through existing ceiling	½ hour
2*	Hardboard sheet 3 mm thick nailed to floorboards. Two Supalux strips 9 mm thick × 50 mm deep fixed to each side of joists with nails. 12 mm Supalux boards laid on top of support strips	½ hour
3*	Plasterboard 9.5 mm thick nailed to joists through existing ceiling with gypsum plaster finish 9 mm thick	1 hour
4*	Hardboard sheet 4.8 mm thick nailed to floorboards. Supalux board 12 mm thick screwed to joists through existing ceiling	1 hour

* See Figure 2.4

Figure 2.1
Upgrading the fire resistance of timber floors

Existing floor construction: plain edge floorboards 22 mm thick on timber joists not less than 38 mm thick. No ceiling		
Detail	Upgrading treatment	Fire resistance
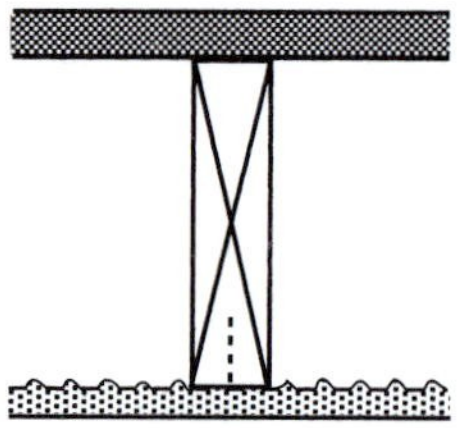	Expanded metal lathing nailed to joists with vermiculite gypsum plaster finish 12.5 mm thick	½ hour
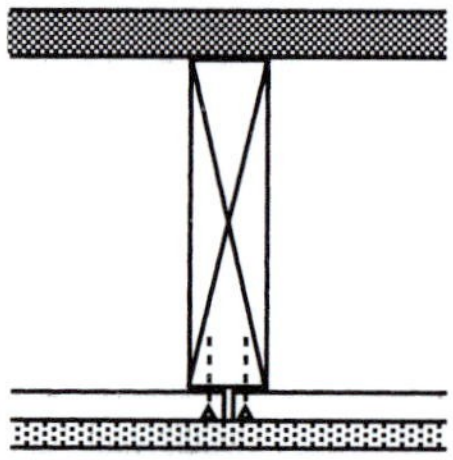	Plasterboard 12.5 mm thick nailed to joists with gypsum plaster finish 12.5 mm thick	½ hour
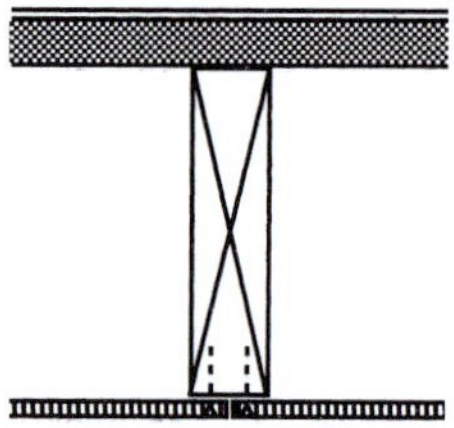	Hardboard sheet 3 mm thick nailed to floorboards Masterboard 6 mm thick nailed to joists	½ hour
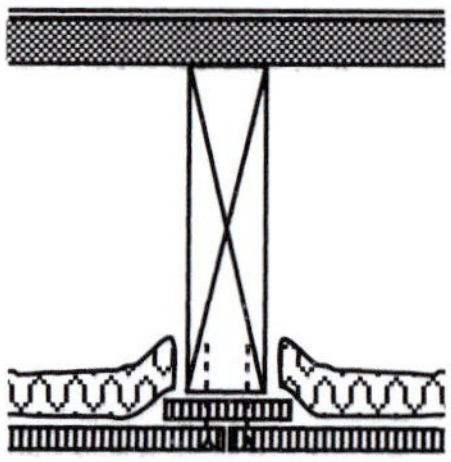	Hardboard sheet 3 mm thick nailed to floor-boards Supalux board 9 mm thick screwed to joists through 80×9 mm thick Supalux fillet overlaid with 60 mm mineral wool mat (density 23 kg/m^3)	1 hour
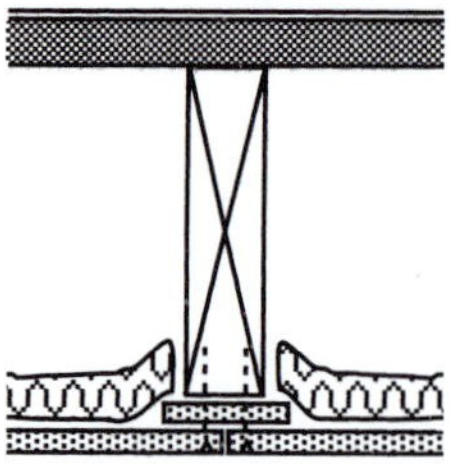	Hardboard sheet 3 mm thick nailed to floor-boards New Tacfire board 9 mm thick nailed or screwed to joists through 75×9 mm New Tacfire cover fillets and overlaid with 40 mm rock wool mat (density 60 kg/m^3)	1 hour

Existing floor construction: tongue-and-groove floorboards not less than 19 mm thick on timber joists not less than 38 mm wide. No ceiling		
Detail	Upgrading treatment	Fire resistance
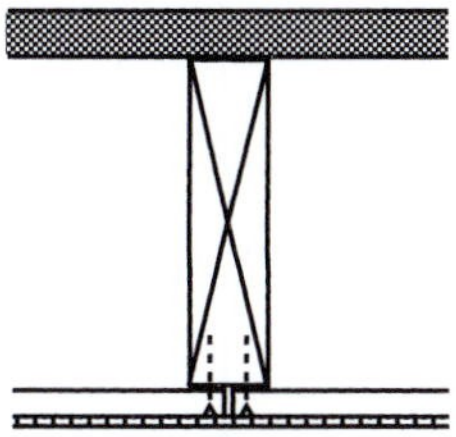	Plasterboard 12.5 mm thick nailed to joists with gypsum plaster finish 5 mm thick	½ hour
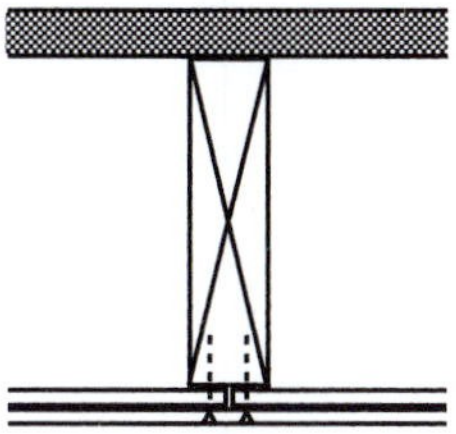	2 layers plasterboard nailed to joists with joints staggered Total thickness 22 mm	½ hour
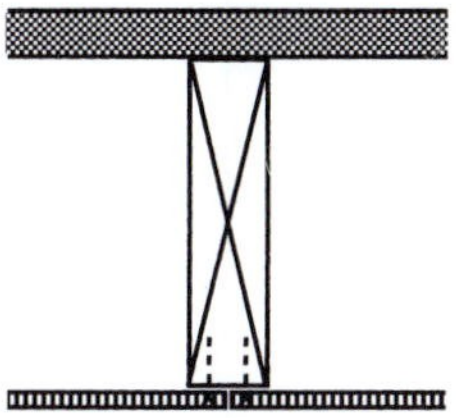	Masterboard 6 mm thick nailed to joists	½ hour
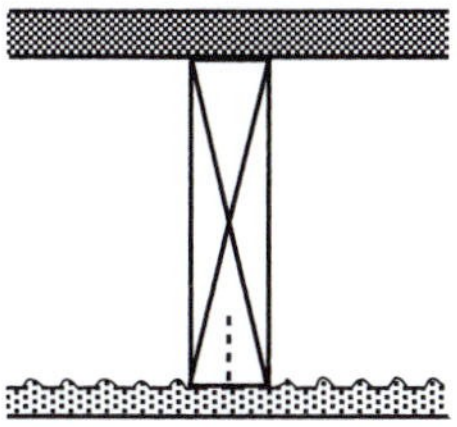	Expanded metal lathing nailed to joists with vermiculite-gypsum plaster finish 12.5 mm thick	1 hour
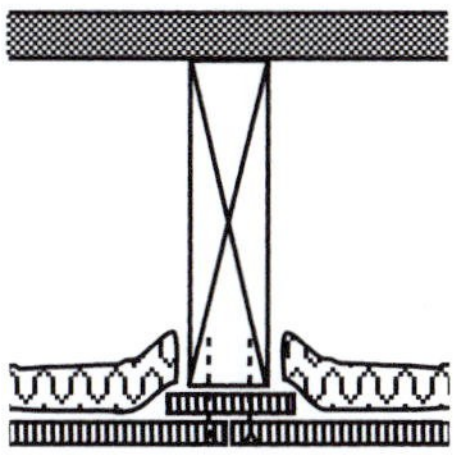	Supalux board 9 mm thick screwed to joists through 80 x 9 mm thick Supalux fillet. Overlaid with 60 mm mineral wool mat (density 23 kg/m^3)	1 hour

Figure 2.2
Upgrading the fire resistance of timber floors

Figure 2.3
Upgrading the fire resistance of timber floors

Existing floor construction: tongue-and-groove floorboards not less than 22 mm thick on timber joists not less than 175 × 50 mm. No ceiling		
Detail	Upgrading treatment	Fire-resistance
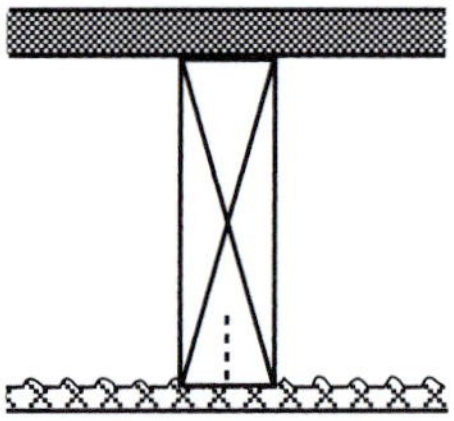	Expanded metal lathing nailed to joists with Sprayed Limpet Mineral Wool–GP Grade finish 13 mm thick	½ hour
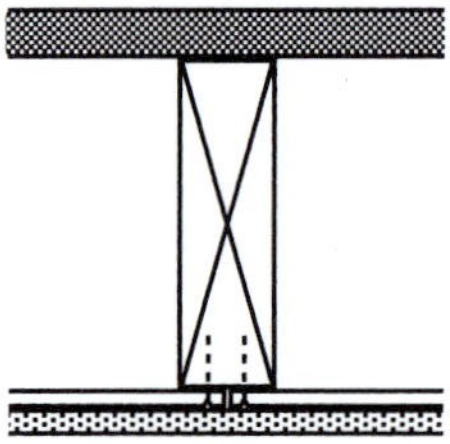	Plasterboard 9.5 mm thick nailed to joists with gypsum plaster finish 12.5 mm thick	½ hour
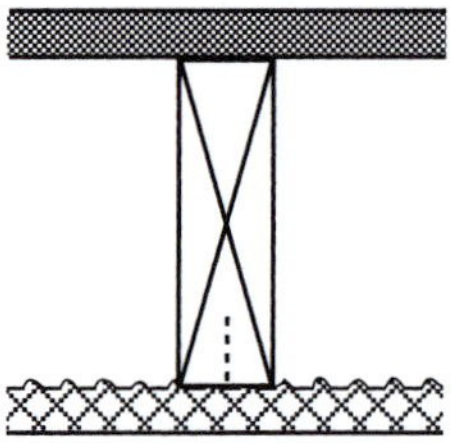	Expanded metal lathing nailed to joists with Sprayed Limpet Mineral Wool–GP Grade finish 22 mm thick	1 hour
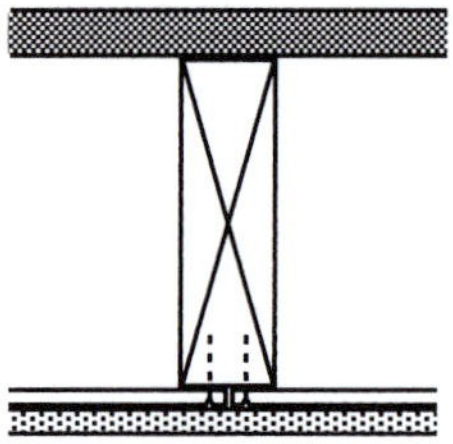	Plasterboard 9.5 mm thick nailed to joists with vermiculite gypsum plaster finish 12.5 mm thick	1 hour
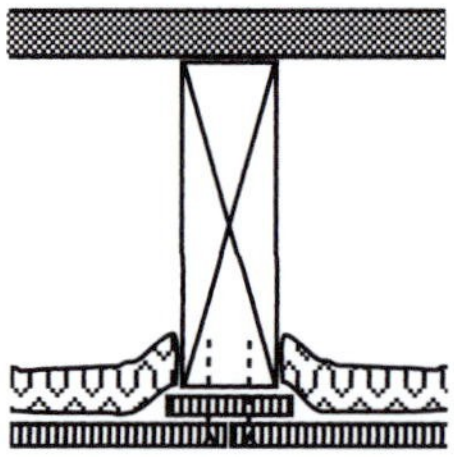	Supalux board 9 mm thick screwed to joists through 80 x 9 mm Supalux fillets. Overlaid with 60 mm mineral wool mat (density 23 kg/m^3)	1 hour

Existing floor construction: plain edge floorboards 22 mm thick on timber joists not less than 50 mm wide. 16 mm lath and plaster ceiling		
Detail	Upgrading treatment	Fire-resistance
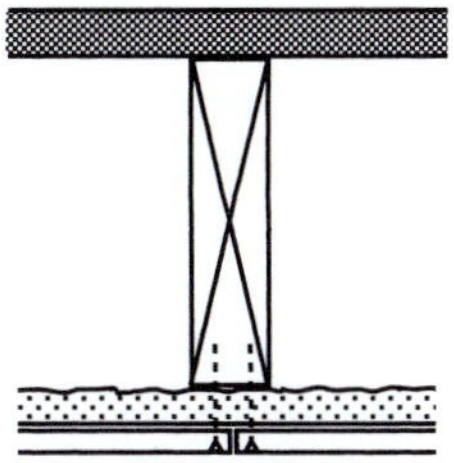	Plasterboard 12.5 mm thick nailed to joists through existing ceiling	½ hour
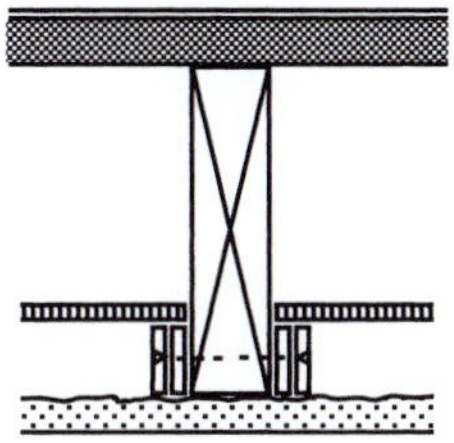	Hardboard sheet 3 mm thick nailed to floorboards 2 Supalux strips 9 mm thick × 50 mm deep fixed to each side of joists with nails 12 mm Supalux boards laid on top of support strips	½ hour
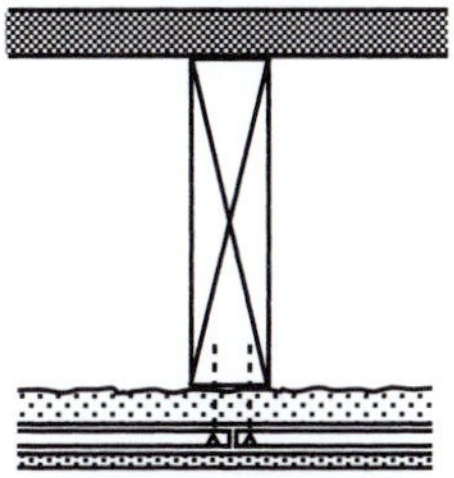	Plasterboard 9.5 mm thick nailed to joists with gypsum plaster finish 9 mm thick	1 hour
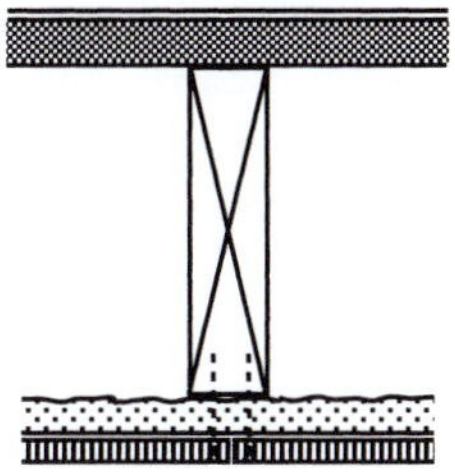	Hardboard sheet 4.8 mm thick nailed to floorboards Supalux board 12 mm thick screwed to joists through existing ceiling	1 hour

Figure 2.4
Upgrading the fire resistance of timber floors

Insertion of a new fire-resisting material into the void between the existing floorboards and ceiling

Intumescent materials

Intumescent materials are applied to surfaces in very thin coatings that, when exposed to fire, undergo a chemical reaction that protects the material to which they have been applied from intense heat. A detailed description of the process is given in Figure 2.8.

Nullifire System W (www.nullifire.com) is a proprietary intumescent material designed specifically for upgrading the fire resistance of timber elements. The waterborne, intumescent coating has a high flame-retardant content that provides spread of flame protection to natural timber surfaces, delaying the onset of charring of the timber by an average of 17 minutes. With a thickness of less than 1 mm, the profile of the timber is unaltered. In contact with heat, the intumescent coating expands, forming a non-combustible insulation barrier, delaying the contact of the heat source and adding to the inherent fire-resistant character of the existing timbers.

It is possible to upgrade existing timber and mouldings to be preserved in refurbishment work. Floors with exposed joists can have their fire rating increased to half-hour fire resistance. The natural look of the timber is largely unaffected as the top seal can be clear (matt or satin). Intumescent basecoats pigmented in any of the full BS 4800 colour range can be used and applied with brush, roller or spray to give the required decorative finish.

The key feature of many of the fire-resisting methods described above is that where a ceiling finish already exists, it will be covered, and therefore concealed, by the new fire-resisting layer applied to its underside. It is quite likely, particularly in older buildings, which have been listed as being of historic or architectural importance, that some ceilings may themselves be protected from alteration, for example if they comprise ornate historic plasterwork. In such cases, upgrading of fire resistance by the addition of a new layer beneath the existing ceiling would not be permitted under any circumstances and, therefore, some other upgrading technique would have to be employed. The addition of a fire-resisting layer to the floor surface above the ceiling is generally inappropriate, as the main protection should be to that part of the floor structure beneath the floorboards, and thus, the only means of providing the required fire resistance is by inserting a new fire-resisting material into the void between the existing floorboards and ornate ceiling.

A proprietary material specifically developed for this purpose is Tilcon Foamed Perlite. This is a site-mixed material, produced from expanded perlite lightweight aggregate, a water-based aerated foam, inorganic hydraulic binders and special additives, that can be pumped into the floor void to upgrade the fire resistance of existing timber floors. The material is produced on site, using a suitable mixer and foam generator, and pump-injected directly into the void after the removal of selected floorboards. The foamed perlite, which sets and cures to a solid light-grey honeycombed matrix, bonds itself to the sides of the joists and is further held in place by metal brackets.

Official fire tests have shown that existing lath and plaster ceilings, even where they have heavy ornate mouldings, give less than half an hour fire resistance. The insertion of Tilcon Foamed Perlite into the floor cavity, to a depth of 175 mm, will increase the fire resistance of such a floor to one hour (load-bearing capacity, integrity and insulation). The upgrading of an existing timber floor using this technique is illustrated in Figure 2.5.

Figure 2.5
(facing page) Upgrading the fire resistance of timber floors

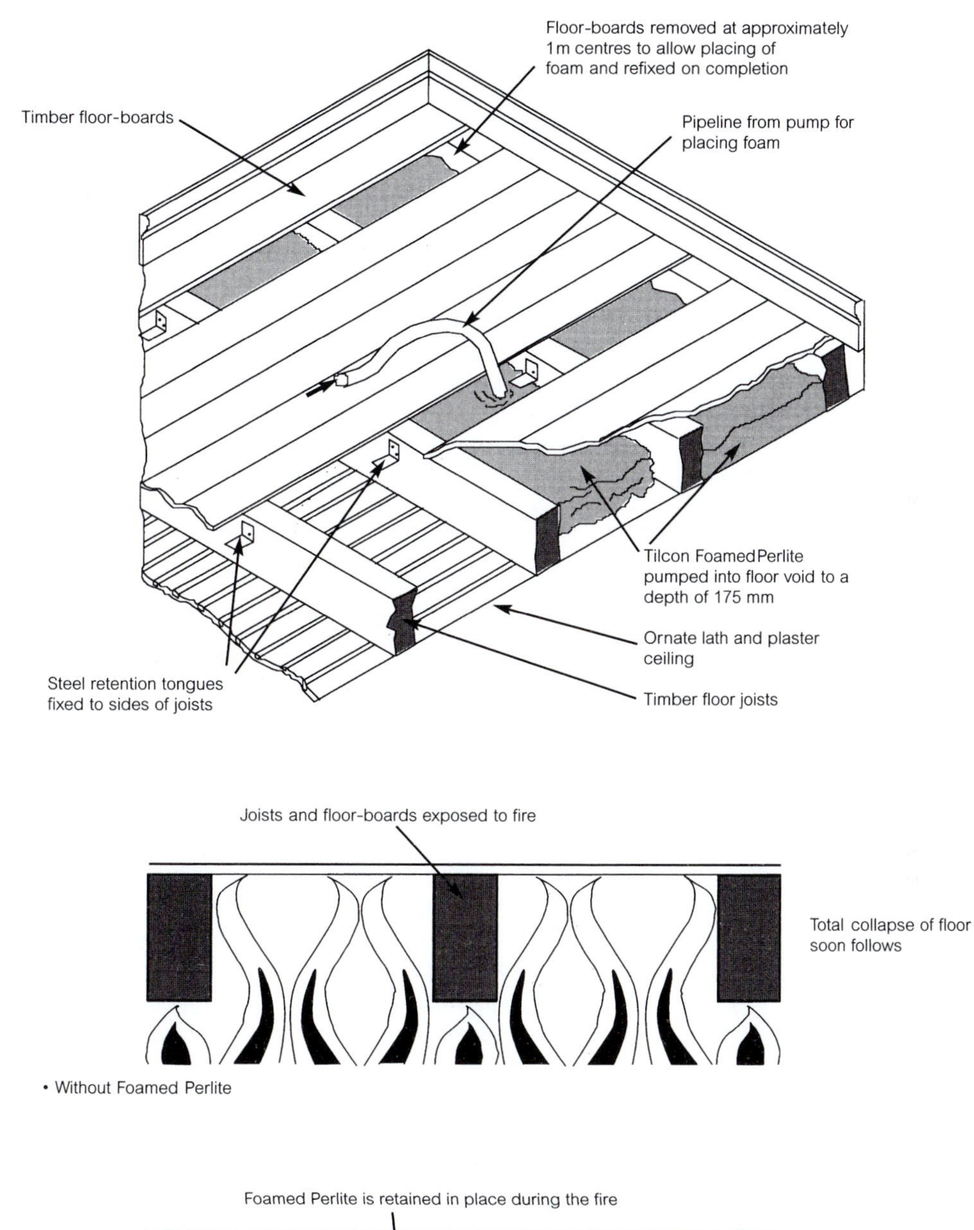

• Without Foamed Perlite

Foamed Perlite is retained in place during the fire

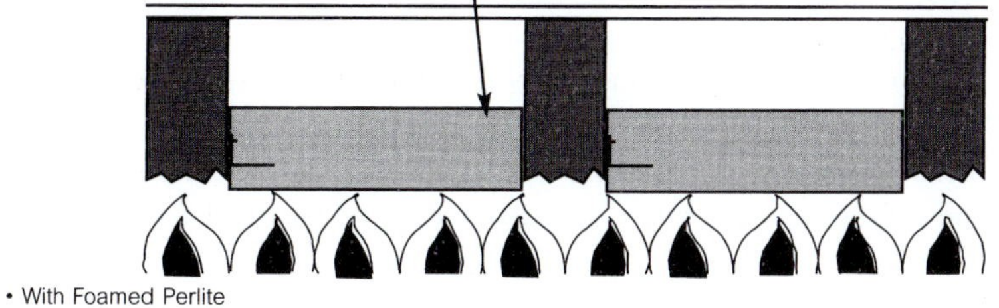

Only a narrow strip of joists is exposed to the fire and the structure is protected

• With Foamed Perlite

Foamed perlite can, of course, be used to upgrade any type of timber floor with a fillable void between the floor surface and ceiling, and the material can also be used for a range of other fire-resisting applications. Further advantages of this method are that scaffolding access to the underside of existing floors, needed with other methods, is not required, and existing light fittings are not disturbed.

2.5 Upgrading the fire resistance of wrought-iron, cast-iron and steel elements

It should be noted that the upgrading techniques and specifications are provided for guidance only, and that the product manufacturers should be consulted for detailed specifications and instructions.

Although wrought and cast iron and steel are non-combustible materials, they are at least as vulnerable as timber to building fires; indeed large-section timber beams are often better at withstanding fire thanks to natural, protective sacrificial charring of their outer layers, which offers protection from further damage. Wrought and cast iron and steel, while being incombustible, are, nevertheless, highly vulnerable to fire, sustaining considerable strength loss and distortion at relatively low temperatures in the evolution of a typical building fire. It should also be noted that wrought and cast iron are less predictable than steel when exposed to fire because of their inherent susceptibility to cracking.

As stated earlier, a large numbers of old buildings that are suitable for refurbishment and alteration contain exposed structural beams and columns of iron or steel, and it is often convenient and economical, particularly in the lower-key categories of refurbishment, to retain these elements. However, as with timber floor structures, the fire resistance of exposed iron and steel work does not approach the standards that are required today, and some form of upgrading will therefore be necessary. In many of our older industrial buildings, structural frames comprising ornate, circular cast-iron columns and wrought-iron main beams support timber secondary beams and floor-boarding. It is often desirable to retain the shape and form of the existing columns as an architectural feature in the refurbished building. In such cases, the materials used in upgrading fire resistance must be capable of application in the form of a very thin coating if the shape and form of the existing sections are to be preserved.

A wide range of techniques and materials are available for upgrading the fire resistance of exposed iron and steel structural elements and all of them are suitable for 'I'-section beams and columns. The techniques available for upgrading circular columns, however, are more limited, since some of the materials used are not capable of being applied to circular sections. The techniques and materials used vary considerably in their application and use and fall into four basic categories, which are described below and illustrated in Figures 2.6 and 2.7.

The figures give the fire resistances that can be achieved for various thicknesses of protection, but it should be noted that the data are only approximate and intended to be an indication of requirements for sections of 'typical size' ('I' sections: 406 mm × 178 mm × 74 kg/m; circular sections: 190 mm diameter × 8 mm thick). In practice, the actual thickness of protection required will vary and is based on the ratio of the section's exposed surface area to the area of its cross-section. The higher the surface area for a given cross-sectional area, the more heat will be absorbed and thus the lower the inherent fire resistance of the section. The actual thickness of the protection required will therefore increase as the ratio of surface area to cross-sectional area increases, and it is necessary to refer to manufacturers' tables, or to use calculations, in order to arrive at the actual thickness of protection required for any particular section. It should also be noted that the thickness of fire protection required for beams is usually less than that required for columns, because beams are exposed to fire only on three sides, whereas columns are normally exposed on all four sides; this is applicable only where the top flange of the beam is protected by a dense concrete floor slab of at least 100 mm thick.

Solid encasement

Although other materials can be used, solid encasement normally involves casting *in situ* concrete around the sections being upgraded. To obtain a good bond between the concrete and the steel or iron elements, steel mesh wrapping fabric is applied to the sections before erecting the temporary formwork and pouring the concrete. The minimum concrete cover to sections being protected must be 25 mm to allow for the maximum coarse aggregate size, and this will give 2 hours' fire resistance to any steel or cast-iron section. Good-quality control in the production of the concrete is essential since its performance in fire can vary considerably depending on its density, moisture content and the aggregates used.

When considering the use of *in situ* concrete to upgrade existing cast-iron and steel elements, it should be borne in mind that it is a messy operation, and that erection of temporary formwork within the existing building, together with the provision of adequate access into the building for the mixed concrete, may cause considerable difficulties.

Lightweight hollow encasement

Hollow encasement techniques essentially involve 'boxing-in' the columns or beams being upgraded, with fire-resisting materials of various types. Several different materials are available, and a number of them are described below.

Expanded metal lathing and plaster

Expanded metal lathing is wrapped around the column or beam to form the key for a wet plaster finish. The lathing can be fixed to the steel section using wire clips or steel stirrups, or by spot-welding. With 'I'-section columns it is advisable to fix metal angle beads at the arrises (internal angles) to provide additional protection against mechanical damage. Where this method is used for circular columns, the metal lathing follows the column profile, and the protection does not, therefore, form a hollow casing (see Figure 2.6). Fire resistances of up to one and a half hours can easily be obtained using this method, and the thickness of plaster required can be reduced if lightweight vermiculite–gypsum plaster is used. The fire resistances obtainable using this method are given in Figure 2.6.

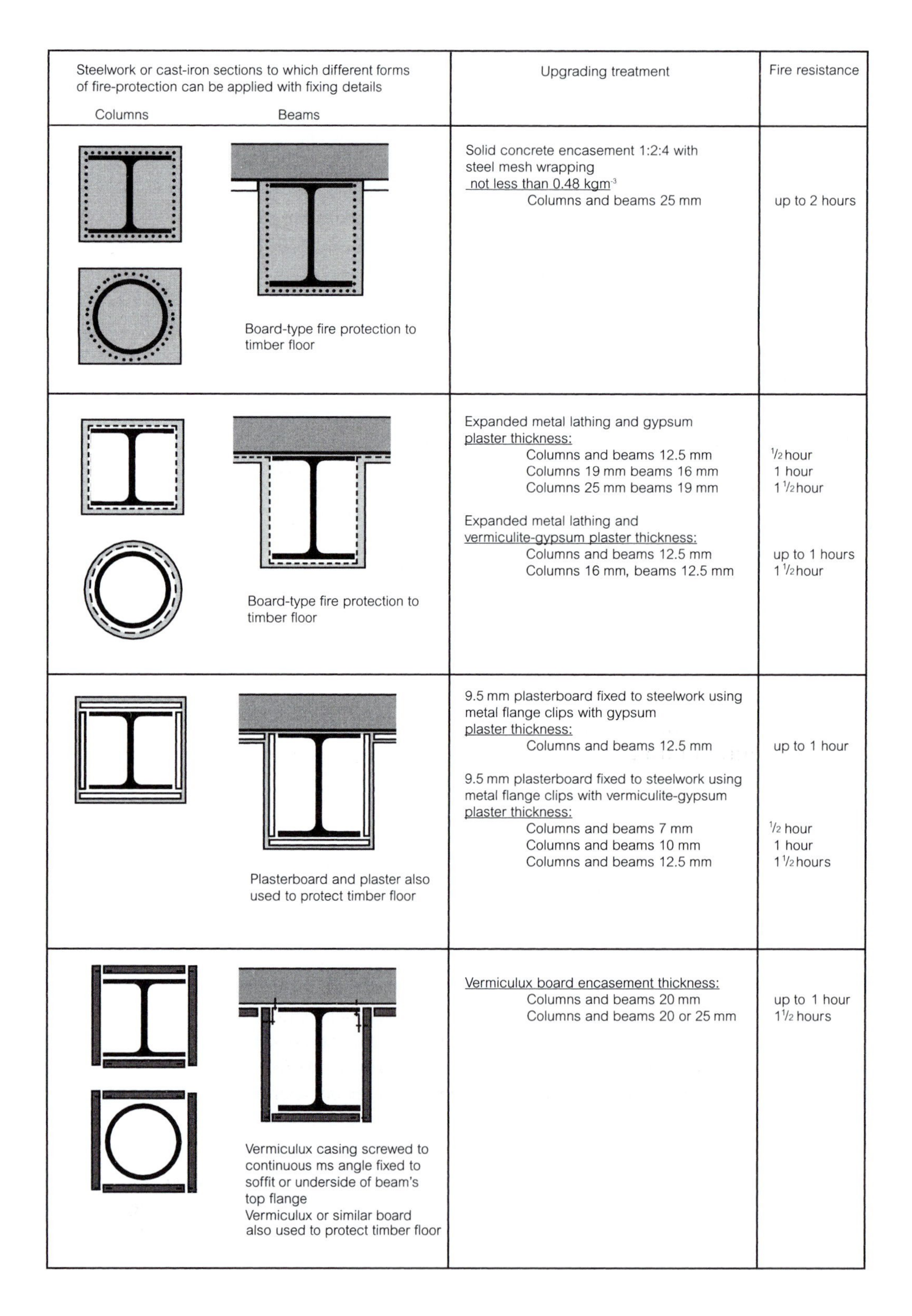

Steelwork or cast-iron sections to which different forms of fire-protection can be applied with fixing details Columns Beams	Upgrading treatment	Fire resistance
Board-type fire protection to timber floor	Solid concrete encasement 1:2:4 with steel mesh wrapping not less than 0.48 kgm^{-3} Columns and beams 25 mm	up to 2 hours
Board-type fire protection to timber floor	Expanded metal lathing and gypsum plaster thickness: Columns and beams 12.5 mm Columns 19 mm beams 16 mm Columns 25 mm beams 19 mm Expanded metal lathing and vermiculite-gypsum plaster thickness: Columns and beams 12.5 mm Columns 16 mm, beams 12.5 mm	 ½ hour 1 hour 1 ½ hour up to 1 hours 1 ½ hour
Plasterboard and plaster also used to protect timber floor	9.5 mm plasterboard fixed to steelwork using metal flange clips with gypsum plaster thickness: Columns and beams 12.5 mm 9.5 mm plasterboard fixed to steelwork using metal flange clips with vermiculite-gypsum plaster thickness: Columns and beams 7 mm Columns and beams 10 mm Columns and beams 12.5 mm	 up to 1 hour ½ hour 1 hour 1 ½ hours
Vermiculux casing screwed to continuous ms angle fixed to soffit or underside of beam's top flange Vermiculux or similar board also used to protect timber floor	Vermiculux board encasement thickness: Columns and beams 20 mm Columns and beams 20 or 25 mm	 up to 1 hour 1½ hours

Plasterboard encasement

This was the first material to be used for lightweight hollow encasement and it is still by no means uncommon in upgrading work. Lengths of 9.5 mm thick plasterboard are cut to size and fixed to the steel sections by means of special metal flange clips. The plasterboard is then finished with a single coat of wet plaster, its thickness depending on the degree of fire resistance required. Because of the technique used to secure the plasterboard casing to the steelwork, this method cannot be used with circular-section columns. As with the previous method, longer periods of fire resistance can be obtained if vermiculite-gypsum plaster is used (see Figure 2.6).

Vermiculux board encasement

Vermiculux is a proprietary (www.promat.co.uk), low-density board material manufactured from exfoliated vermiculite, other non-organic fibres and fillers and a calcium-silicate matrix. The boards (standard size 1,220 × 610 mm and 1,220 × 1,220 mm, in thicknesses from 20 to 60 mm) are cut to size on site and fixed around the steel sections by edge-screwing. Opposite edges of boards are rebated to allow lapped cross-joints to be made without the need for noggings or backing strips. Adhesives are not necessary, and the iron or steelwork needs no cleaning or priming. The completed casing can be finished (after making good countersunk screw fixings, junctions, etc., by filling and sanding) by direct painting, papering or tiling. Alternatively, a wet-skim plaster finish can be applied. Where Vermiculux is used to encase beams beneath an existing floor, it is first necessary to fix continuous mild steel angles to the soffit or beam flange by shot-firing, as shown in Figure 2.6, to enable the boards enclosing the beam sides to be secured. Figure 2.6 also shows the fire resistances that can be achieved in constructions using this material.

Supalux board encasement

Supalux board (www.promat.co.uk), previously described, is too thin to be edge-screwed and is fixed around the members being upgraded by being screwed to continuous steel angle. The steel angle forms a framework around the section. For top flanges of beams the angle framing is shot-fired either to the flange or the floor soffit, as shown in Figure 2.7. Unlike Vermiculux, the Supalux board edges do not have rebates, and thus joints between different lengths of boards require the insertion of Supalux backing strips, to which the ends of the board lengths being jointed are screwed (see Figure 2.7). Fire resistances of up to two hours can be obtained with Supalux constructions, and the material can be finished as for Vermiculux. The fixing methods used for Supalux encasement rule out its application to circular-section columns.

Vicuclad board encasement

Vicuclad (www.promat.co.uk) is a rigid monolithic board material produced from exfoliated vermiculite and inorganic binders. The board has a smooth surface and

Figure 2.6
(facing page) Upgrading the fire resistance of iron and steel beams and columns

Steelwork or cast-iron sections to which different forms of fire-protection can be applied with fixing details Columns Beams	Upgrading treatment	Fire resistance
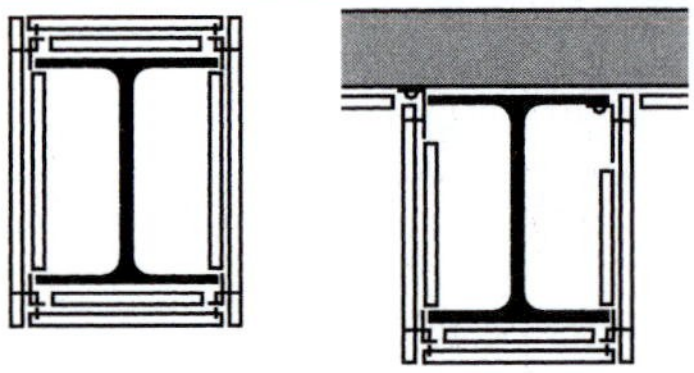 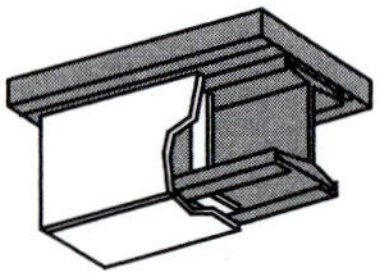 Suplax casing screwed to continuous steel angle frame-work fixed around the section. All transverse joints backed by 75 mm wide suplax backing strips in same thickness as casing. Suplax board also used to protect timber floor	Suplalux board encasement thickness: Columns and beams 6 mm Columns and beams 9 mm Columns 12 mm	½ hour 1 hour ½ hour 2 hours
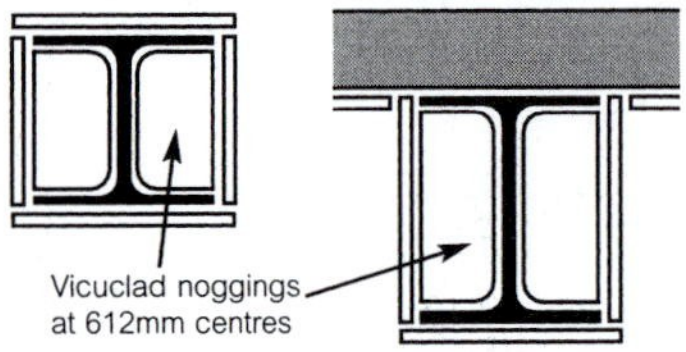 Vicuclad noggings at 612mm centres Vicuclad noggings fixed behind all joints with cement into web of steel section at 612 mm centres. Vicuclad casing fixed with cement to flanges and to noggings. Edge nailing using galvanised nails at all joints. Alternative fixing by screwing to steel angle fixed around existing sections	Vicuclad board encasement thickness: Columns and beams 18 mm Columns and beams 18–30 mm Columns and beams 30–60 mm Columns and beams 45–80 mm	up to 1 hour 1½ hours 3 hours 4 hours
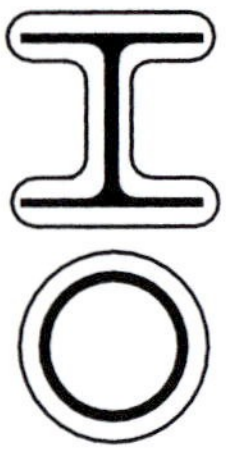 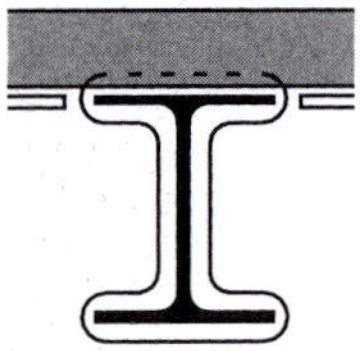 Board-type fire protection to timber floor	Sprayed Limpet mineral wool - GP Grade Columns and beams 10 mm Columns and beams 12–14 mm Columns and beams 23–25 mm	½ hour 1 hour 1½ hours
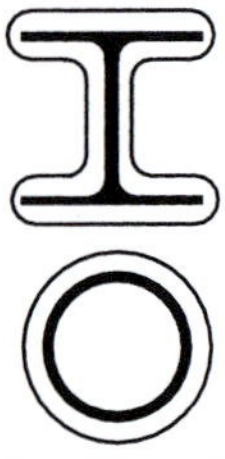 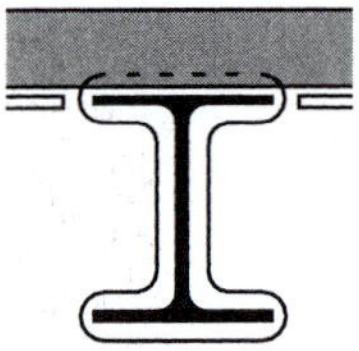 Board-type fire protection to timber floor	Mandolite CP2 sprayed vermiculite - cement thickness: Columns and beams 8–9 mm Columns and beams 13–15 mm Columns and beams 18–20 mm	½ hour 1 hour 1½ hours
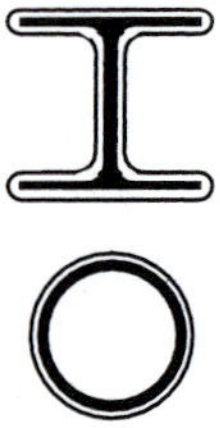 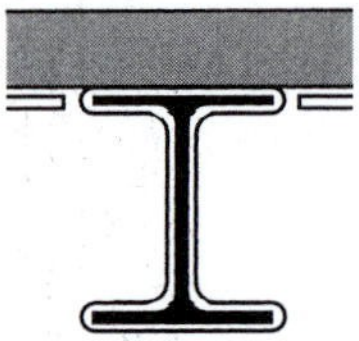 Board-type fire protection to timber floor	Nullifire intumescent coating applied by brush roller or spray Thickness can range from 0.3 mm to 2.0 mm dependent on HP/A calculation and type of section	½ hour to 2 hour

is oatmeal in colour. The boards, available in two grades – 900R (1,000 × 610 × 18–40 mm thick) and 1050R (1,000 × 610 × 45–80 mm thick) – are capable of upgrading the fire resistance of steel columns and beams by up to 4 hours.

The boards are cut to size on site and fixed to the existing sections either using a non-combustible cement, Vicuclad noggings and edge-nailing, as shown in Figure 2.7, or they can be fixed by screwing to galvanised steel angle fixed around the existing sections. Vicuclad can be plastered, tiled, painted, papered (after applying a plaster skim) or finished with a proprietary decorative coating such as Artex.

Spray-applied coatings

The majority of spray-applied materials are sprayed directly onto the surfaces of the beams and columns being upgraded and they therefore follow the existing sections' profiles.

Sprayed Limpet Mineral Wool – GP grade

Sprayed Limpet Mineral Wool (www.thermica.co.uk) is proprietary material, previously described. It is sprayed directly onto the cleaned surface of the section being upgraded in one continuous application, until the desired thickness has been obtained. The thicknesses required for various periods of fire resistance are given in Figure 2.7. Sprayed Limpet Mineral Wool – GP grade can also be used to provide a hollow casing by being sprayed onto expanded metal lathing that has been wrapped around the steel sections.

Sprayed vermiculite-cement

Pre-mixed vermiculite and Portland cement, to which water is added on site, have been previously used to provide fire protection. After mixing, the vermiculite-based render is sprayed directly onto the sections being upgraded, which must be clean and free of any surface impurities that might prevent adhesion. The render or concrete, which is sprayed into position, can be built up to the required thickness by spraying over a series of passes of one or more coats. After drying, the render forms an off-white textured surface that, if required, may be finished with a top-coat paint. Figure 2.7 gives the fire resistances that can be obtained using Mandolite CP2, which is a vermiculite–Portland cement concrete used for fire protection. This method of fire protection used to be quite common, but has reduced with intumescent paints becoming much more popular.

Intumescent coatings

Intumescent materials, which are applied in very thin layers, have unique fire-resisting properties and comprise formulations of resinous binders, pigments, blowing agents and fillers.

These ingredients are stable and unreactive *in situ* at ambient temperatures and do not degrade with time. However, at high temperatures, as in a building fire, the ingredients undergo a well-defined chemical reaction that produces an expanded,

Figure 2.7
(facing page) Upgrading the fire resistance of iron and steel beams and columns

three-dimensional 'meringue-like' char, which has a volume many times that of the original thin-layered coating. The char has low thermal conductivity, giving it good insulating characteristics against the heat and damaging effects of a building fire. The Nullifire S range, S605, S606 and S706 (www.nullifire.com) are typical examples of proprietary intumescent paints; they consist of an intumescent solvent-based basecoat and a decorative topseal coat, which is available in any of the full BS 4800 colour range. The basecoat provides the fire protection, and the topseal, in addition to providing a decorative surface, protects the basecoat from mechanical damage and gives a 'wipe clean' surface. Both the intumescent basecoat and the decorative topseal coat can be applied by brush, roller or spray. Prior to application of the intumescent basecoat, the existing sections being upgraded should be coated with a suitable primer.

One of the principal advantages of using intumescent materials for upgrading existing iron and steelwork is that they are extremely thin (only 1 mm for one-hour fire resistance), and they are therefore ideal where, for architectural reasons, the existing profiles of ornate sections need to be preserved.

Nullifire also produce S607 and S707. These are water-based, spray-grade versions of the S range. Fire ratings for intumscent paints range from 30 to 120 minutes.

Figure 2.7 indicates the fire upgrading obtainable using Nullifire intumescent protection, and Figure 2.8 shows how the process of intumescence takes place.

2.6 Upgrading the fire resistance of doors

Where an existing building is improved or undergoes a change of use, it is likely that some of the existing internal doors will need either to be replaced or upgraded to comply with the requirements of the Building Regulations. Typical locations where fire doors are required include:

- doors separating flats or maisonettes from spaces in common use;
- doors penetrating protecting structures (i.e., fire-resisting enclosures to stairwells, lift shafts, etc.);
- doors penetrating compartment walls (i.e., fire-resisting walls used to subdivide a building into compartments in order to restrict fire spread).

Fire doors will usually need to have a fire resistance of 1 hour, 30 minutes or 20 minutes (measured against the criterion of 'Integrity'; see Section 2.3), and most types of existing door construction are capable of being upgraded to twenty minutes or half an hour with relative ease. However, upgrading an existing door to 1-hour standard is more difficult, and often produces a rather cumbersome result. Where 1-hour fire doors are required, therefore, it is usually preferable to replace the existing doors rather than to attempt to upgrade them.

The techniques used to upgrade the fire rating of existing timber-panelled doors are relatively simple and inexpensive. Several proprietary fire-resisting board materials are available in various thicknesses, and these can be nailed or screwed to the existing door's surface to achieve the standard required. Supalux board (www.promat.co.uk), previously described, is used for a number of fire-upgrading

Figure 2.8
(facing page) The process of intumescence

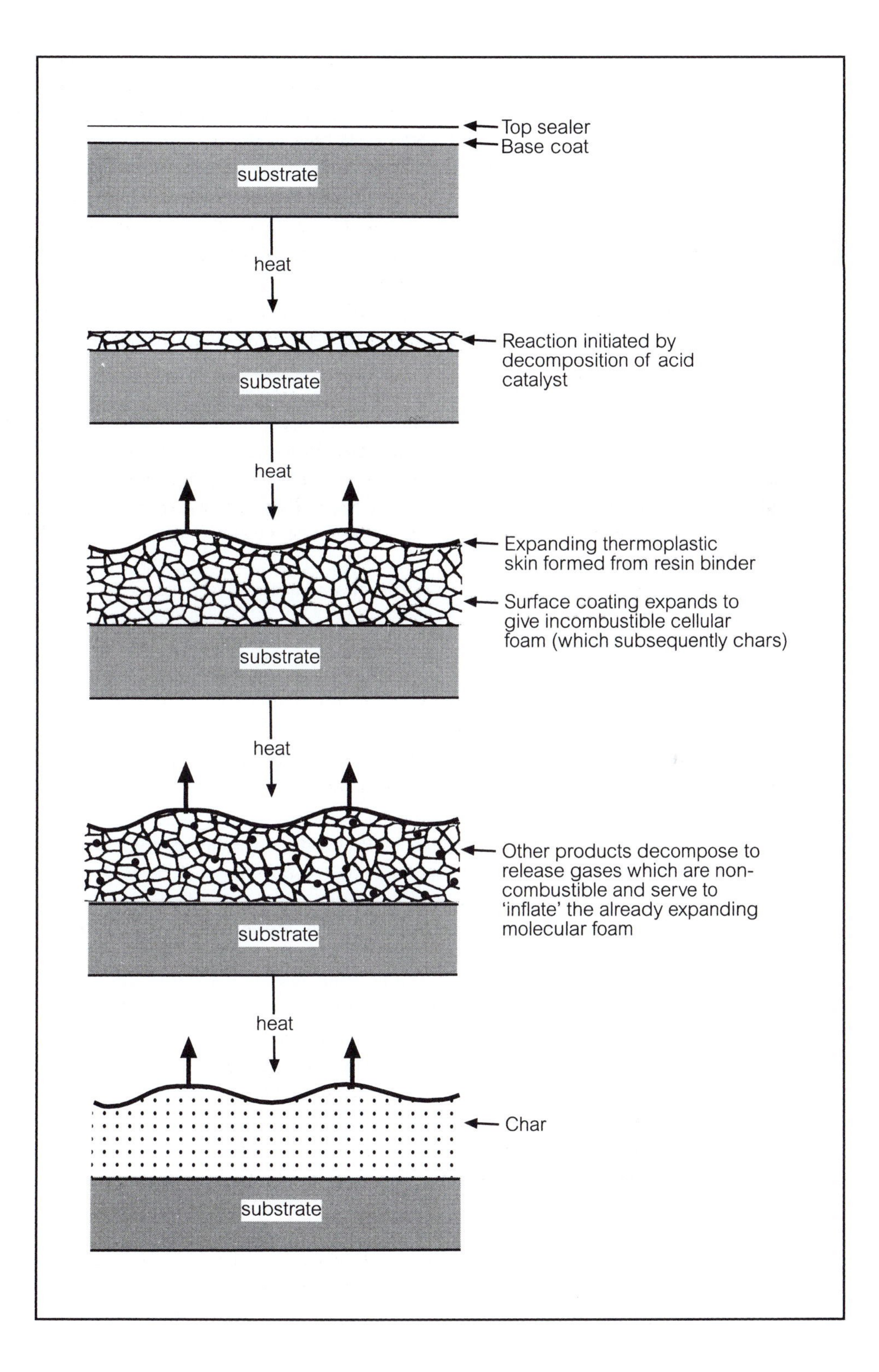
Top sealer
Base coat
substrate
heat
Reaction initiated by decomposition of acid catalyst
substrate
heat
Expanding thermoplastic skin formed from resin binder
Surface coating expands to give incombustible cellular foam (which subsequently chars)
substrate
heat
Other products decompose to release gases which are non-combustible and serve to 'inflate' the already expanding molecular foam
substrate
heat
Char
substrate

Detail	Upgrading treatment	Fire resistance
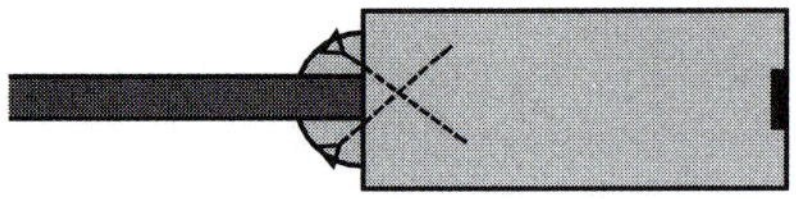	Existing door: Panelled door Minimum 40 mm thick Timber panels replaced with 12 mm Masterboard or 12 mm Supalux set in intumescent mastic and held in place with 11 mm hardwood quadrant beads skew nailed at 200 mm centres. Intumescent strips: 15 × 4 mm thick in both vertical edges and top edge of door. Door stop 12 mm deep.	½ hour
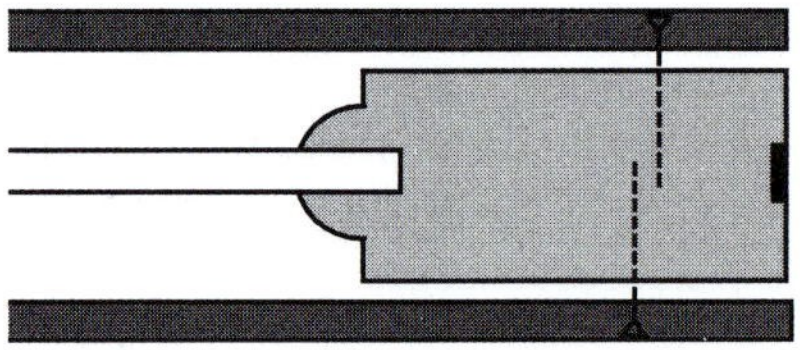	Existing door: Panelled door Minimum 44 mm thick with minimum 6 mm thick panels 6 mm Masterboard or 6mm Supalux screwed to both faces around perimeter and across mid-rail at 150 mm centres. Intumescent strips: 15 × 4 mm thick in both vertical edges and top edge of door. Door stop 12 mm deep.	½ hour

Figure 2.9
Upgrading the fire resistance of doors

applications, and its use in the upgrading of existing doors is illustrated in Figure 2.9. After fixing to the existing door, all screw/nail holes are filled, and the new Supalux surface is painted to complete the upgrading of the door. It will be seen from Figure 2.9 that, in addition to treating the surface of the door, it is also necessary to insert intumescent strips along the edges in order to obtain the required fire rating. In the event of fire, the intumescent strips expand and seal the gaps around the door's edge, preventing the passage of smoke and flames. A further aspect requiring attention is the existing doorstops, which must be at least 12 mm deep.

As an alternative to the above methods, panelled doors can be upgraded to twenty minutes' fire resistance using Nullifire System W (www.nullifire.com) intumescent basecoat and topseal. However, for panelled doors to be upgraded to twenty minutes' fire resistance they should have minimum thicknesses of solid timber as follows:

- **all rails and stiles** **35 mm**
- **panels** **8 mm**
- **maximum gap door-frame** **3 mm**

Intumescent strips along the door edges are also required, as for the other upgrading methods described above.

2.7 Upgrading the fire resistance of walls

In the majority of older buildings, the existing walls are normally masonry or brickwork, both of which have excellent fire resistance and are, therefore, unlikely to need upgrading. For example, an unplastered 100 mm thick brick wall will give a fire resistance of two hours, which is more than adequate in virtually all circumstances. It is quite possible, however, that some of the internal walls within an existing building represent more recent additions that may be of less substantial construction, such as concrete blockwork or timber studding. Unplastered block walls, 75 mm and 100 mm thick, will give a fire resistance of one hour and two hours, respectively, and are therefore, like brickwork and masonry, unlikely to need upgrading for fire-protection purposes. Many basic timber stud partitions, however, only give half-hour fire resistance and therefore need upgrading if a higher standard is required. One of the simplest ways of upgrading such partitions is to nail an additional layer of 9.5 mm plasterboard to each side, which will give a rating of one hour. If greater periods of fire resistance are required, a wet plaster finish, using vermiculite-gypsum plaster if necessary, may be added, or thicker plasterboard used.

References

Building Research Establishment (1984) *Increasing the Fire-resistance of Existing Timber Floors* (Digest 208), 2nd edn (revised), Watford: BRE.

Department for Communities and Local Government (2006) *The Building Regulations,* Approved Document B (Fire safety) – *Volume 1: Dwellinghouses* (2006 edn), London: HMSO (www.planningportal.gov.uk, accessed 23 July 2007).

Department for Communities and Local Government (2006) *The Building Regulations,* Approved Document B (Fire safety) – *Volume 2: Buildings Other than Dwellinghouses* (2006 edn), London: HMSO (www.planningportal.gov.uk, accessed 23 July 2007).

Department of the Environment, Transport & the Regions (1999) *Manual to the Building Regulations*, London: DETR.

Emmitt, S. and Gorse, C. (2006) *Advanced Construction of Buildings*, Oxford: Blackwell Publishing.

Powell Smith, V., Billington, M. J. and Waters, J. R. (2007) *The Building Regulations Explained and Illustrated*, 13th edn, Blackwell Science, Oxford.

Stephenson, J. (2004) *Building Regulations Explained*, London: E. & F.N. Spon.

Urban Task Force (2005) 'Towards a strong urban renaissance: An independent report by the Urban Regeneration Task Force chaired by Lord Rogers of Riverside', Urban Task Force (www.urbantaskforce.org, accessed 7 July 2007).

CHAPTER 3 Upgrading internal surfaces

3.1 General

A highly effective means of improving the interior of an existing building as part of any refurbishment scheme, even in 'low-key' refurbishments, is to upgrade its internal surfaces. If the existing surfaces are in good condition, the necessary upgrading might only involve providing a new coat of paint, or wall covering. On the other hand, in older buildings, and particularly those that have been neglected and/or disused for long periods, the existing surfaces may be in such poor condition that they need to be completely replaced. Where existing surfaces need to be upgraded, it will usually be the walls that involve most work, the upgrading of ceilings and floors tending to be less complicated.

With the increased level of concern regarding the impact of the built environment on the environment and specifically CO_2 emissions, there is a push to upgrade the existing building stock so that it is more energy efficient. There is a move to make buildings more airtight and thermally efficient. While insulation can be added externally it is not always appropriate, especially if the external façade is listed, so efforts to upgrade thermal performance may concentrate on the internal elements of the structure. When upgrading internal surfaces it may be possible to increase the building's energy efficiency by adding insulation or making the finishes more airtight. Internal insulation can be effective, but care should also be taken to ensure that a situation where interstitial condensation can occur is not created.

3.2 Upgrading wall surfaces

In many buildings suitable for refurbishment, such as utilitarian factories, mills, warehouses, churches and agricultural barns, the structural walls may never have received any applied finishes. Such buildings, therefore, often consist of exposed masonry or brickwork that, in its present condition, is unlikely to be of adequate standard in any high-quality refurbishment scheme. Where plaster finishes have been applied they may have deteriorated beyond repair because of penetrating dampness or general neglect, especially where the building has been unoccupied, vandalised or open to the elements over a long period. In such cases, it will therefore be essential to upgrade the existing wall surfaces to satisfy modern standards, and this can normally be achieved, after suitable preparation, either by applying a new plaster finish, or by applying some form of dry lining to the existing wall.

Plaster finishes

Provided the existing wall is not suffering from dampness and is in good condition, a normal plaster finish can be applied. Any existing surface finish should be removed and the brickwork or masonry joints well raked out to provide a suitable key for the new plaster. The wall should then be thoroughly brushed down to remove any dust, efflorescence salt or loose particles before applying the new finish. It is quite possible, particularly in older buildings, that the existing wall surfaces may be so uneven that a normal two-coat plaster finish will not be sufficient to give a true surface. In such cases, the application of two undercoats (a render and a floating coat) will usually be sufficient to fill out the deeper depressions. If this is not possible, then a dry-lining system may be the only solution.

Special plasters – specifically for replastering applications in refurbishment and upgrading work – have been developed, although most of the multipurpose-use and durable plasters achieve similar results nowadays (www.british-gypsum.com). Thistle Renovating plaster was one such product that was designed specifically for renovation work. This plaster was designed to have a greater resistance to efflorescence, which is often present in older buildings. The undercoat plaster is for general replastering applications and, with a final coat of Thistle Renovating Finish, provides a smooth, virtually inert, high-quality surface with earlier surface drying and a higher than normal resistance to efflorescence. Thistle plaster is a lightweight, retarded hemihydrate, premixed gypsum undercoat plaster containing special aggregate and additives. The plaster has a controlled set that eliminates costly waiting time between coats and produces a finish that is free from shrinkage cracks. Where the wall to be replastered is damp, replastering should be delayed as long as possible to allow the background to dry out, and any source of dampness must be identified and eliminated. Any salts brought to the surface of the background must also be carefully removed. Thistle Renovating plaster can be applied where only residual moisture is present, for example after most of the moisture in the wall has dried out following the insertion of a new damp-proof course or other remedial works to prevent laterally penetrating dampness. Low-suction backgrounds, such as engineering bricks or very dense masonry, should be treated prior to plastering with a neat coat of water-resisting bonding aid based on EVA or SBR latex, which should be plastered over while still 'tacky'.

A proprietary plastering system specifically developed for application to walls following the installation of a new damp-proof course or remedial action to prevent penetrating dampness is Thistle Dri-Coat (www.british-gypsum.com). This is a premixed, lightweight cement-based undercoat plaster, incorporating expanded perlite aggregate and containing special additives that resist the passage of efflorescent and hygroscopic salts. When dry it also resists the passage of moisture but does not form a barrier to water vapour, therefore allowing any residual dampness to dry from the building structure. With a final coat of Thistle Board Finish or Thistle Multi-Finish, Thistle Dri-Coat provides a smooth, high-quality, virtually inert surface to walls containing residual dampness. It is essential, however, that the source of rising and/or penetrating dampness has been totally eliminated, and, ideally, the replastering should be delayed as long as possible. As with Thistle Renovating plaster, low-suction backgrounds, such as engineering bricks or very dense masonry, should be treated prior to plastering with a neat coat of water-resisting bonding aid based on EVA or SBR latex, which should be plastered over while still 'tacky'.

Further information can also be obtained from Property Repair Systems (www.dampness-info.co.uk).

Dry linings

Dry linings are a suitable and popular alternative to wet plaster as a means of upgrading wall surfaces in refurbishment work. If the existing wall is suffering from dampness, its source should be eliminated prior to applying the lining, since basic dry-lining systems are not suitable for use on damp backgrounds. It is possible, however, with certain modifications, to use dry linings on damp backgrounds, and this application is considered later. A further useful application for dry linings is in upgrading the thermal properties of external walls and this, also, is dealt with later.

Providing a dry-lining wall finish involves the fixing of gypsum plasterboard to the existing wall surface by one of a number of different techniques. Gyproc WallBoard (www.british-gypsum.com), a plasterboard designed to receive direct decoration, is used, thereby eliminating the need for a wet skim coat of plaster. This gives the advantage of allowing the whole finishing process to be dry, involving no wet operations apart from jointing the board edges. Gyproc WallBoard is available in three standard thicknesses: 9.5, 12.5 and 15 mm. Board widths are 900 or 1,200 mm, and lengths range from 1,800 to 3,600 mm. Where there is a risk of interstitial condensation following installation of the dry lining, Gyproc Duplex WallBoard should be used. This is a Gyproc WallBoard backed with a vapour-control membrane, available in three thicknesses: 9.5, 12.5 and 15 mm. Board widths are 900 or 1,200 mm and lengths range from 1,800 to 3,000 mm.

Three basic fixing techniques are used to install dry linings. Although a dry lining installed solely for the purpose of upgrading the quality of an internal surface finish would use only plain WallBoard, Figures 3.1–3.8 show the use of thermal boards and other methods of incorporating insulation, as the fixing methods and detailing are similar for both.

Timber batten fixing

Timber battens, 50 mm wide × 25 mm thick, are fixed vertically to the existing wall at 400, 450 or 600 mm centres, depending on the width of the boards used. It is likely that the existing wall surface will be uneven, and it is essential, therefore, that all depressions are packed out with timber or fibreboard pieces in order to achieve correct alignment of the battens. The boards are then nailed or screwed to the battens vertically, with 50 × 25 mm timber noggings inserted to support their horizontal edges. After the fixings and joints have been made good, the WallBoards may be decorated directly by painting, wallpapering or using any other suitable surface finish. Figure 3.1 shows details of a timber batten dry-lining system.

Where insulation is placed on the internal face of an external wall, it is important that there is an effective barrier against interstitial condensation forming behind the insulation (Figure 3.2). Vapour barriers should be applied to the surface of the

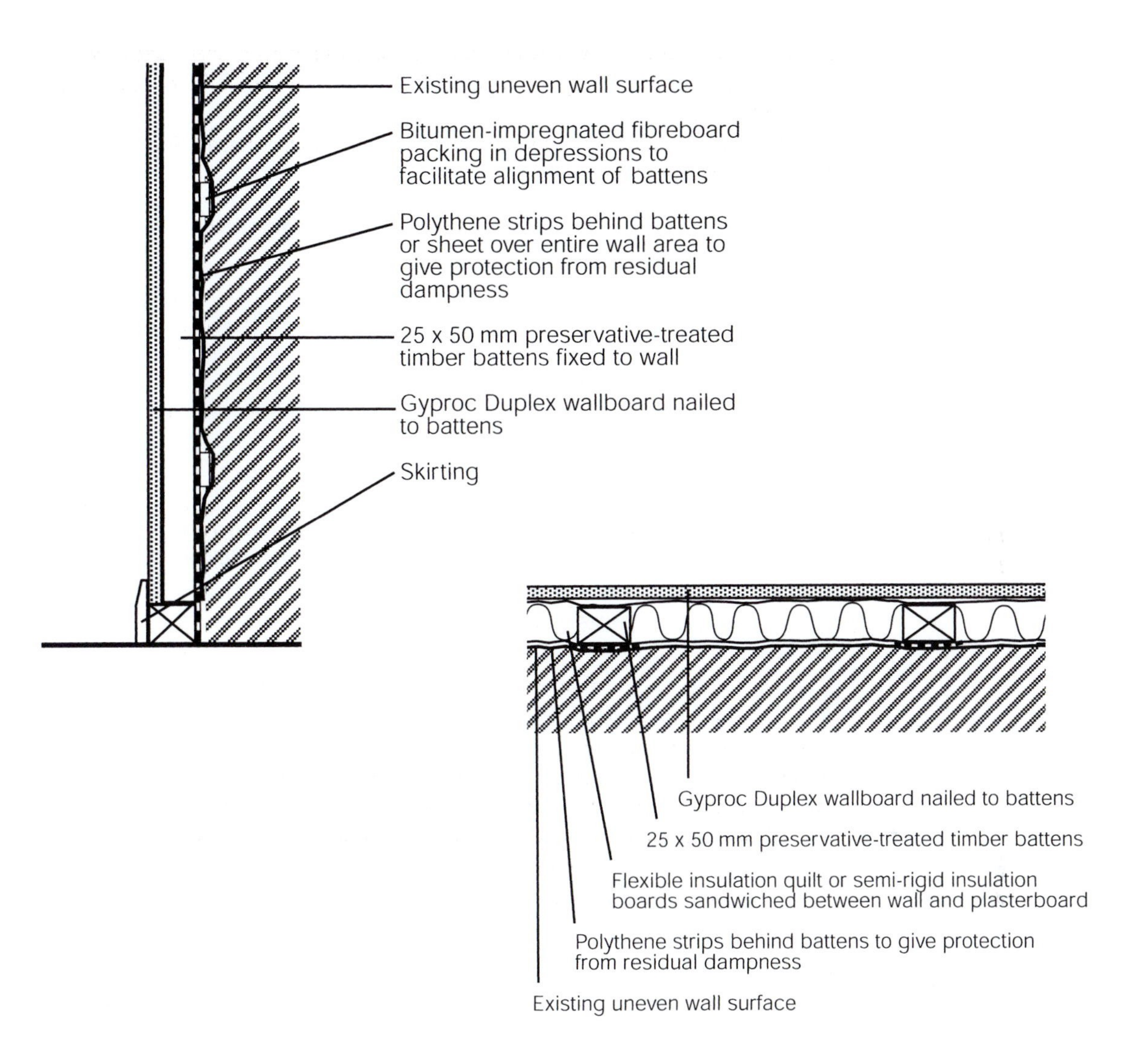

Figure 3.1
Dry linings: timber batten fixing

insulation, ensuring that the warm, moist air from the dwelling cannot penetrate behind the insulation. If warm, moisture-laden air hits a cold surface, the temperature of the air will reduce, meaning that its ability to hold water is reduced. When moisture-laden air hits the cold wall, the air exceeds its water saturation point and has to give up the moisture to the cold surface, causing condensation.

Metal channel fixing

As an alternative to fixing the plasterboard dry lining to timber battens, metal furrings may be used. The Gypframe (www.british-gypsum.com) system uses various

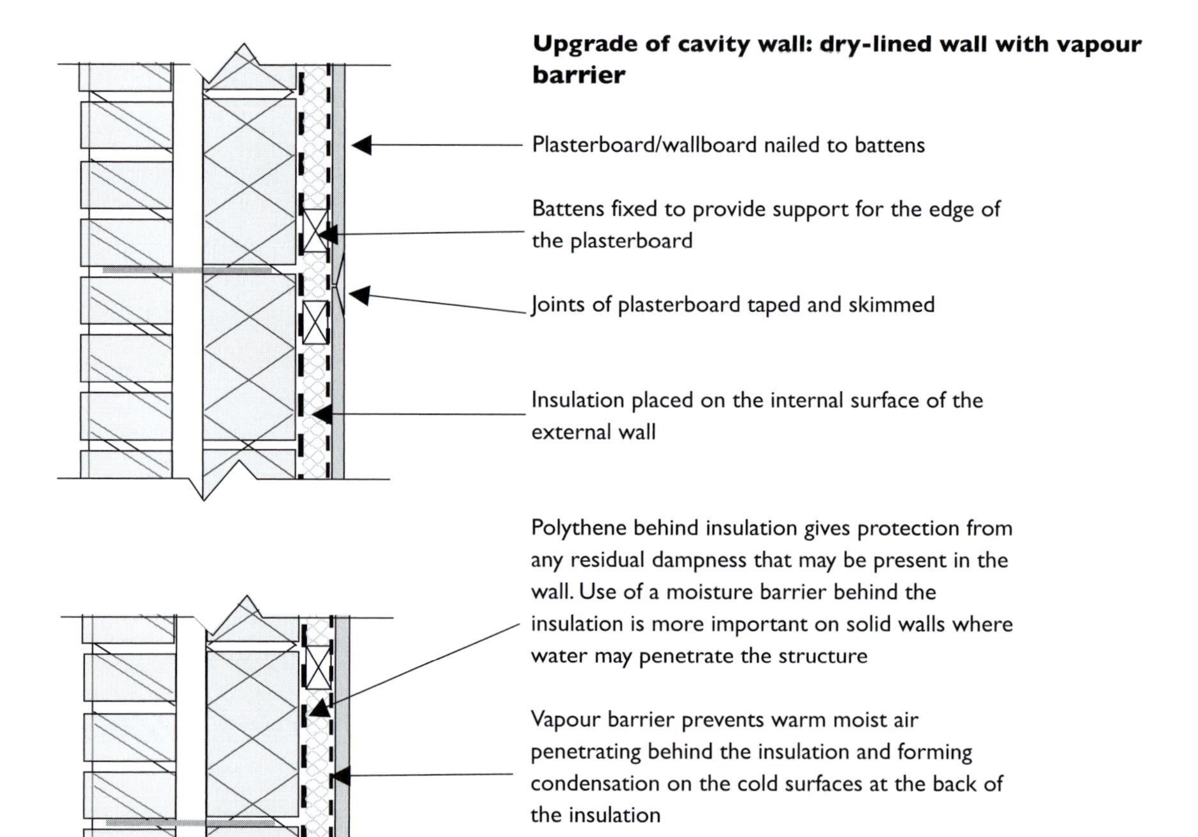

Figure 3.2
Section through dry lined wall with vapour barrier

sized, deep zinc-coated mild-steel channels, which are bonded vertically to the existing wall using special gypsum adhesive. Dabs of Gyproc Dri-Wall adhesive, 200 mm long, are applied to the wall at 600 mm centres, in vertical rows where each channel is to be fixed. The channels are then pressed on to the adhesive and aligned. Horizontal channels are fixed 30 mm below the ceiling and 15 mm above the finished floor level. The system is capable of accommodating irregularities in the background of up to 25 mm, by using the adhesive to fill out any depressions that exist. When the adhesive has set, each 12.5 mm WallBoard is screwed to its three vertical channels using a powered screwdriver. The fixings and joints are then made good, and a suitable decorative finish is applied to complete the upgrading

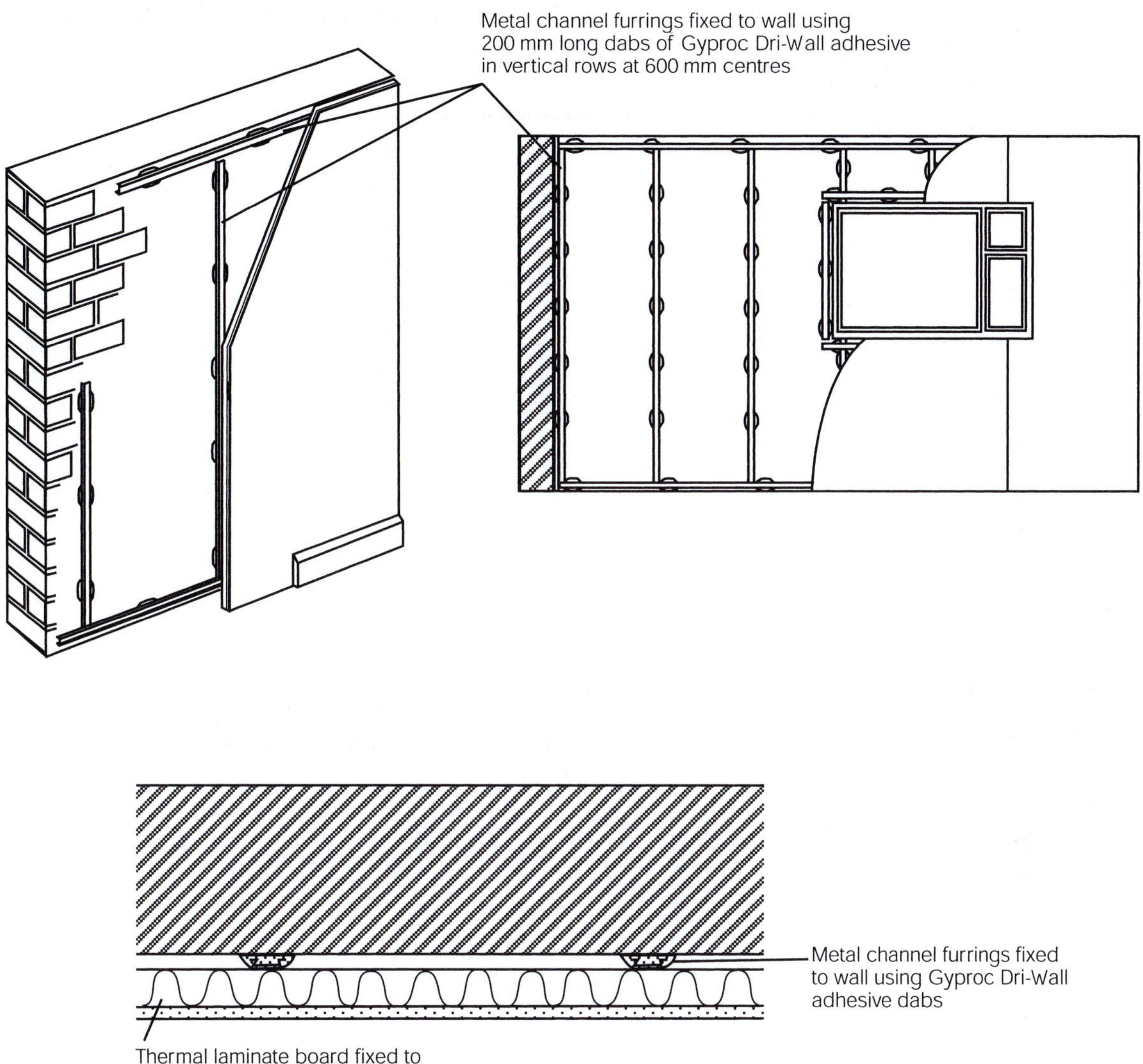

Figure 3.3
Dry linings: metal channel fixing

of the wall. Where thermal laminates rather than plain WallBoards are used to provide the new internal surface, they are fixed to the channels with a special gun-applied Gyproc sealant as an adhesive, with screws as secondary fixings to the vertical board edges. Details of the Gyproc dry-lined wall system are shown in Figure 3.3.

The Gyproc Gypliner Wall Lining system, similar in principle to the Dri-Wall MF system described above, uses lightweight metal track, channel and fixing brackets to form a framework for the fixing of plasterboards or thermal laminates. The framework is secured to the wall with metal brackets that are fixed to the wall, not by adhesive, but by Gypliner Anchors (pre-plugged nails). A major advantage of this system is that the metal fixing brackets allow a variable cavity depth behind the framework, of 25 to 130 mm, to suit specific requirements such as the incorporation of services.

The system is installed by fixing metal track along the floor and ceiling to give the required stand-off (cavity), followed by anchoring of the metal fixing brackets to the wall. The vertical channels are then friction-fitted into the horizontal track and screwed to the fixing brackets. The new wall lining is completed by screwing Gyproc plasterboard or thermal laminate to the framing members.

Direct adhesive fixing

Plasterboard dry linings can be bonded direct to the existing wall surfaces using special adhesive, thereby ruling out the need to install timber battens or metal furrings. A widely used proprietary system is Gyproc dry lining, which employs a special Gyproc Dri-Wall adhesive to bond plain WallBoards or Gyproc Thermal Laminates direct to the existing wall surface. First, a continuous fillet (or ribbon) of adhesive is applied to the existing wall perimeter, followed by dabs of adhesive

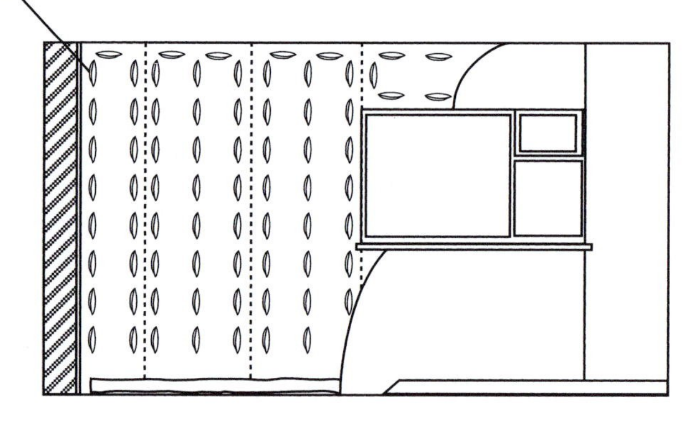

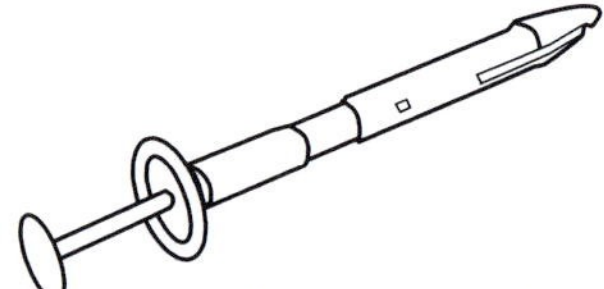

Thermal laminate board fixed to wall
using Gyproc Dri-Wall adhesive dabs

Nailable plugs provide secondary
mechanical fixing

Figure 3.4
Dry linings: direct adhesive fixing

A continuous ribbon of plaster prevents air flowing between wall and plasterboard. This seals off the cavity between the plasterboard and wall and prevents water laden air penetrating. If the air is allowed to circulate behind the plasterboard condensation will form on the cold surface of the wall
Plaster dabs
Sheets of plasterboard
Skirting board
All joints taped and sealed

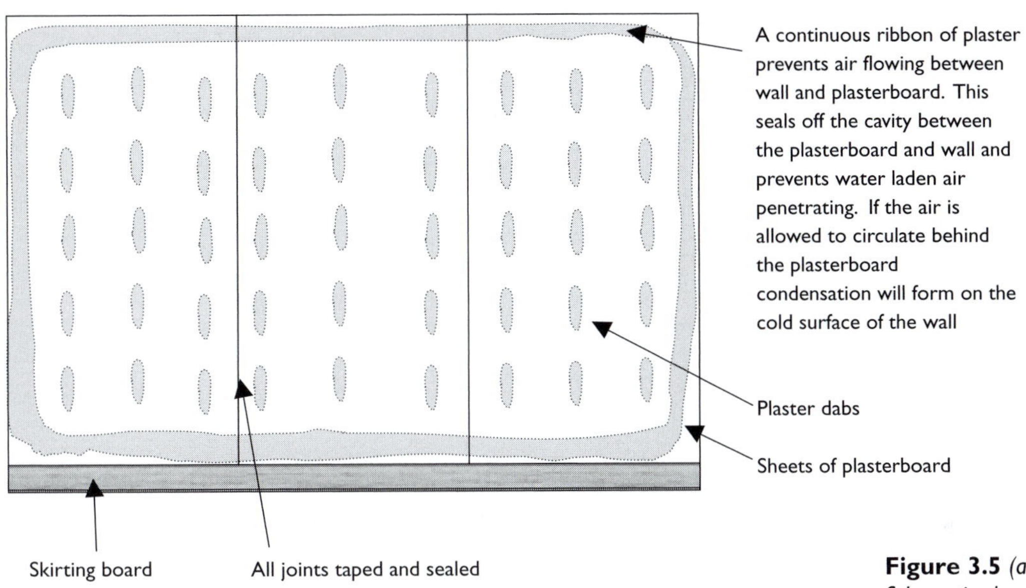

Figure 3.5 *(above)* *Schematic elevation: application of dry lining, showing continuous ribbon sealing plasterboard*

Section through dry-lined wall

Fixing adhesive runs continuously around perimeter of plasterboard to prevent air movement. Improves airtightness and reduces the possibility of interstitial condensation
Edges of plasterboard taped and skimmed
All existing joints should be roughly repointed or skimmed over to prevent air penetration through the brickwork
Continuous plaster ribbon runs along the bottom, top and sides of plasterboard

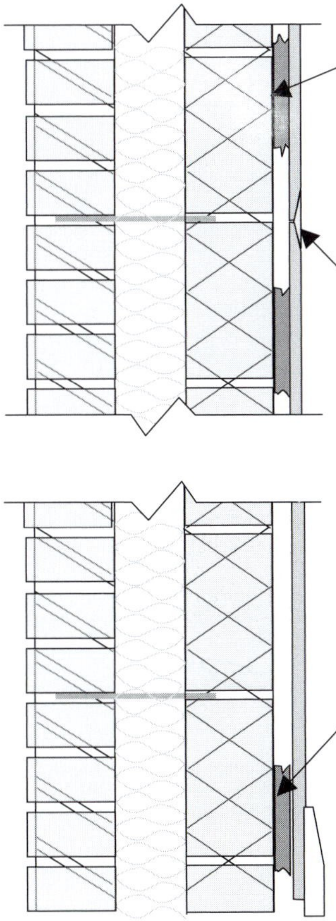

Figure 3.6 *(left)* *Section through dry-lined wall, with continuous ribbon of plaster preventing air circulation*

Figure 3.7
Section of dry-lined wall with thin layer of plaster (pargin) roughly applied to seal surface of porous blockwork

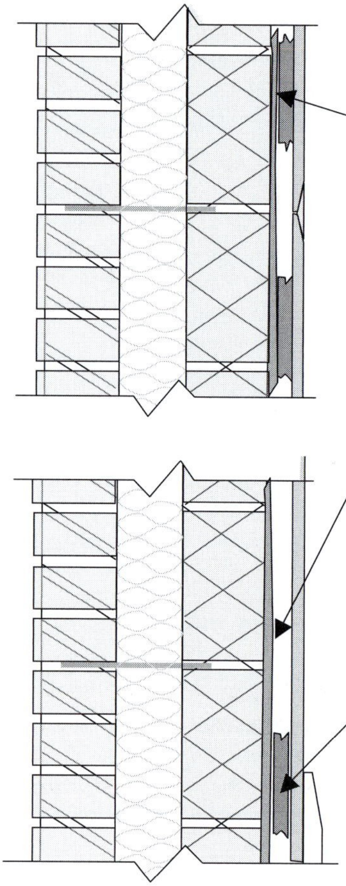

Section through dry-lined wall

Thin layer of plaster roughly applied to the surface of the blockwork. This helps to seal the porous blocks and any gaps in the blockwork joints. Sealing the blockwork prevents the gap behind the plasterboard connecting with the cavity and allowing the passage of air

Continuous ribbon runs along the bottom, top and sides of plasterboard

All mortar joints filled with mortar to prevent air movement

Figure 3.8
(below) Dry linings: direct adhesive fixing

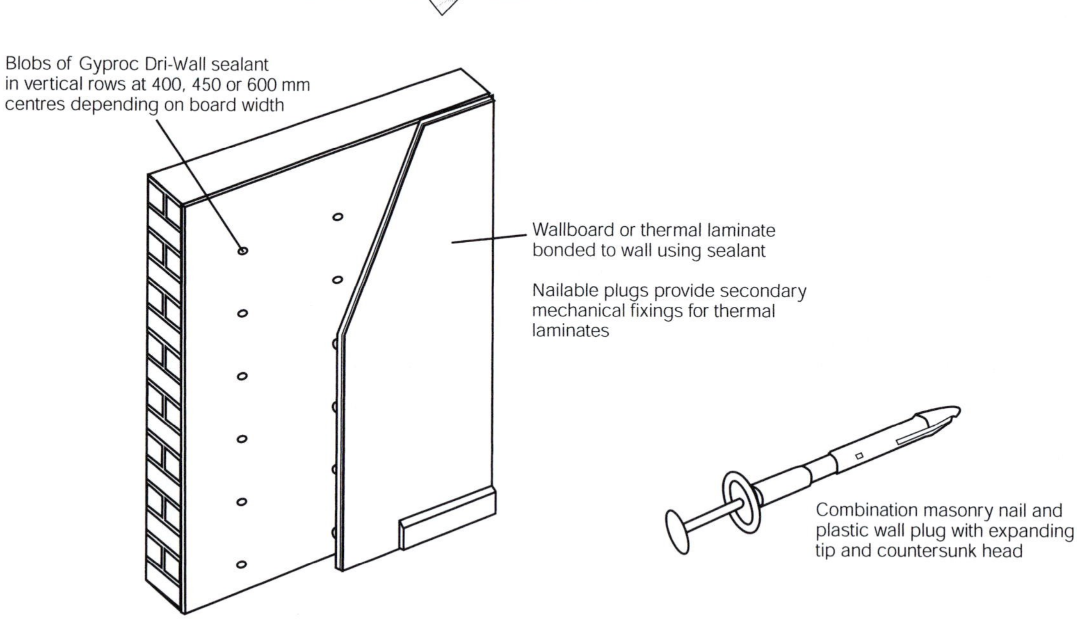

applied in vertical rows 400, 450 or 600 mm apart, depending on board width, with intermediate horizontal dabs at ceiling level. The board is offered up to the wall with the lower edge resting on packing strips, and a footlifter is used to lift the board tight to the ceiling. Where thermal laminates are used rather than plain WallBoards, nailable plugs are installed to provide secondary mechanical fixings. Figure 3.8 shows details of the Gyproc Dri-Wall TL system using thermal laminate plasterboard to form the dry lining.

As an alternative to the Gyproc Dri-Wall system, the Dri-Wall RF system provides a method of fixing boards directly to solid walls, including plastered walls in refurbishment work, using blobs of Gyproc sealant. The blobs of sealant are gun-applied to the wall surface in vertical rows, as shown in Figures 3.3 and 3.4. Around the edges, a continuous ribbon of adhesive or sealant should be run to ensure that the void behind the plasterboard does not act as a passage for air (see Figure 3.4). The board is offered up to the wall and fixed in the same way as the Dri-Wall adhesive system, described above, nailable plugs being used to provide secondary fixings where thermal laminates are used.

To improve airtightness of the structure and prevent air circulating around the back of the plasterboard, it is advisable to ensure that a continuous ribbon (or fillet) of plaster adhesive is placed around the edge of the plasterboard (Figures 3.5, 3.6 and 3.7). If the walls of the structure are particularly porous, it may be necessary to apply a thin skim of rough plaster across the face of the wall (parging); this will help fill any small holes and seal the walls (Figure 3.7). If walls are structurally sound, but porosity is a problem, it may be more advantageous to use wet plaster rather than dry lining. The application of a parge coat and then dry lining seems quite labour intensive. Normally, dry lining is selected over wet plaster because it is more economical and quicker. As wet trades are reduced with dry lining, the drying out time is reduced. However, if it is necessary to skim the walls with parging, then the advantages of using dry lining over wet, two-coat plaster are reduced. Wet plaster is often much more durable and airtight than general dry-lining systems.

3.3 Upgrading ceiling surfaces

The upgrading of existing ceiling finishes is generally less complicated than upgrading walls. Provided the structure above is sound, any conventional ceiling finish can be applied. In older buildings with timber floor structures, this will usually involve removal of the existing ceiling and its replacement with a new plasterboard ceiling. A useful alternative, particularly in buildings with high ceilings and where new service installations need to be concealed, is to install a suspended ceiling system, which can be very effective in reducing the room height and providing a generous services void above. Figure 3.9 shows a general suspended ceiling arrangement, and Figure 3.10 shows the different types of grid that are used to provide different aesthetic appeal.

In certain buildings the upgrading of existing ceilings may be closely associated with other upgrading work. For example, upgrading the fire resistance of timber floors usually involves treatment to the existing ceiling (see Chapter 2) and it can therefore be designed to satisfy both upgrading requirements in one single operation.

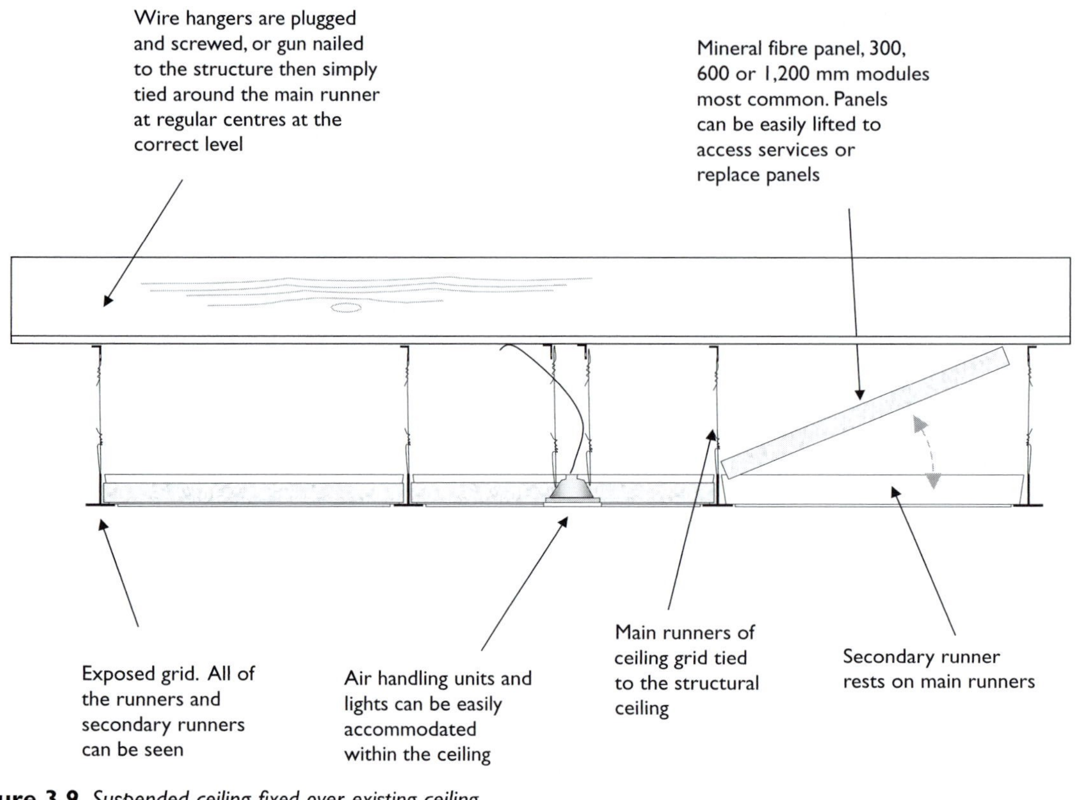
Wire hangers are plugged and screwed, or gun nailed to the structure then simply tied around the main runner at regular centres at the correct level
Mineral fibre panel, 300, 600 or 1,200 mm modules most common. Panels can be easily lifted to access services or replace panels
Exposed grid. All of the runners and secondary runners can be seen
Air handling units and lights can be easily accommodated within the ceiling
Main runners of ceiling grid tied to the structural ceiling
Secondary runner rests on main runners

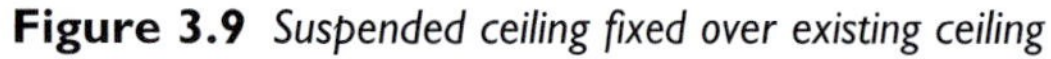

Figure 3.9 *Suspended ceiling fixed over existing ceiling*

Exposed grid. All of the runners and secondary runners can be seen

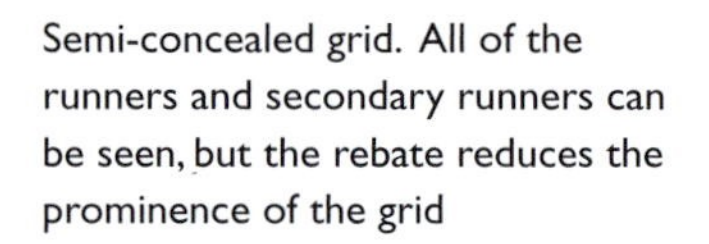

Semi-concealed grid. All of the runners and secondary runners can be seen, but the rebate reduces the prominence of the grid

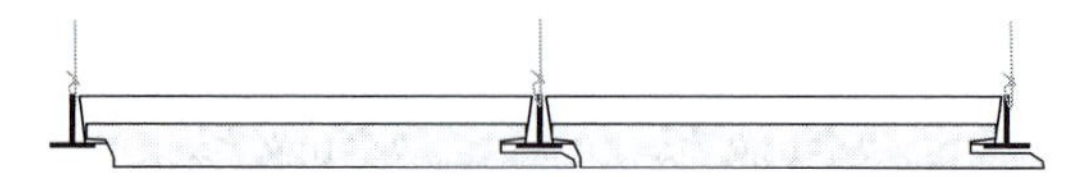

Concealed grid. All of the runners hidden beneath the ceiling panels

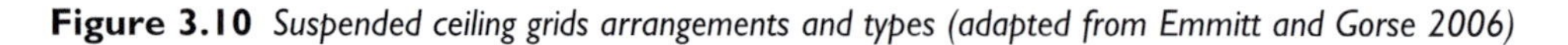

Figure 3.10 *Suspended ceiling grids arrangements and types (adapted from Emmitt and Gorse 2006)*

3.4 Upgrading floor surfaces

Resurfacing with timber

A very wide range of options is available for the upgrading of existing floors, and, provided the structure is sound, any suitable finish can be applied. In many older buildings, the existing floors may be uneven, and the first operation will involve resurfacing the sub-floor in order to provide a level base for the new finish. With timber floors, the simplest solution is to overlay the existing floorboards with hardboard, or, where the surface is very uneven, plywood. In order to achieve a level surface, it may be necessary first to sand down high points and pack out depressions before fixing the new sheeting. In extreme cases, where the existing boarding is so uneven that levelling with a new surface is impossible, or where decay has occurred, it will be necessary to remove the boarding completely and replace it with new material. Existing concrete floors are less likely to need upgrading, apart from their surface finishes. However, if necessary, timber grounds or metal fixings can be inserted to receive a new wood surface.

Rescreeding

Where an existing concrete floor surface has deteriorated, an alternative to providing a new timber surface is to rescreed it, or add a new screed where one did not previously exist. This will require an extensive amount of preparation, which may involve breaking up and removing an existing, deteriorated screed and treating the exposed concrete surface to provide a suitable key for the new screed. It is also a time-consuming, labour-intensive task, and it may, therefore, be worth considering applying one of the proprietary floor-levelling compounds of the type described below.

Proprietary floor-levelling compounds

A widely used alternative to the above methods, where an existing floor surface is in poor condition and needs upgrading, is to apply a proprietary, synthetic resin-based floor-levelling compound. Evo-Stik Floor Level and Fill system (www.evo-stik.co.uk) is a ready-mixed filler and self-levelling compound for preparing old or uneven floors prior to laying new coverings. It can be used over concrete, cement–sand screeds, ceramic tiles and asphalt. The material sets to an extremely durable, smooth surface that prevents the unsightly appearance and possible damage caused to the new floor-covering by any irregularities in the existing floor. Existing uneven floors are treated in three stages, as follows.

Preparation

Any loose or damaged areas in the floor should be chased out, and dirt and grit should be brushed out from holes or cracks. The floor should then be thoroughly cleaned. The whole floor must be sound and not subject to rising damp or any other structural defect.

Filling

The paste-like compound is applied with a trowel directly from its container, in layers up to 20 mm thick, to fill any depressions in the floor surface. Each layer must be allowed to dry before the next is applied. The surface is then smoothed with a float, a perfect surface not being necessary if it is intended to level over the top.

Levelling

Any holes deeper than 3 mm should be filled as previously described. The level-and-fill compound, which is diluted with water for the levelling operation to improve its flow capabilities, is poured on to the floor and spread evenly with a float to give a continuous layer, 1 mm thick and no more than 3 mm at any point. The self-levelling properties allow trowel marks to flow out.

A thin coat will take light foot traffic after drying overnight with good ventilation, and the new floor-covering can be laid after 24 hours.

References

Building Research Establishment (1998) *Replacing Plasterwork* (Good Repair Guide 18), Watford: BRE.

Emmitt, S. and Gorse, C. (2006) *Advanced Construction of Buildings*, Oxford: Blackwell Publishing.

Upgrading the thermal performance of existing elements

CHAPTER 4

4.1 General

An extremely important consideration, both in new construction and refurbishment work, is the provision of good thermal insulation to the external envelope in order to minimise heat loss, reduce heating costs, conserve fuel resources, reduce environmental pollution and maximise thermal comfort. Since the 1970s, the importance of providing adequate thermal insulation has been reflected by several changes to the Building Regulations, which now demand considerably greater standards of thermal insulation than were required earlier. It is therefore likely that most buildings that undergo refurbishment today will not, as they stand, comply with current Building Regulations regarding thermal insulation.

The construction of many older buildings is not conducive to the retention of heat within their interior spaces. In buildings with thick masonry external walls, heat is quickly absorbed into the walls. Because of the dense masonry construction, which absorbs the heat, brick or stone buildings can take more energy to heat the internal environment. If the buildings are not properly insulated, heat energy simply passes into and through the structure and is therefore lost. With heavy brick and stone buildings, satisfactory heating can be difficult to achieve, particularly where intermittent heating cycles are operated. However, the heavy internal walls within the structure do have advantages: they can act as a heat store, taking in heat energy during the warm day and giving it out during the cold night. Although there are benefits to heavy masonry construction, attention still needs to be given to the external envelope so that heat cannot flow straight through the building.

In buildings with thinner, solid masonry or brick walls, heat loss through the structure can be considerable, and some form of thermal upgrading will be necessary if heating costs are to be kept within reasonable limits. It should also be noted that untreated cavity walls do not reach the thermal standards required by current Building Regulations, and thus thermal upgrading of cavity walls is often included in the refurbishment of more recently constructed buildings.

In addition to upgrading the external walls, it will be prudent also to attend to the existing roof, which in most cases will not be insulated to current standards, and in some cases not insulated at all. A wide range of thermal upgrading techniques can be applied to existing roofs, the methods used depending principally on the nature of the existing roof structure.

4.2 Statutory requirements

The Building Regulations, Approved Document, Part L1B – Conservation of fuel and power in existing dwellings (Office of the Deputy Prime Minister, 2006) states that: reasonable provision shall be made for the conservation of fuel and power. The document states the need to limit the heat gains and losses through the building fabric, building services and use of those services. Where buildings are of special historic or architectural value, special considerations may apply if compliance with the energy efficiency requirements would unacceptably alter the appearance or character (further information can be found later in this chapter).

When a person intends to renovate a thermal element of a building, it is necessary to ensure that the thermal upgrade complies with the Building Regulations Part L.

Refurbishment schemes are required to comply with specified parts of the Building Regulations, including Part L ('Conservation of Fuel and Power', which includes thermal insulation requirements), if the works fall within the definition of a 'Material Change of Use'. The 'Material Change of Use' of existing buildings is defined and brought within the scope of the Building Regulations by Regulation 5, which gives its definition, and Regulation 6, which specifies the relevant requirements. Regulation 5 gives ten cases defining 'Material Change of Use':

- where the building is used as a dwelling, where previously it was not;
- where the building contains a flat, where previously it did not;
- where the building is used as a hotel or boarding-house, where previously it was not;
- where the building is used as an institution, where previously it was not;
- where the building is a public building, where previously it was not;
- where a building is not exempt from control, where previously it was exempt under Schedule 2 of the Regulations;
- where a dwelling contains a greater or lesser number of dwellings than it did previously;
- where the building contains a room for residential purposes, where previously it did not;
- where the building, which contains at least one room for residential purposes, contains a greater or lesser number of such rooms than it did previously;
- where the building is used as a shop, where previously it was not.

Typical examples of material changes of use would therefore include the refurbishment and conversion of:

- a redundant church into flats;
- a riverside warehouse into a hotel;
- a disused agricultural barn or stable into a house;
- a redundant railway engine shed into a community centre.

Where a material change of use occurs under any of the cases defined above then the Approved Document must be applied. The guidance provided by the

Approved Document differs depending on whether the work involves a thermal element, whether any of the thermal elements are being renovated or whether the thermal elements are just being retained. Thermal elements include any part of the building that forms the external element, for example roof, walls, windows, doors and floors. The following provides examples of how the Approved Document L: Conservation of Fuel and Power would apply:

- If the roof structure is being substantially replaced, thermal insulation should be provided to ensure a U-value of 0.35 W/m²K or lower (see Table 4.1).
- Where the roof is not being replaced but is retained, additional insulation should be provided to upgrade the U-value to 0.16 W/m²K (for insulation at ceiling level) and 0.20 W/m²K (for insulation between rafters) (see Table 4.2).
- If the ground-floor structure is being substantially replaced, thermal insulation should be used to provided a U-value not exceeding 0.25 W/m²K.
- If a new floor is installed in an extension, the U-value should be no greater than 0.22 W/m²K (see Table 4.1).
- When substantially replacing complete external walls, a U-value of 0.35 W/m²K should be achieved.
- If the internal surfaces of external walls are being renovated over a substantial area, their thermal insulation should be upgraded by providing an improved thermal performance of 0.35 W/m²K (for walls other than cavity walls).
- If insulation can be injected into the cavity, the improved U-value should be no greater than 0.55 W/m²K (this high U-value is only acceptable where a cavity can be filled with insulation). Where insulation cannot be inserted in a cavity, external or internal insulation should achieve a U-value no greater than 0.35 W/m²K. If upgrading is achieved using a dry-lining system, the gaps

Table 4.1 Standards for thermal elements for new elements in an extension and replacement elements in an existing dwelling

Building element	*A: U-value standards for new thermal elements (e.g. in an extension), W/m²K*	*B: U-value standards for replacement thermal elements in existing dwellings, W/m²K*
Wall	0.30	0.35*
Pitched roof (insulation at ceiling level)	0.16	0.16
Piched roof (insulation at rafter level)	0.20	0.20
Flat roof or a roof with integral insulation	0.20	0.25
Floors	0.22	0.25

* A reduced provision may be acceptable if meeting the prescribed standard would result in a reduction of more than 5 per cent of the floor area. If upgrading of floors creates significant problems in relation to adjoining floors, an alternative calculation method may be used.

Adapted from Table 4, Approved Document L1B (Office of the Deputy Prime Minister, 2006)

Table 4.2 Standards for the upgrading of retained elements

Building element	*A: Threshold value, W/m²K*	*B: Improved value, W/m²K*
Cavity wall (suitable for cavity wall insulation)	0.70	0.55
Other walls	0.70	0.35
Floors	0.70	0.25
Pitched roof (insulation at ceiling level)	0.35	0.16
Pitched roof (insulation between rafters)	0.35	0.20
Flat roof or a roof with integral insulation	0.35	0.25

Adapted from Table 5, Approved Document L1B (Office of the Deputy Prime Minister, 2006)

Table 4.3 Standards and reasonable provision when replacing or installing new windows and doors

Fitting	*A: New fittings in an extension, W/m²K*	*B: Replacement fittings in an existing dwelling, W/m²K*
Windows, roof lights and roof windows	1.8 or centre pane 1.2 or window energy rating band D	2.0 or centre pane 1.2 or window energy rating band E
Doors with 50% or greater of their internal area glazed	2.2 or centre pane 1.2	2.2 or centre pane 1.2
Other doors	3.0	3.0

Adapted from Approved Document L1B (Office of the Deputy Prime Minister, 2006)

Table 4.4 Flexible design and limiting U-value standards

Building element	*A: Area-weighted average U-value W/m²K*	*B: Limiting U-value W/m²K*
Wall	0.35	0.70
Floor	0.25	0.70
Roof	0.25	0.35
Windows, roof windows, rooflights and doors	2.2	3.3

Adapted from Table 1, Approved Document L1B (Office of the Deputy Prime Minister, 2006)

between the lining and masonry should be sealed at the edges of window and door openings and at wall, floor and ceiling junctions to prevent infiltration of cold, outside air.

- Where replacement windows are installed as part of the refurbishment, they should be draught-stripped and have an average U-value not exceeding 1.8 W/m^{2}K. However, it is recognised that this may be inappropriate in conservation work and in other situations where existing window design needs to be preserved. More detailed options for replacement windows are shown in Table 4.3.

The guidance for renovations is slightly complicated, and thought needs to be taken on whether, given the nature of work undertaken, thermal upgrading is required. Building Control can be contacted for advice in such situations.

New and replacement elements of the building

Where elements of construction are classed as newly constructed thermal elements, for example when constructing walls, floors, roofs etc. as part of an extension, they would need to meet the standards set out in column A in Table 4.1, which is an extract of the standards given in the Approved Document L1B. Where components of an existing building are replaced, the standards set out in column B would apply.

Where the thermal elements of the building are being renovated, the Building Regulations consider that it is necessary to upgrade elements of the building that have U-values worse than those shown in Table 4.2, column A (threshold value); the Regulations require the improved standards to comply with or be better than those set out in column B (improved U-value).

More flexible approaches in the selection of U-values and design of thermal elements are available by substantially improving the U-value in one aspect of the building, so that compensation could be applied to another element of the building. Alternatively, SAP 2005 (DETR 2005) calculations can be undertaken to show that the CO_2 emissions comply with the Regulations. Where compensation methods are used, the weighted average should be no worse than that set out in Table 4.4, column A, and none of the individual elements and components should be worse than those set out in column B.

Thermal upgrading: buildings of special historic interest or architectural value

When applying the Building Regulations Part L, special consideration is given to buildings that have special historic interest or architectural value. If it is considered that the energy efficiency requirements would unacceptably alter the character or appearance of the building, then special considerations would apply. Such buildings should still aim to improve the energy efficiency where it is practically possible. Changes should be made where they do not prejudice the character of the building or pose a risk that could result in the long-term deterioration of the fabric and fittings of the host building. English Heritage provides guidance when designing appropriate energy performance standards for such buildings. Further information can be found on the English Heritage website (a downloadable copy of *Building Regulations and Historic Buildings* (2004) is available at www.english-heritage.org.uk).

Issues relating to work on historic building that warrant sympathetic treatment from the Building Regulations include:

- restoring the historic character of a building that has been subject to inappropriate alteration;
- rebuilding of a former historic building (e.g. following a fire);
- making provisions enabling the fabric of the historic building to breath to reduce moisture and prevent the potential of long-term decay.

The guidance documents of the Building Regulations advise that the local authority's conservation officer should be contacted in such circumstances. The officer can give advice on the balance between historic building conservation and energy efficiency improvements.

The Building Regulations

It can be seen, therefore, that many refurbishment schemes will require thermal upgrading in order to comply with the Building Regulations and, even where it is not a statutory requirement, thermal upgrading will be a highly desirable feature of any refurbishment scheme because of the many benefits that it can provide. The Building Regulations can be accessed through the government Planning Portal (www.planningportal.gov.uk.).

4.3 Upgrading the thermal performance of walls

As explained above, many categories of refurbishment will include the addition or upgrading of thermal insulation to the existing walls. This is usually applied by adding a layer of insulating material to either the inside face or the outside face, or, in the case of cavity walls, by injecting an insulating fill into the cavity. The current Building Regulations U-value requirements for external walls vary depending on whether the element is considered to be new, retained or refurbished element. In most cases, the maximum U-value of 0.35 W/m^2K applies. Whichever method of upgrading is used, the aim should be to achieve lower U-values than those specified, providing an energy efficient building. With solid-walled buildings, the choice between internally applied or externally applied insulation will depend on two main factors: first, whether the building is heated intermittently or continuously; and second, the thermal capacity of the walls. Internally applied insulation is effective where the building is heated intermittently, since it prevents heat being absorbed by, and lost into, the walls, thereby giving a more rapid warm-up period. However, the addition of internal insulation will cause the existing wall structure to be colder, increasing the risk of condensation on or within it. It is therefore vital that an efficient vapour barrier be provided on the warm side of the new insulation to minimise this risk.

Externally applied insulation is effective where the building has thick walls of high thermal capacity and is heated continuously. Heat is absorbed into, and retained by, the walls and 'given back' to the interior, and the risk of condensation is minimised. Externally applied insulation is also effective with thinner walls of low thermal capacity, regardless of whether the heating is intermittent or continuous.

The relative merits and disadvantages of externally and internally applied thermal insulation are given in Table 4.5.

Internally applied insulation

The application of thermal insulation to the internal surfaces of external walls is often combined with the upgrading of their surface finishes, and one of the most convenient ways of achieving this is to incorporate a layer of thermal insulation into a dry-lining system as follows.

Separate sandwich insulation

With basic timber batten dry linings, a separate layer of insulation can be sandwiched into the space that has been created between the surface of the existing wall and the plasterboard, as shown in Figure 4.1. Standard size insulation battens used to be supplied in thicknesses from 25 mm thick. Such small thicknesses are now relatively uncommon when upgrading the thermal resistance of a building. It is more usual to see insulation in excess of 75 mm thick. Suitable insulation materials include flexible quilts, or semi-rigid batts of glass mineral wool or rockwool.

Isover timber frame batts of glass mineral wool, 60 mm thick, used in conjunction with a timber batten dry-lining system, will upgrade the U-value of an existing 220 mm solid plastered brick external wall from 2.17 to 0.52 W/m^{2}K. To achieve the Building Regulations standard of 0.35 W/m^{2}K, 100 mm thick batts would be required. It is becoming more common to use a steel frame stud wall rather than timber frame; again 100 mm thick batts would be inserted between the steel studs to achieve a U-value of 0.35 W/m^{2}K (further information can be found at Isover, www.isover.co.uk).

Rockwool 'flexi' flexible insulating slabs, 50 mm thick, placed between the timber battens of a dry-lining system, will upgrade a 220 mm solid plastered brick wall to 0.54 W/m^{2}K, and 75 mm slabs will upgrade the same wall to 0.41 W/m^{2}K. The slabs are available in a range of thicknesses. The thickness required to upgrade to existing Building Regulations would be dependent on existing construction: an uninsulated solid masonry wall would normally require 90 to 100 mm of insulation to meet 0.35 W/m^{2}K (further information can be found at www.rockwool.co.uk).

Linings with pre-bonded insulation

The modern alternative to incorporating sandwich insulation is to apply a dry lining comprising plasterboard with a layer of rigid insulation pre-bonded to it. Examples of proprietary systems are described below.

Kingsan Kooltherm insulated dry-lining board is available in a variety of sizes. The plasterboard thickness is 12.5 mm, with insulant thicknesses varying from

Table 4.5 The relative merits and disadvantages of externally and internally applied insulation

	Externally applied insulation		*Internally applied insulation*
1	Building can be almost totally wrapped in insulation so areas of cold bridging are significantly reduced	1	Cold bridging – where internal walls and floors abut the façade – is eliminated, therefore further reducing heat loss and surface condensation
2	Existing wall is kept warm and dry, thus increasing its insulation value and heat storage capacity	2	Application is not affected by weather
3	Need external scaffolding system, although few services, fittings and fixings need to be relocated to provide access to the surface	3	Access to surfaces being treated is easier; unless ceilings are high there is no need for scaffolding
4	Where external appearance is shabby and in need of a face lift, can be used to improve aesthetics	4	Has no effect on the external appearance of the building
5	The risk of interstitial condensation, within the thickness of the wall, is reduced	5	Cheaper than externally applied insulation
6	Existing wall is protected from the external environment	6	Existing wall is not protected
7	No internal work involved	7	Does not eliminate structural cold bridging
8	Avoids disruption to, or masking of, existing interior wall finishes, which may have to be preserved if the building is listed	8	Masks existing interior wall finishes
9	If occupants continue to use the building, the scaffolding should afford proper protection to the building users	9	Causes serious internal disruption
10	Less disruption to the occupants	10	Eliminates surface condensation
11	No loss of floor space	11	Perimeter floor space is reduced
12	Easier to apply around doors and windows	12	Difficult to apply around doors, windows and internal fittings. Wall mounted services and finishes need to be remounted
13	Provides an improved external finish to buildings whose appearance has deteriorated because of weathering and atmospheric pollution	13	Can be applied selectively to various parts of the building
14	Has a significant effect on the external appearance of the building and is therefore unsuitable for certain historic buildings where the existing appearance must be preserved	14	Can produce interstitial condensation risk. Vapour barriers should be used to reduce the risk of interstitial condensation
15	Provides an extra barrier from the weather and environment – reducing the impact of seasonal variation on the structure (expansion, contraction and frost attack)	15	Reduces heat protection of outside wall
16	Thickness of insulation only limited by fixing device and strength of existing structure	16	Practical limitations on thickness
17	More expensive than internally applied insulation	17	Interstitial condensation risk with some insulants
18	Can be applied with almost no limit of thickness – will accommodate future standards	18	Vapour barriers must be used to prevent interstitial condensation
19	Can correct adverse dew point situation	19	Fire risk with some insulants
20	Produces similar savings in heat loss and energy consumption (up to 50%) to internally applied insulation	20	Produces similar savings in heat loss and energy consumption (up to 50%) to externally applied insulation

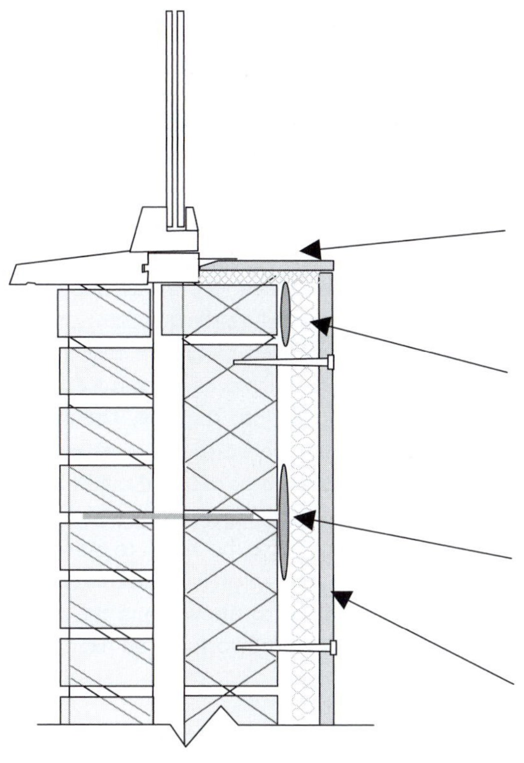
Upgrade of cavity wall: dry-lined wall with plasterboard/insulation laminate

Narrow sections of laminated plasterboard (20–25 mm insulation) used around window and door reveals. Fixed with plaster adhesive

Ribbon of plaster placed around the top and bottom of board to give good fixing and seal the void behind the laminate board

Dot and dab adhesive bonds the insulation plasterboard laminate to the wall

Minimum 57 mm laminate board (12.5 mm plasterboard and 55 mm insulation) to provide U-value of 0.35 W/m²k

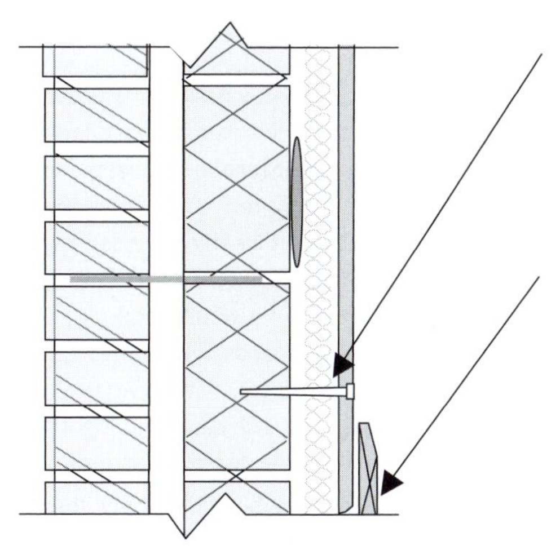
Nailable fixing plugs fixed after the adhesive has set. Holds the boards securely in position, fixed at the top, middle and bottom of each board

Edges of the plasterboard and junctions (e.g. between plasterboard, skirting and floor) should be sealed to ensure airtightness

Figure 4.1 *Section through dry-lined wall: composite plaster/insulation board fixed to timber battens*

20 to 85 mm thick in increments of 5 mm. When applied to a brick and block cavity wall, a 57.5 mm thick board (45 mm insulation, 12.5 mm plasterboard) achieves a U-value of 0.35 W/m^2 K, and a 77.5 mm board (65 mm insulation and 12.5 plasterboard) achieves a U-value of 0.27 W/m^2 K. Note that these figures make no allowance for windows etc.

The insulated plasterboard is fixed using plaster adhesive, bonded directly to the brickwork or blockwork (www.insulateonline.com). A ribbon of decorator's corking or silicone sealant is run around the skirting, ceiling and edges of the board, as well as at window and door openings, providing a continuous seal around the edges of the board. Plaster adhesive dabs are applied to the wall, and the laminate insulated board is offered to the adhesive. When refurbishing buildings, it is essential that all loose material and decoration be removed to ensure that a good bond between the substrate and the insulation board is achieved. To ensure a good bond, fixing plugs can also be fixed through the board into the wall. The nailable plugs are fixed at the top, middle and bottom of the board (three plugs per board are fixed after the plaster adhesive has set). Narrow widths of board can be used around window and door reveals to reduce cold bridging (Figure 4.1).

The insulation/plaster laminated board can also be fixed to timber or steel studwork. Where the surface is particularly uneven or it is difficult to get a good bond to the surface, timber battens can be fixed to the wall and the plasterboard nailed directly to the boards. Alternatively steel studwork can be fixed to the walls (Figure 4.2).

Styroliner: Styroliner (see Figure 4.3) comprises plasterboard, 9.5 mm thick, factory-bonded to a backing of Styrofoam closed-cell extruded polystyrene (www.styroliner.co.uk). The product is available with a tapered-edged manila-faced plasterboard for direct decoration or, alternatively, with square-edged grey-faced plasterboard for finishing with a skim coat of plaster.

The closed-cell structure of the Styrofoam (www.dow.com) prevents capillary absorption of moisture, and, in addition to thermal insulation, Styroliner LK provides an effective moisture barrier and vapour check. Styroliner boards are 1,200 mm wide × 2,438 mm long and are available in seven overall thicknesses, from 21.5 mm to 55.5 mm. Styroliner LK can be fixed directly to most common forms of wall construction, including plastered surfaces, masonry and brickwork, using a suitable proprietary adhesive, such as Ardurit X7 (www.ardex.co.uk), applied either to the wall itself, or to the reverse side of the boards. The adhesive is applied to form a continuous band around the board perimeter, with a vertical central band of adhesive. The boards are pressed against the adhesive and tightly against the ceiling. Secondary fixings are recommended in the form of at least four Tapcon self-tapping masonry screws per board, or any other proprietary fixing system.

As an alternative to direct adhesive fixing, thermal boards can be fixed to timber battens. The boards are fixed to the battens using galvanised plasterboard nails or screws. This method is more suited to very uneven backgrounds, fibreboard packing being used to fill any depressions in order to ensure correct alignment of the battens. The following table (Table 4.6) gives an indication of the possible U-values that can be achieved using Styrofoam.

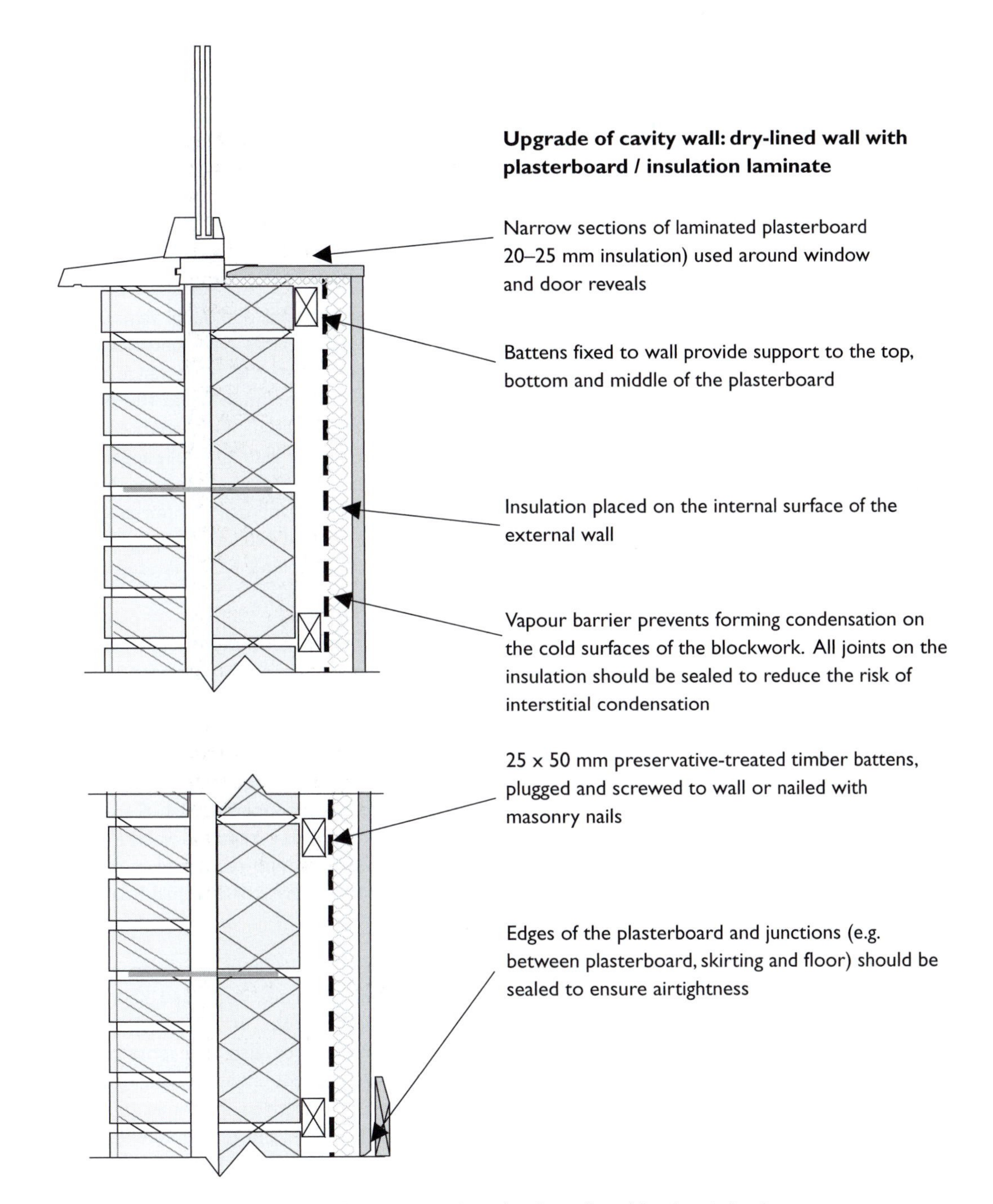

Upgrade of cavity wall: dry-lined wall with plasterboard / insulation laminate
Narrow sections of laminated plasterboard 20–25 mm insulation) used around window and door reveals
Battens fixed to wall provide support to the top, bottom and middle of the plasterboard
Insulation placed on the internal surface of the external wall
Vapour barrier prevents forming condensation on the cold surfaces of the blockwork. All joints on the insulation should be sealed to reduce the risk of interstitial condensation
25 x 50 mm preservative-treated timber battens, plugged and screwed to wall or nailed with masonry nails
Edges of the plasterboard and junctions (e.g. between plasterboard, skirting and floor) should be sealed to ensure airtightness

Figure 4.2 *Section through dry-lined wall: composite plaster/insulation board fixed to timber battens (www.insulateonline.com)*

Gyproc thermal board: The Gyproc range of plasterboard and insulation laminated together includes: Gyproc ThermalLine basic, plus and super (www.british-gypsum.bpb.co.uk). The basic plasterboard thermal laminate does not include an integral vapour check. Where a vapour check is not included, it is strongly advised that one is included in the overall construction to prevent interstitial condensation. The plus and super range both include integral vapour checks. Even though vapour checks are integral, attention must be paid to the joints, edges, top and bottom of the wall construction to ensure that the vapour check is continuous and sealed. To ensure vapour resistivity across the whole wall Gyproc recommends that all WallBoards are taped and filled; the application of two coats of Gyproc drywall sealer once the WallBoards are in position provides additional vapour resistance. The plus and super ranges of thermal plasterboards have better water-vapour resistivity and thermal-insulation properties than the basic range. Where the required standard of thermal upgrading is high or humidity in the property is a potential problem, the plus and super boards would be more appropriate than the basic range. Interstitial condensation is always a risk with internally applied insulation; care should be taken to ensure all service entries and other features that penetrate the boards are sealed. The plasterboard thermal laminate comprises 9.5 mm thick Gyproc WallBoard, factory-bonded to a backing of extruded polystyrene insulating board or phenolic foam (Phenol formaldehyde foam), which is used in the super range.

Gyproc Thermal Board Plus: Gyproc Thermal Board Plus is composed of 9.5 mm thick Gyproc WallBoard factory-bonded to a backing of closed-cell extruded polystyrene. With the exception of the insulant, the laminate uses expanded polystyrene as the insulating material. Board sizes are 2,700 and 2,400 mm long × 1,200 mm wide and are available in five overall thicknesses of 27, 35, 42, 50 and 55 mm. The thermal performance of closed-cell extruded polystyrene is superior to that of expanded polystyrene, and its inherent high resistance to the passage of water vapour rules out the need for a separate vapour-check layer to be incorporated into the laminate where vapour resistance is required. The U-value comparisons below illustrate the superiority of Thermal Board Plus over standard Thermal Board.

Gyproc Thermal Board Super: Gyproc Thermal Board Super is similar in all respects, including fixing methods, to the thermal laminates described above, with the exception of its insulant, CFC-free phenolic foam, which is superior to both expanded and extruded polystyrene. The boards also incorporate an integral vapour-control layer as standard. The more efficient phenolic foam insulant enables high standards of insulation to be achieved by a relatively thin laminate with minimal encroachment on floor space. Board sizes are 2,700 and 2,400 mm long × 1,200 mm wide, available in three overall thicknesses of 50, 60 and 65 mm. Tables 4.7 and 4.8 illustrate the relative efficiencies of Gyproc Thermal Board.

Externally applied insulation

Two of the most important factors that can determine whether externally applied insulation is used in preference to internally applied insulation are the condition

Table 4.6 U-values achieved when using Styrofoam applied to solid walls

Styrofoam thickness (internal insulation), mm	*U-value achieved, W/m²K*
110	0.25
90	0.28
80	0.30
70	0.35

Source: building.dow.com

Table 4.7 Thermal resistance for Gyproc ThermalLine Plus (9.5 mm WallBoard + extruded polystyrene)

Thickness, mm	*R-values (thermal resistance), m²K/W*	*U-value (for unit only), m²K/W*
27	0.63	1.287
35	0.88	1.156
40	1.07	2.857
45	1.25	0.800
55	1.57	0.630

Source: www.british-gypsum.bpb.com

Table 4.8 Thermal resistance for Gyproc ThermalLine Super (9.5 mm WallBoard + phenolic foam insulation)

Thickness, mm	*R-values (thermal resistance), m²K/W*	*U-value (for unit only), m²K/W*
50	1.79	0.55
60	2.32	0.43
65	2.55	0.39

Source: www.british-gypsum.bpb.com

Table 4.9 Thermal resistance for Gyproc Platinum (12.5 mm WallBoard + expanded polystyrene)

Thickness, mm	*R-values, (thermal resistance), m²K/W*	*U-value (for unit only), m²K/W*
28	0.59	1.695
40	0.89	1.123
50	1.26	0.793
60	1.58	0.633

Source: www.british-gypsum.bpb.com

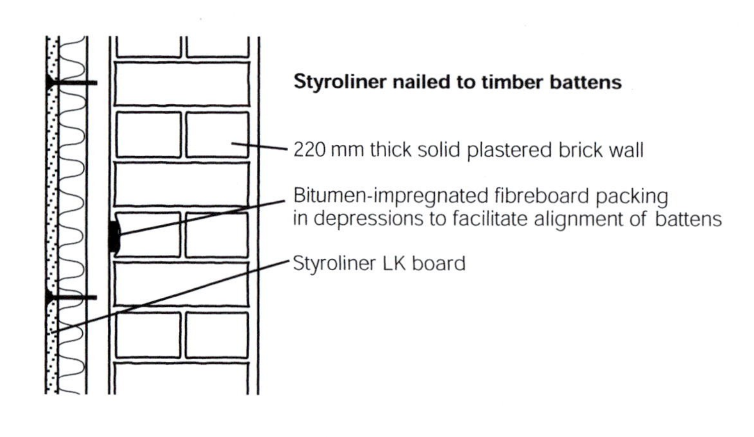

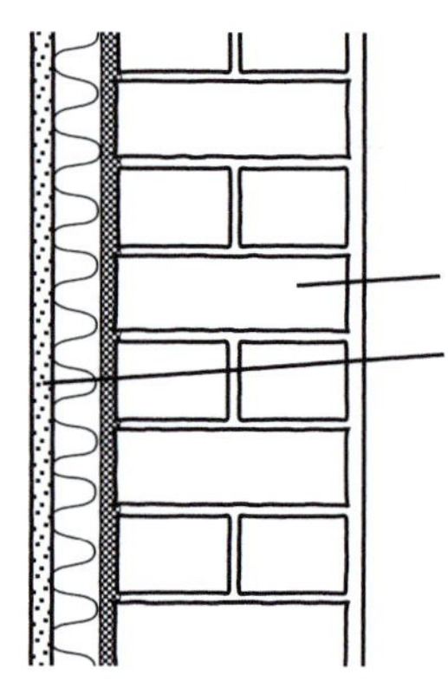

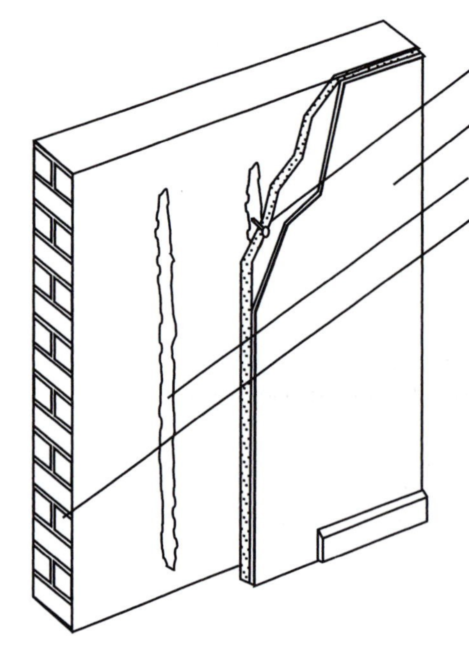

Figure 4.3
Thermal composite (plasterboard and dry lining)

and appearance of the existing external wall surfaces. For example, if the walls are constructed from fine, ornate masonry and they possess architectural and aesthetic merit, then masking them with externally applied insulation will not be appropriate. On the other hand, if the walls are uninteresting, in poor condition or affected by dampness, then externally applied insulation may be the ideal solution, since, in one operation, their thermal performance, appearance and weather protection can be significantly improved. A number of other factors must also be considered before deciding whether or not to opt for external insulation, and these have been listed in Table 4.5.

Although the use of externally applied insulation systems rules out the need to disrupt the interior of the building, certain modifications to the exterior will be necessary to allow for the increase in the thickness of the wall. This may involve the extension of windowsills, removal and repositioning of rainwater and waste pipes, provision of metal flashings, the accommodation of airbricks and other work to facilitate addition of the new insulation layer.

Where the existing wall has a tile or weatherboard external cladding, it may be possible to insert a layer of rigid or flexible insulation behind it. This will involve removal and replacement of the cladding, but has the advantage of maintaining the original exterior appearance of the building. It should be noted that, where this method is employed, provision should be made for adequate ventilation behind the cladding.

Four basic forms of externally applied thermal insulation are described here.

Expanded polystyrene boards with render finish

A polystyrene external wall insulation system can be used (see Figure 4.4; see also Table 4.10 for U-values). The system shown comprises 1,220 × 610 mm stipple-coated expanded polystyrene insulation, with a density of 15 kg/m^3 applied to the existing external wall surface and overlaid with expanded metal lathing (for example, Rendalath reinforced render carrier; www.brc-special-products.co.uk) to act as a key for the render finish. The insulation boards and lathing are secured mechanically to the wall using polypropylene, nylon or stainless steel fixing pins, which are tapped home into holes drilled through the lathing and insulation and into the existing brickwork or masonry. The render consists of an 8–10 mm undercoat, 6–8 mm topcoat and finish, applied two days after the undercoat. The undercoat comprises ordinary Portland cement, sand and a dry plasticiser/waterproofer with dry polymers that are factory-batched, needing only the addition of clean water on site. The factory-batched topcoat, either traditional or with polymer additives, is finished with decorative pre-washed and graded aggregates, or a range of acrylic decorative textured finishes. The insulation board is available in a range of thicknesses from 15 to 100 mm. A polystyrene, externally applied system, using 50 mm thick insulation, will upgrade the U-value of a 225 mm solid plastered brick external wall from 2.17 to 0.50 W/m^{2}K. To achieve the Building Regulations standard of 0.35 W/m^{2}K, 80 mm boards would be required.

Rockwool (www.rockwool.co.uk) and Alumasc (www.alumasc-exteriors.co.uk) both supply insulation systems that are capable of carrying render.

As an alternative, the Terratherm external wall insulation system (see Figure 4.5) comprises 1,000 × 600 mm expanded polystyrene boards fixed to the existing

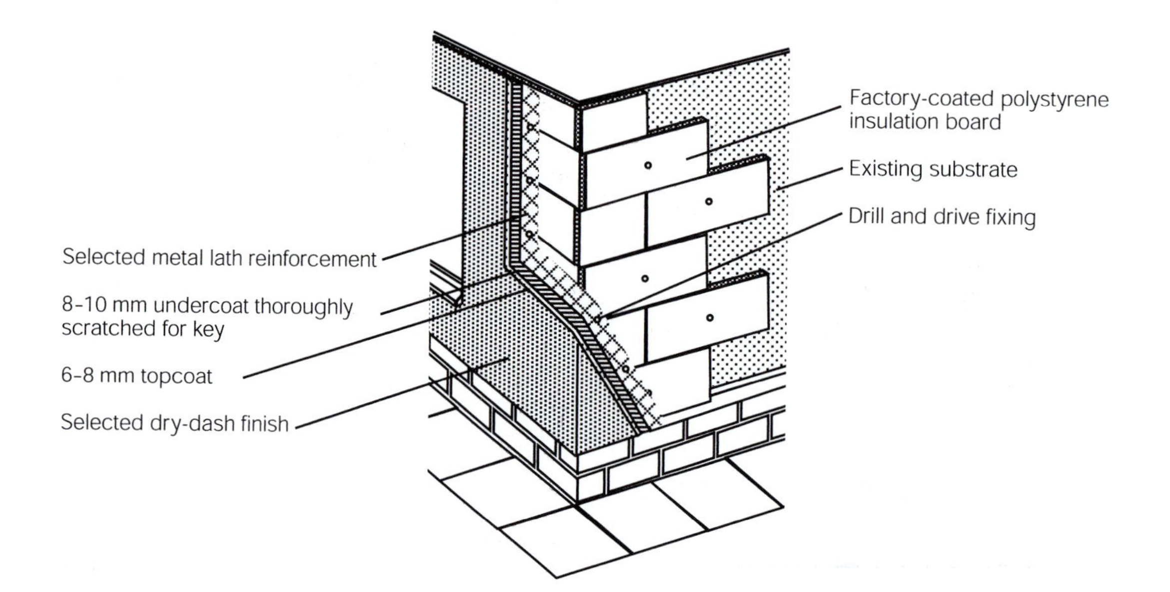

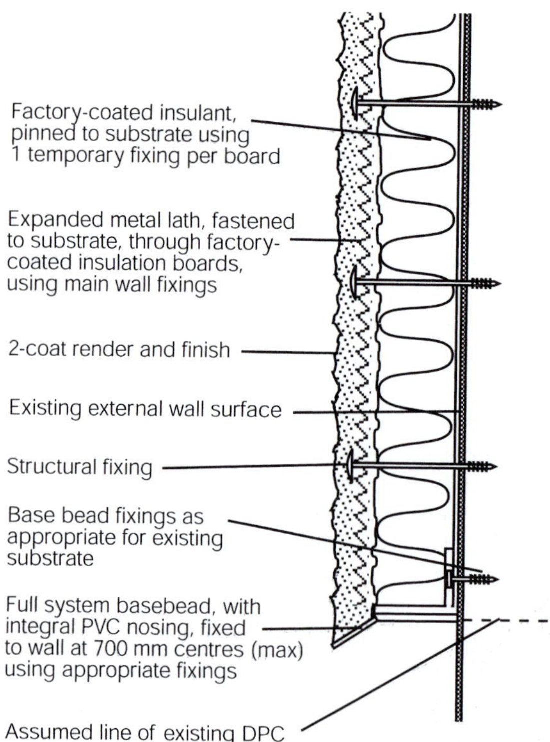

Figure 4.4
Polystyrene, externally applied insulation with render carried on render reinforcement

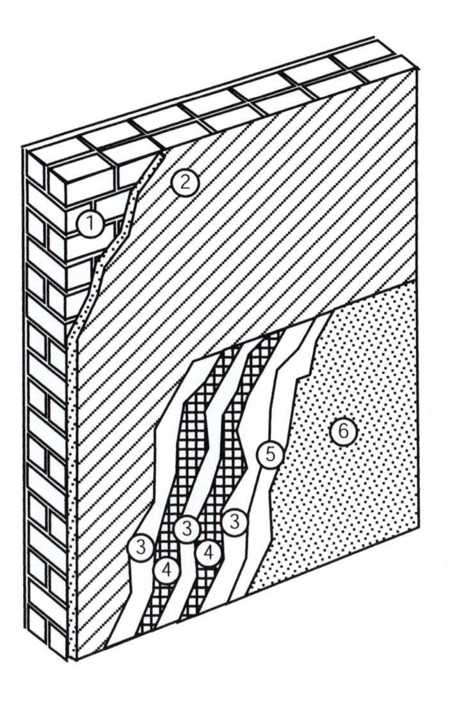

1. Existing wall surface
2. Expanded polystyrene board fixed with adhesive
3. Terratherm basecoat
4. Glass-fibre reinforcing embedded in basecoat
5. Primer applied by roller
6. 2 mm thick textured or 3 mm thick fine aggregate finish applied by trowel

Figure 4.5 *Terratherm PSB externally applied insulation*

Table 4.10
Thermal insulation values for standard Expolath insulation applied to a 225 mm brick external wall with 9.5 mm plasterboard

Insulation thickness, mm	*U-value, W/m²K*
15	0.88
20	0.79
30	0.65
40	0.56
50	0.49
60	0.43
70	0.39
80	0.35
90	0.32
100	0.30

wall surface using adhesive, or a mechanically fixed rail system, respectively. Two coats of basecoat mortar are then applied to the insulation, each with glass-fibre reinforcing mesh embedded into it. The adhesive and basecoat mortar consist of cement mixed with Terratherm Basecoat in the proportions 1:2.5 and 1:3, respectively. A primer is applied by roller to the final basecoat, and the insulation system is finished with either a coloured textured finish, 2 mm thick, or a coloured fine aggregate finish, 3 mm thick, applied by trowel. Additional mechanical support, in the form of stainless steel and polypropylene fixings, supplements the adhesive in securing the expanded polystyrene boards to the existing wall.

A range of new multipurpose renders are available at Weber (www.weber-sbd.com). The insulation boards are available in a range of thicknesses up to 100 mm.

Polyisocyanurate (PIR) insulation foam boards with finish secured on timber battens

Celtex is another alternative and offers single insulation boards with thicknesses up to 200 mm (Figure 4.6). Typically, insulation thicknesses range from 35–90 mm;

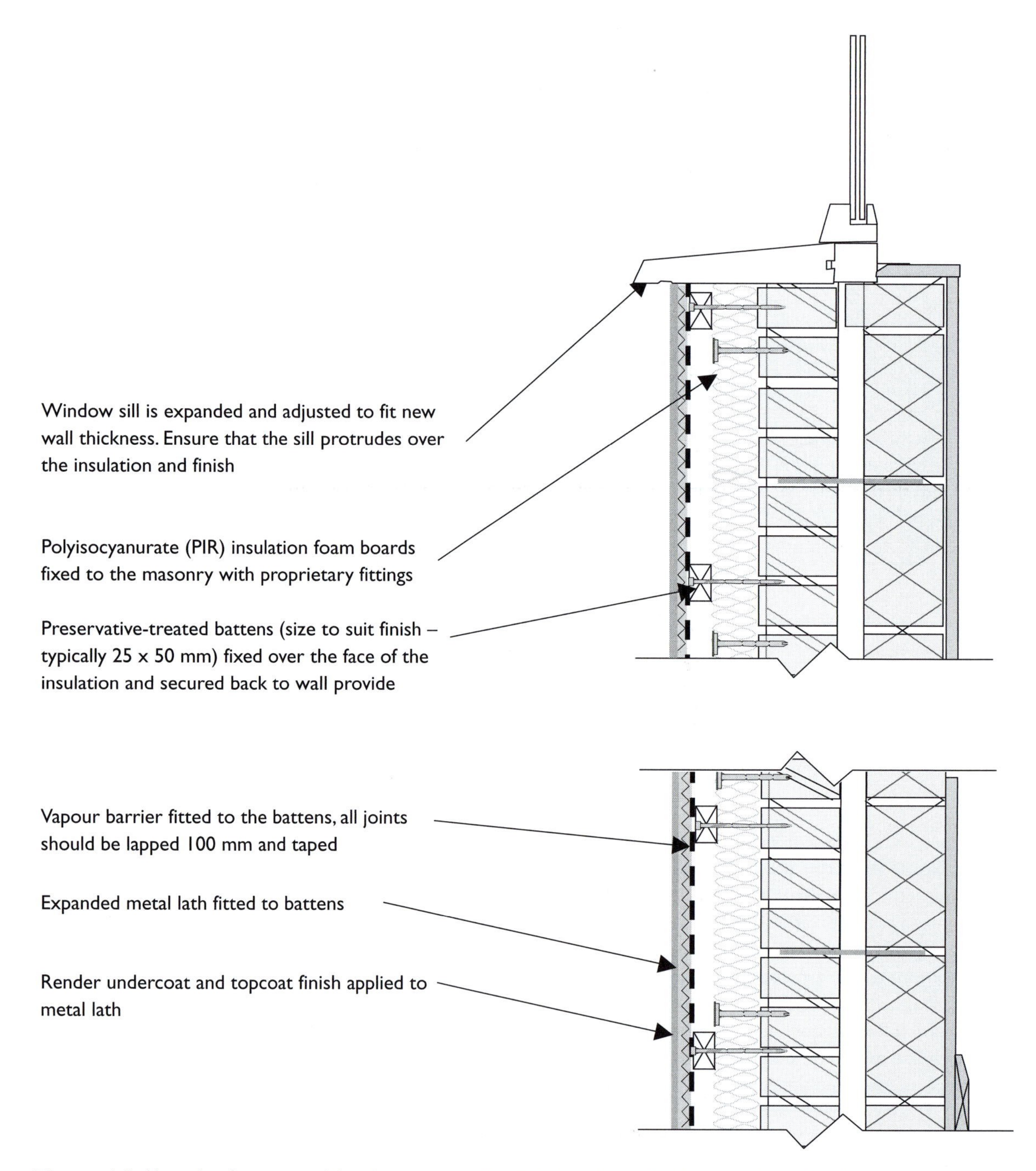
Window sill is expanded and adjusted to fit new wall thickness. Ensure that the sill protrudes over the insulation and finish
Polyisocyanurate (PIR) insulation foam boards fixed to the masonry with proprietary fittings
Preservative-treated battens (size to suit finish – typically 25 x 50 mm) fixed over the face of the insulation and secured back to wall provide
Vapour barrier fitted to the battens, all joints should be lapped 100 mm and taped
Expanded metal lath fitted to battens
Render undercoat and topcoat finish applied to metal lath

Figure 4.6 *Upgrade of cavity or solid wall using externally applied insulation; finish secured on timber battens (adapted from www.celotex.co.uk)*

however, it is likely that the 150 mm and 200 mm range offered by Celtex will become much more popular. The Celtex system is fastened to the substrate using proprietary fasteners that are fixed into predrilled holes. Unlike the systems previous described, the finish to the Celtex system is secured on timber battens rather than directly to the face of the insulation. The preservative-treated timber battens are applied vertically over the face of the insulation and secured to the masonry. A breather membrane is fixed over the battens, with 100 mm laps. Various claddings can be fitted to the battens; commonly used finishes include render or tile hanging. For the render finish, metal lathing is fixed to the battens and reinforcing ribs. For tile hanging, cross battens are fixed to the vertical battens, allowing tile hanging.

Mineral wool slabs with render finish

The Rockwool RockShield rigid slab external wall insulation system (see Figure 4.7) comprises mineral wool slabs, 1,000 × 500 mm, fixed to the existing wall surface using a combination of adhesive and mechanical fixings. The slabs receive an 8 mm or 12 mm thick render incorporating a reinforcing mesh to strengthen the system. The thicker render finish offers a higher tolerance to uneven wall surfaces. The render receives either Silcoplast ready-mixed finish, available in a stippled texture, or Liteplast finish, available in either stippled or dragged texture. Liteplast is supplied as a dry mortar, requiring the addition of clean water on site. Both topcoat finishes are supplied in a wide range of colours. RockShield rigid slabs are available in thicknesses from 30 mm to 140 mm in 10 mm increments: 30 mm slabs will upgrade the U-value of a 225 mm solid plastered brick external wall from 2.17 to 0.69 W/m^{2}K; 60 mm slabs will upgrade the same wall to 0.44 W/m^{2}K. Examples of the U-values that can be achieved are provided on the Rockwool web site

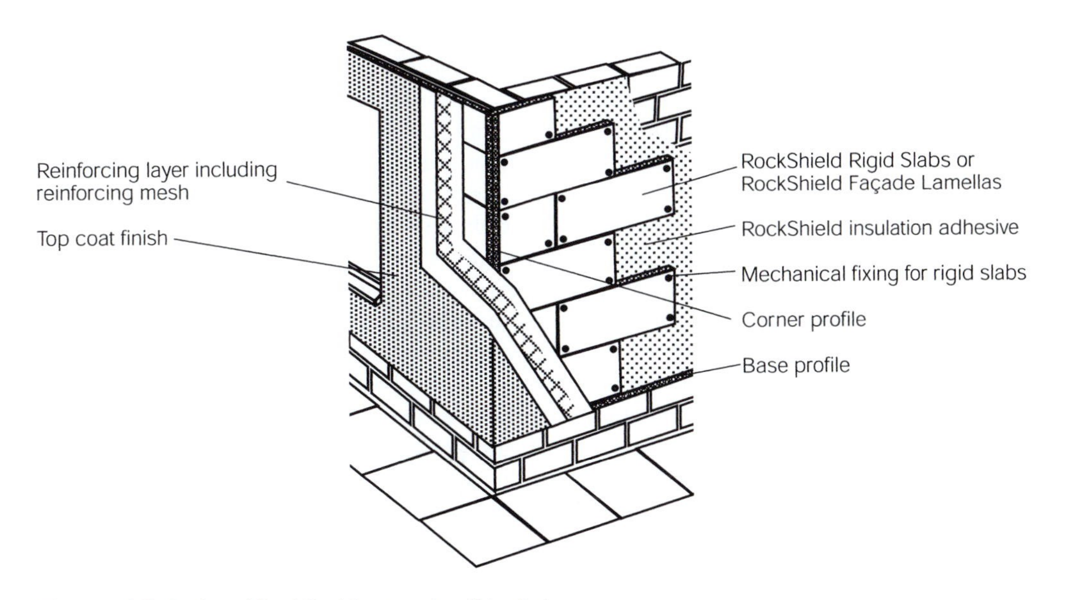

Figure 4.7 *Rockwool RockShield external wall insulation*

(www.rockwool.co.uk). For a 225 mm solid brick wall with 120 mm of RockShield insulation, a U-value of 0.26 W/m^2K can be achieved.

As an alternative to RockShield rigid slabs, RockShield façade lamellas, 1,000 × 200 × 30–150 mm thick, manufactured from the same material and fixed with adhesive only, may be used.

Closed-cell phenolic foam boards with render finish

The reinforced external wall insulation system shown in Figure 4.8 comprises 2360 × 600 mm closed-cell phenolic foam insulation boards incorporating a factory-applied welded mesh render carrier, forming a system panel (further information can be found on the following website: www.brc-special-products.co.uk). The render carrier is formed from either a 1.6 mm diameter galvanised, or a 1.5 mm diameter stainless steel, wire mesh interwoven with a perforated layer of chipboard paper designed to provide uniform suction for the render. The system panels are mechanically secured to the external surface of the wall using stainless steel and polypropylene fixing pins, which are tapped into holes drilled through the panels and at least 50 mm into the wall. The fixings are positioned at 300 mm vertical centres and 600 mm horizontal centres. The panels, once fixed to the wall, receive a 10–12 mm thick cement-based standard or lightweight basecoat render. This is followed by one of a range of finishes, including normal and lightweight cement-based dash receivers for dry spar dash and cement-based coloured topcoat. The combined render basecoat and finishing coat thickness is 20 mm.

The high-performance Thermalath phenolic foam insulation boards are supplied in a range of thicknesses between 25 and 70 mm.

Externally applied insulation: domestic case study

The houses shown (Photographs 4.1–4.8), which have been upgraded with externally applied insulation and render, were originally uninsulated, built with solid concrete block and panel walls and a render finish. Prior to the renovation, the houses looked untidy, and the render was cracked, with some parts of the render damaged and missing (Photograph 4.1). Where cracks were large, they acted as a water trap, providing a risk of damp penetrating the structure. While the external façade was unsightly, the main structure of the buildings was sound.

Owing to the old single-glazed windows and poorly fitting doors, the houses were draughty. None of the walls was insulated, and thus the buildings were cold and expensive to heat. By using an insulation system that was externally applied, a thermal upgrade was carried out with minimum disruption to the tenants. The tenants occupied the buildings throughout the upgrade and refurbishment.

The decision to improve the thermal performance by upgrading using externally applied insulation was relatively straightforward. Upgrading insulation internally causes considerable disruption to the occupants, and it is difficult to achieve an effective thermal encasement. When the insulation is applied externally, the whole structure can be encased in the insulation, with few cold bridges. Insulation that is applied to the inside skin of the wall tends to be disrupted by floors, ceilings, services, built-in fittings and other obstacles; cold bridges are often present owing

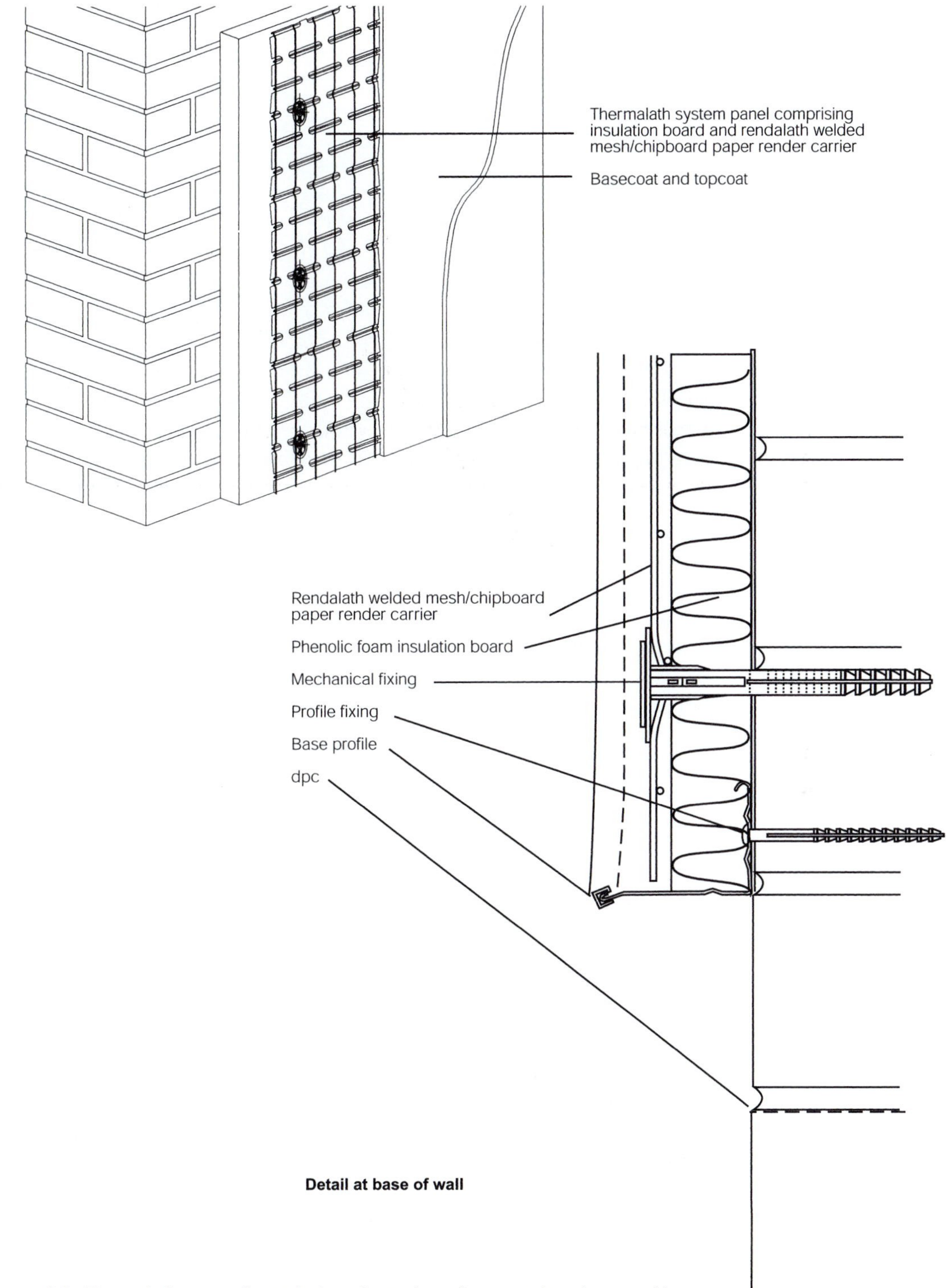

Figure 4.8 *Thermalath externally applied insulation (www.brc-special-products.co.uk)*

Photograph 4.1
Existing house, scaffolding erected in preparation for external insulation

to the gaps in the insulation. Where the aesthetics of the building are not considered important or, as in this case, are in need of a face lift, external insulation coupled with a new finish provides a more effective solution. Photograph 4.1 shows the house prior to the renovation.

Where the render had cracked or where there were breaks in the external envelope, epoxy resin and foam filler were applied to seal the structure. All loose parts of the structure and existing renders were removed. Most manufacturers of external insulation and render materials have their own 'dubbing' (a special render mix) for filling the gaps in the existing finish and treating surfaces.

Photograph 4.2
Insulation is inserted on to fixing brackets and clipped in place

Photograph 4.3
Insulation trimmed around window openings and ventilation and cut to make corners

As part of the thermal upgrade, UPVc replacement double-glazed windows were installed. Replacement cavity ties were installed to ensure that the skins of the cavity remained stable.

The insulation is fixed to the studs and trimmed at the corners. Careful attention is given to the openings in the building. Insulation is neatly cut around window and door openings. Holes are cut for ventilation ducts that are present, extra fixings are used to ensure insulation is properly secured around the openings, and trims are fitted to provide a neat finish.

Photograph 4.4
Glass-fibre scrim (fabric reinforcement) helps the render to bond to the insulation and reduces cracks at insulation joints

In a relatively short period of time, the insulation can be cut and fitted around the building. Where possible, independent scaffolding is used, enabling good access and easy fitting of the insulation to the façade.

Around the wall openings that allow for doors, windows, etc. extra fittings are normally specified to ensure that the insulation is securely fixed. Photograph 4.3 shows the additional mechanical fixings used around the window and at the corner of the building.

Prior to the positioning of the insulation, all cables, outlets, services and features are removed and they are replaced once the insulation and render have been applied.

Glass-fibre scrim (Photograph 4.4) is used to provide a fabric reinforcement over the whole area of the insulation. The scrim bridges the joints in the insulation

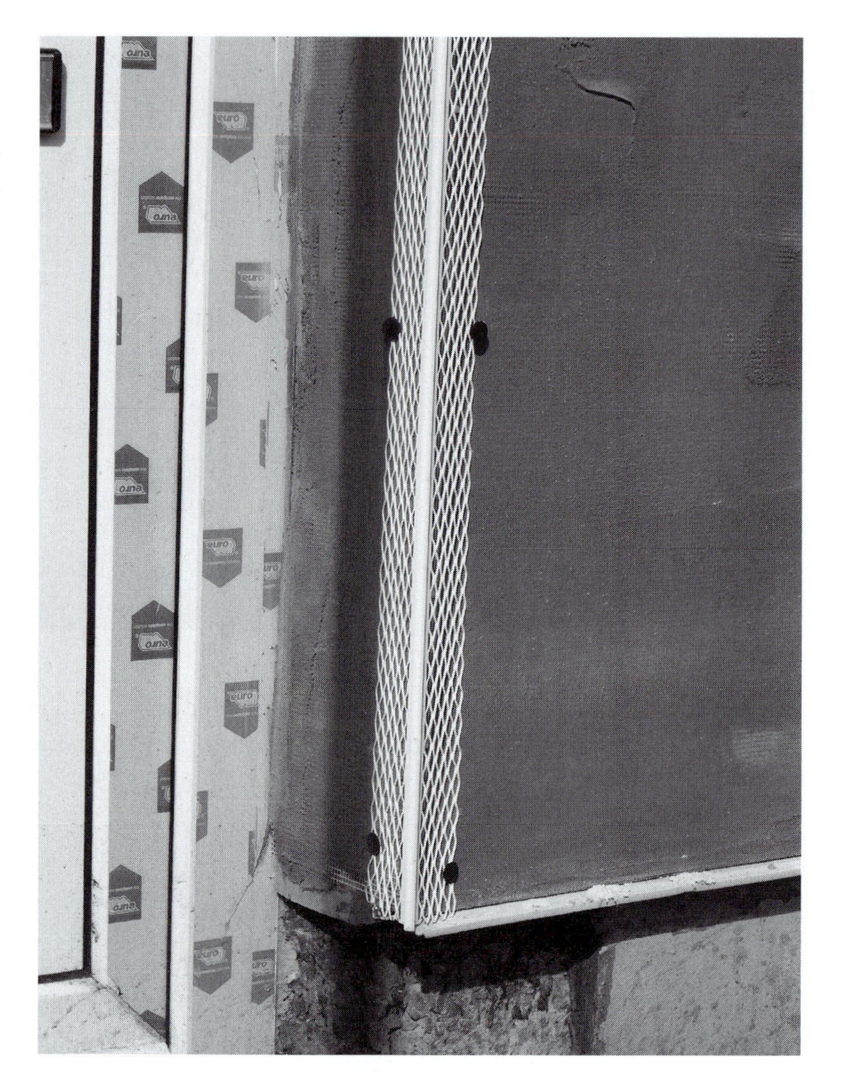

Photograph 4.5
Polyester-coated galvanized steel render corner beads and edge trims fitted to provide a neat edge for the final coat of render (base coat trowelled on the pebble dash finish is to be applied)

and ensures that cracks do not appear in the render. Scrim adhesive was trowelled onto the insulation, and the reinforcing scrim was offered to the adhesive and smoothed into place. The scrim must be completely covered in adhesive and properly bonded to the surface of the insulation. The scrim overlaps 75–100 mm at joints, depending on manufacturer's instructions. Additional patches of scrim reinforcement were used around window and door openings, ensuring additional strength in the render at points of greatest exposure and weakness. The extra patches (500–600 × 250–300 mm) of the scrim can be placed diagonally at the corners.

Photograph 4.5 Polyester-coated galvanised steel render corner beads and edge trims fitted to provide a neat edge for the final coat of render (base coat trowelled onto the pebble dash finish is to be applied).

The base coat of the render is smoothly trowelled onto the scrim in thicknesses 5–8 mm thick, depending on manufacturer's instructions. Most base coats normally need a minimum of three days' drying time, in good drying conditions. Manufacturers' drying times must be strictly adhered to before topcoats are applied.

Polyester coated galvanised steel render beads are used at the external corners, window reveals, door heads and jambs. Movement, expansion and contraction joints must be applied in accordance with the manufacturer's instructions. Movement joints are normally formed out of the same material is the render beads, in this case polyester-coated galvanised steel.

The appearance of a flat brick lintel was achieved using a specialist brick finish render (see Photographs 4.6 and 4.7). The imitation brick, formed out of coloured render, looks like it was constructed with concrete bricks. Once finished, the house looks as if it is a new build (Photograph 4.8).

Externally applied insulation: three-storey apartments case study

The following photographs show blocks of flats that have been upgraded in the same manner as described above (see Photographs 4.9–4.12).

Once the scaffolding is in place, and fixtures and fittings are removed from the face of the wall, the insulation can be applied quickly.

The condition of the substrate to which the insulation is to be fixed must be checked. A full survey must be undertaken on the existing structure prior to the insulation being fitted. Any minor defects in the existing building's fabric and finish must be made good in accordance with the insulation supplier's recommendations. Ensure that a good bond can be achieved between the insulation and existing wall.

Where render has broken away from the substrate, it is normal to remove all of the loose material, make a cut 100 mm clear from the edge of the broken render and then remove the render. Between the cut line and the broken area the render is removed, exposing substrate. This ensures that the remaining render is stable and secure, and provides a neat area to repair (Figure 4.9).

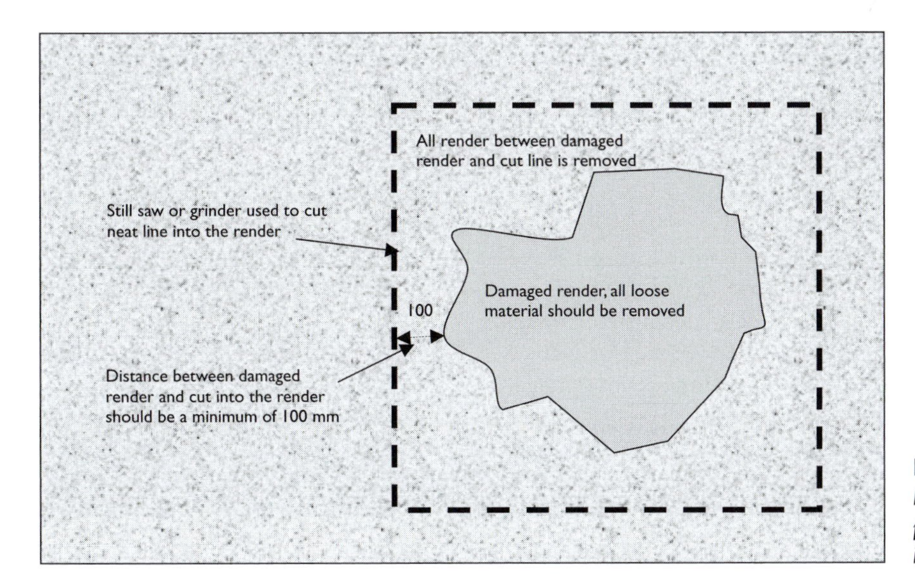

Figure 4.9
Repairing render finish to provide substrate for thermal insulation

Photograph 4.6
Head of window prepared to provide a render brick finish

Photograph 4.7
Brick detail formed out of render. It is difficult to recognise the difference between brick render and bricks

Photograph 4.8 *Finished building, fully insulated without major disruption to occupants*

Photograph 4.9 *External insulation bonded to existing wall*

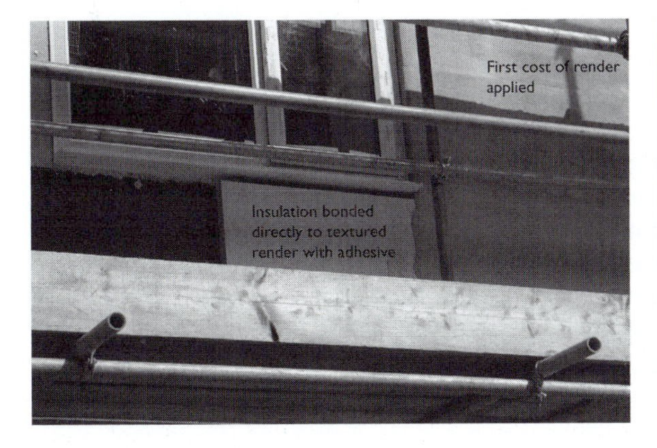

Photograph 4.10
First coat of render applied directly to the external insulation

Photograph 4.11
Existing textured render provided a strong substrate with which to bond the insulation

Photograph 4.12
External upgrade complete: finished flats with new roof covering, windows and a coat of pebble dash render applied to the external insulation

High-rise external insulation: case study

The following multi-storey flats were upgraded with Swislab insulated render system, which is specifically designed for refurbishment projects (www.alumasc-exteriors.co.uk). The insulation slabs chosen for this project were high-density rock fibre, with resins that aid bonding and repel water. Together with the base and decorative render coat, the external insulation protects the façade from the weather. The insulation thicknesses range from 40 to 140 mm.

When upgrading external insulation, the windows are normally upgraded so that the sills protrude beyond the new layer of insulation. Sills are normally required to protrude a minimum of 50 mm beyond the surface (façade) of the building (Photograph 4.13).

The number of fittings required to secure the insulation will depend on the type of insulation fitted, the exposure of the building and the strength of the substrate. Each manufacture has slightly different requirements; however, *in situ* tests should also be undertaken to ensure that all fittings and the insulation are securely fixed. A pull-out test can be carried out on site to determine the size and the number of fittings required per board. In the example shown, two fittings have been installed in the centre of each insulation slab (Photographs 4.14 and 4.15). To ensure a good bond, 6 mm of bedding adhesive is applied immediately before the insulation slabs are applied and should still be wet, ensuring a good bond (Photograph 4.16). Once the insulation is positioned, a glass-fibre scrim coat is fixed in place using 3 mm of scrim adhesive, which is combed or trowelled across the surface of the insulation. Angle, window, door and movement beads must be positioned. The render base coat is then applied in 8 mm layers using a wooden or plastic float and then allowed a minimum of 72 hours to dry (in good weather conditions) before the silicone topcoat and finish are applied (Photograph 4.17).

Photograph 4.13
(facing page) *The block of flats with new windows installed*

SHERBURN COURT

Photograph 4.14
Insulation slabs fixed chemically with bedding adhesive and mechanically with two fixings per slab

Photograph 4.15
Insulation fixed and render applied in bays up to the horizontal and vertical movement joints

Photograph 4.16
Glass-fibre scrim reinforcement fixed in place using 3 mm of scrim adhesive and 8 mm base coat render applied

Photograph 4.17
Completed building

Injected cavity-fill insulation

This form of insulation was developed specifically for improving the thermal properties of existing cavity-walled buildings in order to reduce energy consumption and fuel costs. The insulation, which is injected into the cavity via holes drilled through the outer leaf, may be selected from a number of different proprietary materials. Injected cavity-fill materials currently in use include expanded polystyrene beads and mineral fibres, both of which are capable of greatly reducing the U-value of a standard cavity wall.

Where a cavity-walled building requires thermal upgrading, this method is the most preferable solution, since it can be applied without affecting the existing exterior or interior surface, and with minimal disruption.

Rockwool EnergySaver cavity wall insulation (see Figure 4.10; www.rockwool.co.uk) is a fully dry system that uses granulated rockwool blown into an existing external wall cavity to a density of 30–50 kg/m^3 via holes drilled through the outer leaf. The holes are 18 or 25 mm diameter at 1.5 m centres in a staggered 'W'-pattern, normally located at mortar joint positions. Additional holes may be required, for example at windowsills, at the tops of walls, around airbricks or under gables.

When blown mineral wool is injected into the 50 mm cavity of a standard brick outer-leaf/plastered dense blockwork inner-leaf cavity wall, it will improve the U-value from 1.41 W/m^{2}K to 0.55 W/m^{2}K. If the wall has a 65 mm cavity, the improved U-value will be 0.45 W/m^{2}K. Depending on whether the blockwork was dense or lightweight, a cavity of 75 mm can achieve a U-value of 0.32–0.42 W/m^{2}K (further information on U-values can be found at www.rockwool.co.uk). The ability of the injected cavity systems to improve thermal performance is dependent on the width of the cavity and the cleanliness of the cavity. Any obstructions within the cavity may prevent the blown mineral wool filling the void. Reduced width of the cavity caused by irregular brickwork will reduce the potential effectiveness

Figure 4.10
Rockwool EnergySaver cavity wall insulation

of the insulation material. However, where a cavity exists, cavity fill often provides an efficient and effective way of reducing heating costs.

4.4 Upgrading the thermal performance of roofs

Loss of heat through the roof can represent up to a quarter of a building's total heat loss, and thus thermal upgrading of the existing roof structure will be essential if future heating costs are to be minimised. Several techniques are available for use in upgrading roofs, and the work can usually be carried out with the minimum of disruption to the building and its occupants.

The current Building Regulations limit a roof's U-value to 0.25 W/m^{2}K for new build. The requirements for refurbishment, replacement and upgrading are more stringent. For an extension where the fabric of the extension is newly constructed, a pitched roof with insulation at ceiling level should be 0.16 W/m^{2}K, a pitched roof with insulation at rafter level should be 0.20 W/m^{2}K and a flat roof or roof with integral insulation should be 0.20 W/m^{2}K. The values are the same when replacing elements of a building, e.g. replacing the roof, with the exception of a flat roof or roof with integral insulation, which has a slightly lower value of 0.25 W/m^{2}K.

Thermal upgrading of pitched roofs

Pitched roofs may be upgraded by inserting an additional insulating layer, either at ceiling level, or at rafter level, immediately below the roof covering. Where a ceiling exists beneath the roof space, and provided the roof space is not intended for use, the insulation can be inserted at ceiling level to reduce heat loss into the void above, and ultimately to the exterior (see Figure 4.11a).

If, however, the roof space is to be converted into accommodation in the refurbishment/alteration scheme, it will be necessary to provide the new insulation at rafter level (see Figure 4.11b). The new insulation may also have to be inserted at rafter level in buildings where no ceiling exists and where the accommodation extends into the roof space. Typical examples of this include redundant churches and agricultural barns, where the existing open roof space is retained to preserve the original character; and old factories and warehouses, where ceilings were not normally provided beneath the roof space (see Figure 4.11c).

The various methods that can be used to upgrade the thermal performance of pitched roofs are described below and illustrated in Figure 4.11.

Insulation added at the ceiling level creates cold surfaces in the roof space. When warm air, which is laden with water, passes over the cold surfaces and subsequently cools the air, the cool air is no longer capable of holding the water and gives it up as condensation on the cold surface. Cold roofs are susceptible to condensation forming on the rafters, especially where there is no ventilation to help evaporate any condensation that forms. Ventilation should be used to stop the voids becoming stagnant and to carry condensation and water vapour outside the building. Unsealed loft hatches or breaks in the ceiling allow warm moist air from kitchens, bathrooms and living areas to leak into the cold roof. Adequate ventilation must be incorporated in roofs when adding insulation. When adding insulation to create warm or cold roofs, ensure that ventilation requirements are met; see Figure 4.12.

Pitched roof ventilation

Insulation mats at ceiling level

Where a ceiling exists beneath the roof space, the simplest and most common solution is to lay the new insulation immediately above the ceiling, between the joists or the lower ties of the roof trusses. Flexible insulation mats, supplied in rolls of varying widths and thicknesses, are a cheap and efficient way of upgrading thermal insulation using this method (see Figure 4.11a).

When insulation is provided directly above the existing ceiling in this way, it is essential to ensure that the roof space above the insulation is properly ventilated. Providing better insulation at ceiling level has the effect of reducing the temperature within the roof space, which, in turn, increases the risk of condensation. Effective cross-ventilation is therefore essential in order to minimise the condensation risk and avoid potential damage to the roof structure. This can be achieved by forming ventilation openings at the eaves and inserting airbricks in gable ends.

Rockwool Roll takes the form of lightweight, rolled wool insulation mats available in thicknesses of 80, 100, 150 or 170 mm × 400, 600 or 1,200 mm wide × up to 5 m long, depending on thickness. Thicknesses greater than 170 mm are obtained by using double layers of insulation. The insulation is also supplied in precut batts, which are available in widths of 400 and 600 mm.

The U-value of an existing, uninsulated, tiled 30 degree pitch roof, with sarking felt and a plasterboard ceiling with 38 × 97 mm joists at 600 mm centres, can be improved from 2.0 W/m^{2}K to 0.19 W/m^{2}K using 180 mm (100 + 80) thick Rockwool Roll.

To obtain a U-value of 0.11–0.16 W/m^{2}K, Rockwool recommends the use of a 100 mm thickness placed between the joists and another layer of 170 mm Rockwool roll laid across the joists, at 90 degree to the first layer of insulation. All of the layers of insulation should be tightly butted up, with no gaps, and each layer of insulation should be placed on top of the next avoiding air pockets. Possible U-values achieved using Rockwool are shown in Table 4.11; further examples are available at www.rockwool.co.uk.

Table 4.11 Possible U-values achieved using Rockwool roll between and laid on ceiling joists

Insulation thickness, mm	*Joist spacing 400 mm Joist size 47 × 100 mm*	*Joist spacing 600 mm Joist size 47 × 100 mm*
100	0.22 W/m^{2}K	0.21 W/m^{2}K
150	0.17 W/m^{2}K	0.17 W/m^{2}K
170	0.16 W/m^{2}K	0.16 W/m^{2}K
200	0.13 W/m^{2}K	0.12 W/m^{2}K
300	0.11 W/m^{2}K	0.11 W/m^{2}K

Source: Adapted from www.rockwool.co.uk

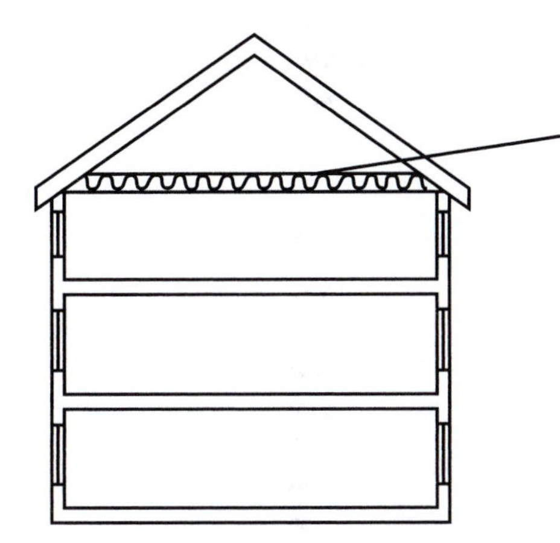
(a) Existing roof void not intended for use
New thermal insulation inserted at ceiling level between or directly beneath ceiling joists

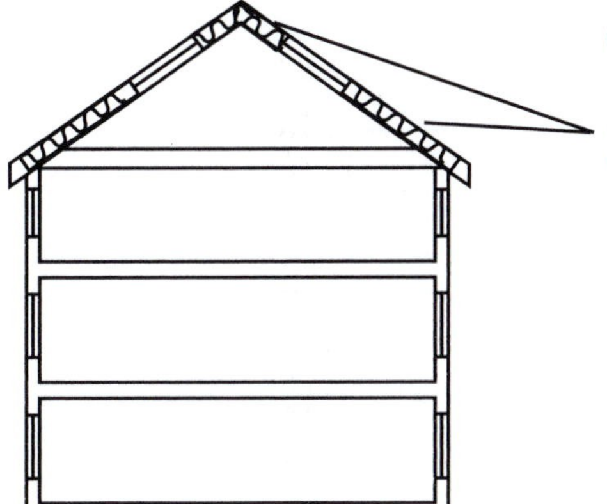
(b) Existing roof void converted to provide additional accommodation
New thermal insulation inserted at rafter level between or directly beneath rafters

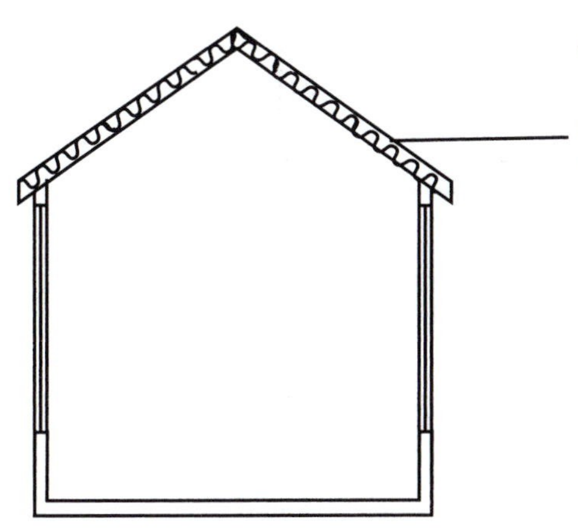
(c) Existing and proposed accommodation extends into roof space (e.g. redundant church)
New thermal insulation inserted at rafter level between or directly beneath rafters

Figure 4.11
Upgrading the thermal performance of pitched roofs

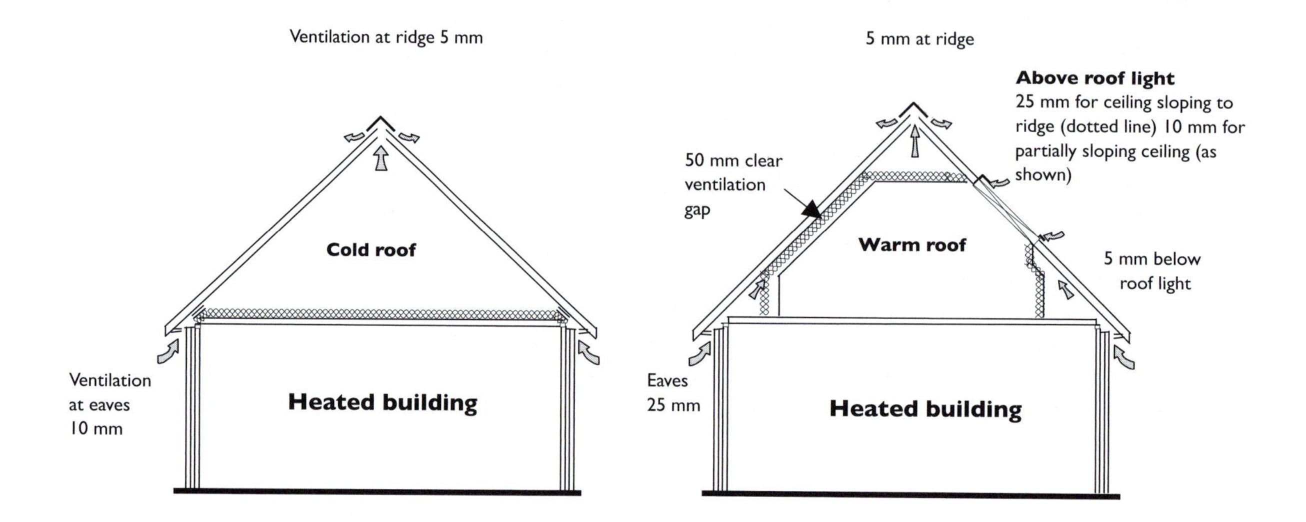

(A) Cold roof, insulation at ceiling level. Roof pitch exceeds 35° or with a span in excess of 10 m

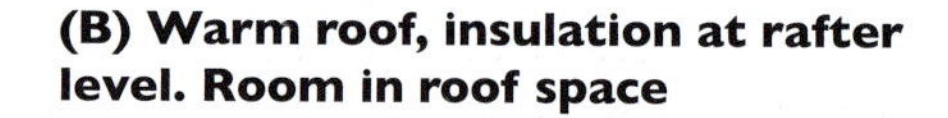

(B) Warm roof, insulation at rafter level. Room in roof space

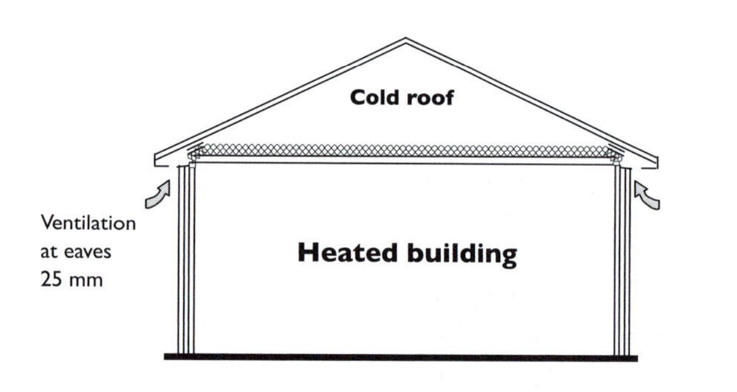

(C) Cold roof, insulation at ceiling level. Roof pitch less than 15°

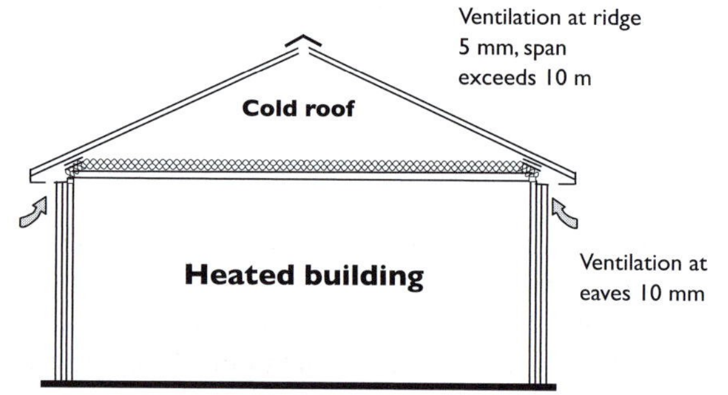

(D) Warm roof, insulation at rafter level. Roof pitch greater than 15°

Figure 4.12 *Hot and cold roofs: ventilation requirements (adapted from Emmitt and Gorse, 2005)*

Loose-fill materials at ceiling level

As an alternative to laying insulating quilts above the existing ceiling, loose-fill insulating materials may be used. These include blown mineral wool, expanded vermiculite and expanded polystyrene beads, all of which can be inserted in varying thicknesses according to the degree of thermal upgrading required (see Figure 4.13b). One disadvantage of using these materials is that their loose, lightweight nature can cause problems such as leakage via ventilation openings and inadvertent blocking of services ducts, flues and so on. As with insulating quilts, it will also be necessary to ventilate the roof space in order to minimise the risk of condensation.

Rockwool EnergySaver blown loft insulation comprises granulated Rockwool blown through a delivery hose directly between and over the existing ceiling joists to the required thickness. The U-value of an existing, uninsulated, tiled 30 degree pitch roof, with sarking felt and a plasterboard ceiling with 38 × 97 mm joists at 600 mm centres, can be improved from 2.0 W/m^{2}K to 0.22 W/m^{2}K using a 175 mm thickness of Rockwool blown loft insulation. Blown insulation in roof spaces is not used as much as the rolls and batts. The rolls and batts are easier to handle and do not suffer from being disturbed by wind.

Thermal boards at ceiling level

Where the existing ceiling is in poor condition and is beyond economic repair, its replacement with a proprietary thermal board such as Gyproc Thermal Board or Styroliner (see Section 4.3) will be a viable solution (see Figure 4.13c). In this way, a new plasterboard ceiling, capable of receiving direct decoration, and an additional insulating layer are provided in a single operation. A further advantage is that thermal boards can be obtained with an integral vapour barrier, sandwiched between the plasterboard and the factory-bonded insulation, therefore significantly reducing the risk of condensation in the roof space above.

Insulation mats or boards at rafter level

Thermal insulation must be provided at rafter level where the roof is to be converted into usable accommodation or where no ceiling exists and the upper rooms extend into the roof space. In most cases, a rigid internal lining of plasterboard or similar material will be fixed to the rafters, and the insulation can therefore be sandwiched between the lining and the roof covering. The insulating material will need to be held in position to prevent settlement or slipping down the roof slope, which could result in gaps opening between sections. This rules out the use of loose-fill materials, the most suitable being rigid insulation boards supported at regular intervals down the roof slope by cross-battens between the rafters. Suitable materials include insulation mats of glass wool or rock fibre and rigid boards of expanded polystyrene or polyurethane foam, all of which are available in a range of sizes and thicknesses to suit individual requirements.

Where this method is used, it is essential that a ventilated airspace of at least 50 mm is provided between the insulation and the sarking felt in order to minimise the risk of condensation. The condensation risk can also be reduced by using

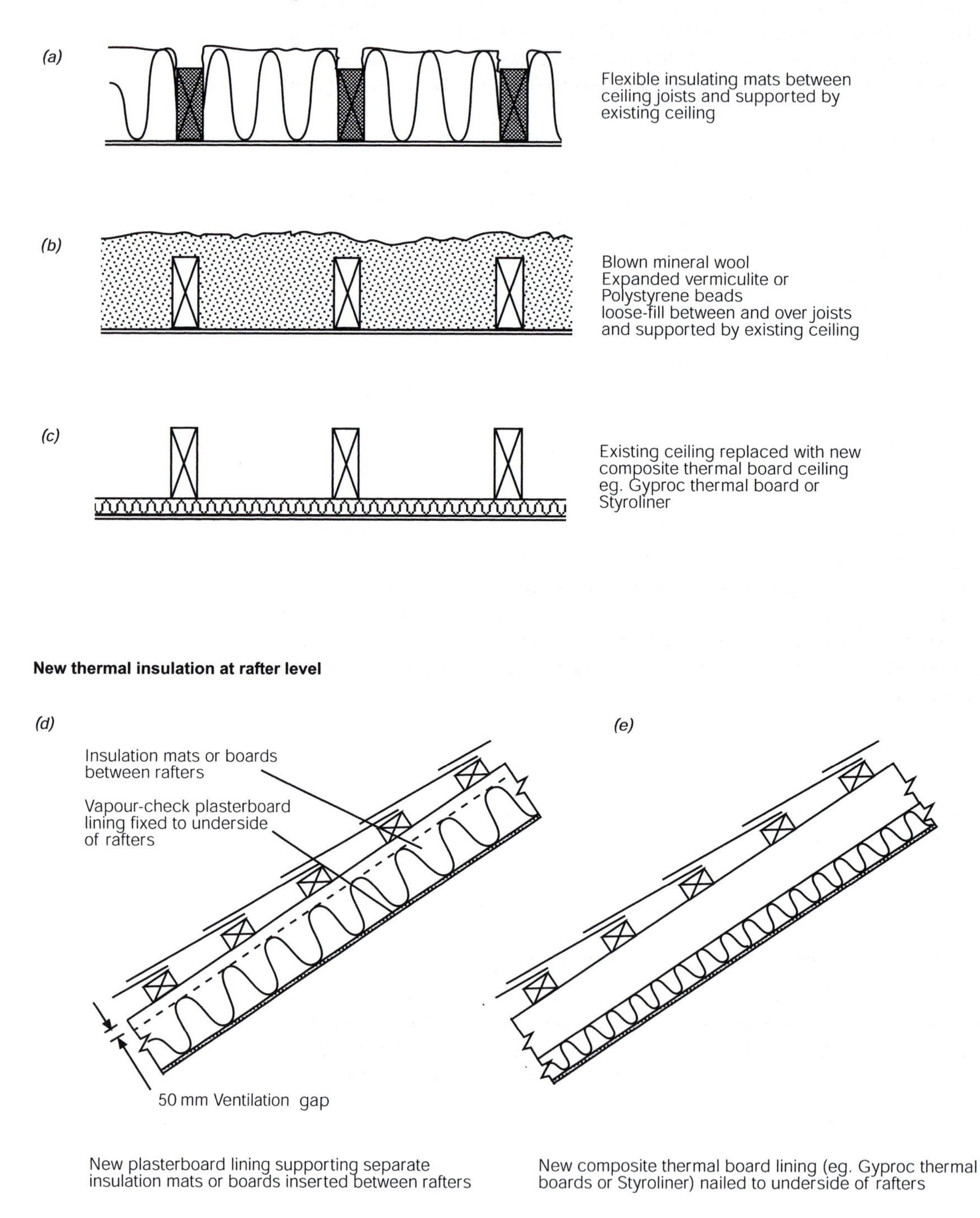
New thermal insulation at ceiling level
(a)
Flexible insulating mats between ceiling joists and supported by existing ceiling
(b)
Blown mineral wool
Expanded vermiculite or
Polystyrene beads
loose-fill between and over joists and supported by existing ceiling
(c)
Existing ceiling replaced with new composite thermal board ceiling eg. Gyproc thermal board or Styroliner
New thermal insulation at rafter level
(d)
(e)
Insulation mats or boards between rafters
Vapour-check plasterboard lining fixed to underside of rafters
50 mm Ventilation gap
New plasterboard lining supporting separate insulation mats or boards inserted between rafters
New composite thermal board lining (eg. Gyproc thermal boards or Styroliner) nailed to underside of rafters

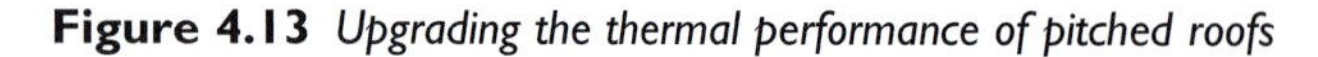

Figure 4.13 *Upgrading the thermal performance of pitched roofs*

vapour-check plasterboard (see Figure 4.13d). A suitable proprietary system of this type is Isover General Purpose Roll glass-wool insulating mat (www.isover.co.uk), available in a range of thicknesses from 60 to 200 mm and widths of up to 1,200 mm, with an internal lining of Gyproc WallBoard Duplex plasterboard (www.british-gypsum.bpb.co.uk). The U-value of a typical, uninsulated pitched roof, covered with slates or tiles and with a sarking felt underlay, can be upgraded from 2.0 W/m^{2}K to 0.15W/m^{2}K. The U-value can be achieved using 200 mm thick Isowool General Purpose Roll lined between rafters and 100 mm glass wool fixed under main rafters between counter-battens, with Gyproc WallBoard Duplex finish. In order to fit this thickness of insulation between the rafters and maintain a 50 mm airspace, it will be necessary to increase the rafter depth by fixing additional battens to the face of the existing rafters, depending on the size of the existing rafters.

A suitable alternative to using a plasterboard lining and separate sandwiched insulation is to use one of the proprietary thermal boards described earlier. Styroliner or Gyproc Thermal Board fixed to the underside of the rafters provides an internal lining, thermal insulation and a vapour check in one operation (see Figure 4.13e).

The thermal upgrading achieved using this method can be improved further by the addition of Isowool General Purpose Roll above the thermal board and between the rafters. For example, 50 mm Thermal Board Super, fixed to the undersides of the rafters and overlaid with 60 mm thick Isowool General Purpose Roll, will upgrade the U-value of a typical, uninsulated pitched roof from 2.0 W/m^{2}K to 0.25 W/m^{2}K. Table 4.12 gives various combinations of insulation between and below rafters that can easily achieve current U-values, providing there is sufficient roof space. Bearing in mind the depth of insulation that is required to meet with current Building Regulations, it may be necessary to remove the roof covering and insulate above the rafters.

Insulation on top and between rafters

Rockwool's Rockfall system, which can be applied on top of, and/or between, the rafters can achieve significant reductions in heat lost through the roof. With the

Table 4.12 U-values achieved using the Rockfall system above and between rafters

Amount of insulation required above and between rafters		*U-value achieved, W/m^{2}K*	
Insulation thickness over the rafter	*Insulation thickness below the rafter*	*47 mm width rafter 400 mm spacing between rafters*	*47 mm width rafter 600 mm spacing between rafters*
40	150	0.22	0.20
50	200	0.17	0.16
80	150	0.18	0.17
80	220	0.13	0.13

Source: Adapted from rockwool.co.uk

addition of insulation over the top of the rafters, a significant reduction in thermal bridging is achieved.

Thermal upgrading of flat roofs

The method used to upgrade an existing flat roof will depend to some extent on its construction. Concrete flat roofs can be upgraded only by adding the new insulation either beneath the slab at ceiling level or on top of the slab. With timber flat roofs, a third option is available, this being to insert the insulation within the void between the ceiling and the roof covering. It should, however, be borne in mind that, where possible, upgrading methods that produce a 'cold roof' should be avoided, as this would involve providing adequate ventilation to remove moisture vapour. A number of methods to upgrade flat roofs are described next.

Thermal boards at ceiling level

The simplest and most cost-effective means of upgrading a flat roof is to provide thermal boards at ceiling level. Proprietary boards, such as Gyproc Thermal Board or Styroliner, can either be fixed directly beneath the existing ceiling or be used to replace the existing ceiling if it is in a poor state of repair (Figure 4.13). The use of thermal board for upgrading rules out the need to gain access to the void (in the case of timber roofs), does not involve external work, and provides a new ceiling capable of direct decoration. However, it does involve some internal disruption and inconvenience to occupants, and it may also be inappropriate where the existing ceiling has ornate plasterwork.

The provision of additional insulation on the underside of the roof structure in this way produces a 'cold roof', that is, the temperature of the roof structure is at, or near, that of the outside air. This condition significantly increases the risk of condensation within the roof construction, and it is therefore essential that precautions are taken to reduce this risk. The use of thermal board with an integral vapour barrier sandwiched between the plasterboard and insulation will minimise the occurrence of condensation, and, if the roof is of timber, the void should also be ventilated.

Insulation mats or boards at ceiling level

This method is only applicable to timber flat roofs and involves inserting the new insulation in the void between the ceiling and roof covering. It is therefore appropriate only in cases where the existing ceiling is in poor condition and needs replacing, and involves inserting insulation mat or rigid insulation boards between the roof joists prior to fixing a new plasterboard ceiling. This method has the advantage of allowing a much thicker layer of insulation to be provided, since the void between the ceiling and roof covering will be at least 100 mm. As with the previous method, however, a cold roof will result, and it will therefore be necessary to use a vapour-check plasterboard and to ventilate the roof void.

Rigid insulation boards on top of existing roof structure

In some cases, it may be necessary to remove and replace the exterior roof covering if it has deteriorated beyond repair, and in this event, new, rigid insulation can be provided before laying the new covering. The Rockwool Hardrock range of insulation boards can be applied to existing timber, concrete and metal roofs to improve their thermal performance (www.rockwool.co.uk). Hardrock Standard Roofing Board sized 1,200 × 600 mm is available in thicknesses from 30 mm to 150 mm, and the U-value of an existing timber-joisted flat roof with plasterboard ceiling can be upgraded to 0.19 W/m^{2}K using 180 mm thick boards (two boards of 85 and 95 mm used to make up the overall thickness). A U-value of 0.13 W/m^{2}K can be achieved using a total thickness of 270 mm Hardrock Standard Roofing Board (2 × 135 mm boards).

Although providing the new insulation on top of the existing roof structure involves external work, there are a number of significant advantages:

- Adding the insulation externally produces a 'warm roof', which minimises the risk of condensation within the roof construction.
- There is no disruption to interior finishes and fittings.
- The occupants of the building undergo minimal inconvenience since no interior work is necessary.
- The acoustic insulation is also increased.

Rigid insulation boards on top of the existing roof covering

If the existing roof covering is still waterproof and in good physical condition, the new thermal insulation can be added on top without it being necessary to take up the existing covering. In addition to the advantages of producing a 'warm roof', the lack of disruption to interior finishes and fittings, and minimal inconvenience to occupants, such systems, as they are laid on top of the existing waterproof covering, provide the latter with protection from frost action, high temperatures caused by solar radiation, and mechanical damage.

Roofmate SL-X (also see the LG-X system) is a proprietary closed-cell extruded polystyrene insulation board specifically developed for the external insulation of flat roofs (www.dow.com). The closed-cell structure of Roofmate gives the boards a high resistance to moisture absorption, providing resistance to freeze/thaw cycles and ensuring that the boards retain their insulation properties throughout the life of the building. The Roofmate SL insulation boards are 1,250 mm long × 600 mm wide × 80–130 mm thick with rebated (ship-lap) edges. Roofmate SL can be used to upgrade the thermal insulation of timber, concrete and metal roofs finished with built-up felt, asphalt and other waterproofing systems, the procedure being as follows:

- The existing roof surface is swept clean of any loose gravel chippings. Well-bonded gravel chippings need not be removed but should be covered with a loose-laid cushioning layer such as Ethafoam 222E extruded polyethylene foam sheet (www.dow.com). The existing waterproof layer should be checked prior to laying the new insulation and any necessary repairs should be carried out as Roofmate SL-X is not a cure for membrane failure.

- The Roofmate insulation boards are laid loose in a brick-bond pattern with their rebated edges pushed tightly together.
- The insulation boards are covered with a ballast layer of either gravel (20–40 mm nominal size), 50–90 mm thick, or concrete paving slabs 40 or 50 mm thick. The thickness of the loading-layer will depend upon the thickness of insulation boards selected.

While 90 mm thick Roofmate SL insulation boards with a 75 mm gravel loading-layer will improve the U-value of an uninsulated, 200 mm thick flat concrete roof with a 50 mm screed and plastered ceiling to 0.30 W/m^2K, 120 mm thick boards with 90 mm of gravel will achieve 0.24 W/m^2K. The addition of further insulation will reduce the U-values further.

An alternative to the above, produced by the same manufacturer, is Roofmate LG-X insulation board. This consists of Roofmate closed-cell extruded polystyrene insulation board, self-finished with a 10 mm thick protective layer of modified mortar on the top surface. Because it avoids the need for overall ballasting, it is lighter than the Roofmate SL-X system, being specifically designed for lightweight roofs that cannot support the weight of the ballast required for Roofmate SL-X.

The Roofmate LG-X boards are 1,200 × 600 × 60, 70, 90, 110 or 130 mm overall thickness and have a specially designed tongue-and-groove detail on their long sides to ensure that they interlock when laid: 90 mm (overall) thick Roofmate LG insulation boards will improve the U-value of an uninsulated timber-joisted roof, with three layers of felt on 19 mm wood-wool slabs and a plasterboard ceiling, from 1.7 W/m^2K to 0.34 W/m^2K; 130 mm thick boards will achieve 0.24 W/m^2K. Extra layers of insulation can be added to achieve current Building Regulations.

4.5 Upgrading the thermal performance of floors

Heat loss through the floor of a building is substantially less than that via the external walls and roof, and, for this reason, the floors are often ignored when refurbishing and upgrading buildings. However, as previously indicated, current Building Regulations requirements will have to be met if a ground-floor structure is being substantially replaced. In any event, all energy-conscious refurbishment/improvement schemes should upgrade the thermal performance of ground and other exposed floors where possible. The current Building Regulations require all exposed floors and ground floors to have a U-value of 0.25 W/m^2K or less, for upgrading, replacement and retained floors.

Styrofloor is a factory-bonded laminate of 18 mm thick P5 chipboard and Styrofoam closed-cell extruded polystyrene insulation board (www.panelsystemsgroup.co.uk). This can be used to provide insulation and finish in one unit. The boards are 2,400 mm long × 600 mm wide, with tongue-and-groove edges to facilitate effective jointing. Styrofloor boards can be laid on existing concrete or timber ground floors to upgrade their thermal insulation and provide a new floor finish in a single operation. All skirtings and other fixtures should first be removed. Uneven floors can be levelled using a proprietary levelling compound; and, if particularly damp conditions exist, and in the absence of a damp-proof membrane, 1,000-gauge polythene sheeting should be laid under the Styrofloor. Prior to laying the boards, the floor should be cleaned and all loose material removed. The boards are laid with cross-joints staggered to produce a brick pattern and with a 10–12 mm expansion gap at all wall abutments. The boards are not secured to the floor, but all of the tongue-and-groove edge-joints are bonded with water-resistant PVA adhesive. A suitable compressible foam filler should be fitted around the perimeter of the floor between the boards and walls before the skirting boards are refixed.

Styrofloor boards are capable of upgrading the thermal insulation of existing concrete and timber ground floors to the current Building Regulations standard of 0.25 W/m^{2}K or better. (Note: Styroliner and Styrofloor are registered trademarks of Panel Systems Ltd; www.styrofoam-online.co.uk.)

Alternatively, insulation can be applied independent of the finish, with wood or concrete being laid on top of the rigid insulation. Depending on the type of floor insulation selected, the range of thicknesses varies (www.floormate-online.co.uk). Board thicknesses range from 25 to 150 mm. Floormate, Strofoam and Styrofloor can all be used to insulate the floor.

References

Anderson, B. (2006) *Conventions for U-value calculations*. BRE Report BR 443, Garston: BRE Press.

Building Research Establishment (1990) *Choosing Between Cavity, Internal and External Wall Insulation* (Good Building Guide 5), Watford: BRE.

Building Research Establishment (1993) *Double Glazing for Heat and Sound Insulation* (Digest 379), Watford: BRE.

Building Research Establishment (1994) *Thermal Insulation – Avoiding Risk*, Watford: BRE.

Department of the Environment, Transport & the Regions (2005) *SAP 2005: The Government's Standard Assessment Procedure for Energy Rating of Dwellings*, Watford: BRE.

Emmitt, S. and Gorse, C. (2005) *Introduction to Construction of Buildings*, Oxford: Blackwell Publishing.

Office of the Deputy Prime Minister (2006) *The Building Regulations 2000: L1B Conservation of Fuel and Power in Existing Dwellings, Approved Document*, London: RIBA.

English Heritage (2004) *Building Regulations and Historic Buildings: Balancing the Needs for Energy Conservation with those of Building Conservation: An Interim Guidance Note on the Application of Part L*, London: English Herritage.

Powell Smith, V. and Billington, M. J. (1999) *The Building Regulations Explained and Illustrated*, Oxford: Blackwell Science.

Richardson, B. A. (1991) *Defects and Deterioration in Building*, London: E. & F.N. Spon.

Stephenson, J. (1995) *Building Regulations Explained*, London: E. & F.N. Spon.

Useful web addresses

Alumasc exteriors: cladding, roofing and external thermal insulation.
www.alumasc-exteriors.co.uk

British gypsum: plaster and plasterboard products, insulated plasterboard also used for thermal properties and fire resistance.
www.british-gypsum.bpb.com

Building Research Establishment: U-value links and information on thermal performance.
www.bre.co.uk/uvalues
Building Science.com: provides information on acoustics, thermal control, moisture control, air flow, climate and light.
www.buildingscience.com
Celtex: Information on thermal insulation.
www.celotex.co.uk
CLEAR Comfortable Low Energy Architecture: information on the calculation of U-values and thermal resistance.
www.learn.londonmet.ac.uk/packages/clear/thermal/buildings/building_fabric/properties/resistance.html
Dow: information on energy efficiency, moisture resistance, construction performance and maintenance.
www.styrofoameurope.com
Hukseflux Thermal Sensors: sensors for measurement of heat flux and thermal conductivity.
www.hukseflux.com
Insulated cladding and render association: information on externally applied insulation.
www.inca-ltd.org.uk
Isover: Information on thermal, acoustic and fire safe solutions.
www.isover.co.uk
Kay-Metzeler: expanded polystyrene products.
www.kaymetzeler.com
Kingspan insulation: cladding and insulation for walls, floors and roofs.
www.insultaion.kingspan.com
Knaur: domestic and industrial insulation, eco-friendly solutions.
www.knaufinsulatotion.co.uk
The Oxford Solar Initiative: information on energy efficiency measures for domestic buildings.
http://oxfordsolar.energyprojects.net/
Permarock: renders and external thermal insulation.
www.permarock.com
Pittsburgh-Corning: cellular glass insulation for building and industrial uses.
www.foamglas.co.uk
Planning building and the environment: communities and local government. The Building Regulations and consultation documents can be accessed from this site.
www.communities.gov.uk
Recticel: thermal insulation products.
www.recticelinsulation.com/UK/EN
Rockwool: thermal and acoustic insulation products and information.
www.rockwool.co.uk
S&B EPS: expanded polystyrene insulation products.
www.sandbeps.com
Scottish Executive: Scottish Building Regulations available on this site.
www.scotland.gov.uk
Springvale EPS: expanded polystyrene insulation products.
www.springvale.com
Sundolitt: expanded polystyrene insulation products.
www.sundolitt.co.uk
Styrene: expanded polystyrene insulation products.
www.styrene.biz

Thermal Economics: information on thermal and acoustic insulation.
www.thermal-economics.co.uk
URSA: glass-wool insulation.
www.ursa-uk.co.uk
Vencel Resil: thermal and sound insulation.
www.vencel.co.uk
Whetherby building systems: case studies of externally applied thermal insulation.
www.wbs-ltd.co.uk/refurbishment.htm

More information on thermal conductivities of insulation can be found at the Thermal Insulation Manufacturers and Suppliers Association (www.timsa.ord.uk).

Calculating U-values and heat loss

Building materials: thermal performance and typical thermal conductivities

The following information is useful for estimating U-values and calculating the thermal resistances of building elements.

The thermal conductivity (W/mK), signified by λ, is the property of a material indicating the material's ability to conduct heat. A higher value indicates a better ability of the material to conduct heat. Good conductors of heat are poor thermal insulators.

The thermal resistance (*R*) is a measure of a material's thermal performance taking into account the material's thickness.

Thermal resistance can be expressed as:

$R = d/\lambda$ Thermal resistance = (Units: m^2K/W)

λ = thermal conductivity (W/mK)

d = thickness of material (in metres)

The thermal resistance is calculated by dividing the material's thickness, in metres, by its thermal conductivity. Higher thermal resistance figures amount to a greater resistance to the transfer of heat where thermal transmittance can be expressed as:

$U = 1/\sum R$ = (units are W/m^2K)

Thermal transmittance is more commonly known as U-value. It indicates the rate of heat flow through a complete element of a building, e.g. a wall, roof, window, door or floor. It is the reciprocal of the sum of the thermal resistances of the components that make up the building element. A cavity wall is made up of brickwork, cavity insulation, blockwork and plasterboard. The resistances of all of these components are added together to make up the sum of the resistances. The internal and external surfaces also have resistances; these are also added to the components. The number one is divided by the total of all of the components' resistances.

The reciprocal of the resistance = the U-value.

A lower U-value = greater resistance and better performance.

Total U-value for a building element is:

$$U = {}^{1}/Rsi_{(\text{inside surface resistance})} + \Sigma R_{(\text{sum of components resistance})} + Rso(\textit{outside surface resistance})$$

This is the external and internal surface resistance (*Rsi* + *Rso*) plus the sum of all of the components of the structure that form an element, e.g. a wall:

$$U = {}^{1}/R_1 + R_2 + R_3 \ldots + Rsi + Rso$$

where R_1, and R_2 are the thermal resistances for each component of the building element, e.g. the brickwork, blockwork, cavity, plaster etc.

For example:

$$\text{Total resistance} = \Sigma R_T$$

where

$$\Sigma R_T = R_{(\text{external surface})} + R_{(\text{brickwork})} + R_{(\text{cavity filled with insulation})} + R_{(\text{blockwork})} + R_{(\text{plaster})} + R_{(\text{internal surface})}$$

and

U-value = *1* divided by the sum of all the resistances of all the building materials

$$= 1/\Sigma R_T$$

Tables such as the one shown at the top of p. 101 can be useful for calculating the total resistance and U-values.

$$\text{U-value of cavity wall} = 1/\Sigma R$$

Heat loss through the building fabric $Q_f = \Sigma(UA\,\Delta T)$

where

U = U-value of element

A = area of element

ΔT = difference in temperature

Component/element	*Thickness*	*Thermal conductivity*	*Thermal resistance*
	d	λ	R = d/λ
R-value$_{\text{(external surface)}}$			
R-value$_{\text{(brickwork)}}$			
R-value$_{\text{(cavity filled with insulation)}}$			
R-value$_{\text{(blockwork)}}$			
R-value$_{\text{(plaster)}}$			
R-value$_{\text{(internal surface)}}$			
ΣR			

To calculate the total heat loss of a building, the heat lost through the fabric and heat lost through ventilation need to be calculated and added together:

Heat loss through ventilation $Q_v = C_v NV\,\Delta T/3600$.

C_v = specific heat capacity of air

N = number of air changes per hour

V = volume of building

ΔT = difference in temperature

Total heat loss = heat loss through the building fabric + ventilation heat loss.

*Typical thermal conductivity of building materials: structural and finishing materials**	*Thermal conductivity (W/mK)*
Acoustic plasterboard	0.25
Aerated concrete slab (500 kg/m^3)	0.16
Aluminium	237
Asphalt (1,700 kg/m^3)	0.50
Bitumen-impregnated fibreboard	0.05
Brickwork (outer leaf 1,700 kg/m^3)	0.84
Brickwork (inner leaf 1,700 kg/m^3)	0.62
Dense aggregate concrete block 1,800 kg/m^3 (exposed)	1.21
Dense aggregate concrete block 1,800 kg/m^3 (protected)	1.13
Calcium silicate board (600 kg/m^3)	0.17
Concrete, general	1.28
Cast concrete (heavyweight 2,300 kg/m^3)	1.63
Cast concrete (dense 2,100 kg/m^3 typical floor)	1.40
Cast concrete (dense 2,000 kg/m^3 typical floor)	1.13
Cast concrete (medium 1,400 kg/m^3)	0.51
Cast concrete (lightweight 1,200 kg/m^3)	0.38
Cast concrete (lightweight 600 kg/m^3)	0.19
Concrete slab (aerated 500 kg/m^3)	0.16
Copper	390
External render sand/cement finish	1.00
External render (1,300 kg/m^3)	0.50
Felt–bitumen layers (1,700 kg/m^3)	0.50
Fibreboard (300 kg/m^3)	0.06
Glass	0.93
Marble	3
Metal tray used in wriggly tin concrete floors (7,800 kg/m^3)	50.00
Mortar (1,750 kg/m^3)	0.80
Oriented strand board	0.13
Outer leaf brick	0.77
Plasterboard	0.21
Plaster dense (1,300 kg/m^3)	0.50
Plaster lightweight (600 kg/m^3)	0.16
Plywood (950 kg/m^3)	0.16
Prefabricated timber wall panels (check manufacturer)	0.12
Screed (1,200 kg/m^3)	0.41
Stone chippings (1,800 kg/m^3)	0.96
Tile hanging (1,900 kg/m^3)	0.84
Timber (650 kg/m^3)	0.14
Timber flooring (650 kg/m^3)	0.14
Timber rafters	0.13
Timber roof or floor joists	0.13
Roof tile (1,900 kg/m^3)	0.84
Timber blocks (650 kg/m^3)	0.14
Web of I-stud timber	0.15
Wood-wool slab (500 kg/m^3)	0.10

Insulation materials and thermal conductivity	
Cellular glass	0.038–0.050
Expanded polystyrene	0.030–0.038
Expanded polystyrene slab (25 kg/m^3)	0.035
Extruded polystyrene	0.029–0.039
Glass mineral wool	0.031–0.044
Mineral quilt (12 kg/m^3)	0.040
Mineral wool slab (25 kg/m^3)	0.035
Phenolic foam	0.021–0.024
Polyisocyanurate	0.022–0.028
Polyurethane	0.022–0.028
Rigid polyurethane	0.022–0.028
Rock mineral wool	0.034–0.042

CHAPTER 5 Upgrading the acoustic performance of existing elements

5.1 General

It is now universally accepted that unwanted and intrusive sound is a common environmental problem and one that can be alleviated to a large extent by paying special attention to the acoustic design and construction of buildings. It is also possible, by using a wide range of techniques, to improve considerably the acoustic performance and sound insulation of elements in existing buildings when refurbishment and alteration work are carried out.

5.2 Statutory requirements

Part E of the Building Regulations, 'Resistance to the Passage of Sound', applies only to dwelling-houses, flats, rooms for residential purposes and schools, there being no statutory requirement to provide sound insulation in other commercial buildings (Department of the Environment and The Welsh Office 1992). Both new dwellings and buildings converted to dwellings must comply with the acoustic standards laid down in Part E of the Regulations. Conversion and refurbishment schemes are required to comply with specified parts of the Building Regulations if they undergo a 'Material Change of Use' – Section 4.2 defines situations that constitute a Material Change of Use. With regard to the acoustic performance of buildings, the works must comply with Part E of the Building Regulations ('Resistance to the Passage of Sound') in the cases of the building being used as a dwelling, where previously it was not, and containing a flat, where previously it did not.

Note that Material Change of Use cases are specified in Building Regulation 5 (see Chapter 4), but compliance with Part E is enforced only in the cases referred to above, that is conversion to dwellings and flats. There are, however, many other examples where good sound insulation is important: in hotels, boarding-houses, hostels and so on, where sleeping accommodation is provided, and where intrusive noise is unacceptable to the occupants. Walls and floors separating offices from workshops in industrial buildings, or consulting rooms from waiting areas in medical centres, are further common examples illustrating the importance of good sound insulation. Despite the fact that the Building Regulations do not enforce acoustic requirements in such cases and the multitude of other situations where good sound insulation is desirable, it is essential that all high-quality refurbishment/conversion schemes incorporate acoustic upgrading where necessary to ensure satisfactory environmental conditions for their occupants. Designers are therefore advised to use Part E of the Building Regulations as a guide for all refurbishment/conversion schemes where sound insulation is a key factor.

In order to meet the sound-insulation requirements of Part E of the Building Regulations, conversions of buildings to dwellings or flats must comply with either Approved Document E: Resistance to the Passage of Sound, section 2, which gives examples of wall types that should achieve the set performance standards; section 3, which gives examples of floor construction to achieve performance standards; section 4, which gives guidance on dwelling houses and flats formed by material change of use; or section 5 of Approved Document E, which permits the repetition of constructions that have been built and tested in a building or a laboratory.

The Building Regulations make special allowances for buildings that have been recognised as having historic value (BSI, 1998). Where buildings are of historic value and there is a need to conserve the special characteristics of the building, it may not be possible to meet the acoustic performance standards set by the Regulations. When working with such buildings, the aim should be 'to improve the sound insulation to the extent that is practically possible'. Sound insulation should be undertaken without prejudicing the character of the historic building or increasing the long-term deterioration of the building fabric or fittings. When making alterations to such buildings, consultation should be undertaken with the local planning authority's conservation officer.

Under the Building Regulations, historic buildings include: listed buildings; buildings in conservation areas; buildings that are of architectural and historical interest; buildings of architectural and historical interest within national parks, areas of outstanding natural beauty and world heritage sites; and vernacular buildings of traditional form and construction.

It is possible that the existing wall and floor constructions already meet the requirements for sound insulation without the need for upgrading. This can be demonstrated by *either* showing that the existing wall or floor is generally similar to one of the acceptable constructions for new buildings, given in sections 1 and 2 of Approved Document E (for example, its mass is within 15 per cent of the section 1 and 2 constructions), *or* carrying out a field test on the construction in accordance with the method specified in section 6 of Approved Document E.

It should be understood that floors and walls may need to be resistant to the passage of one or both of the following types of sound (see Figure 5.10):

- Air-borne sound: air-borne sources, such as speech, musical instruments and audio speakers, create vibrations in the surrounding air that spread out and, in turn, create vibrations in the enclosing walls and floors (elements). These vibrations spread throughout the elements and into connecting elements, forcing the air particles next to them to vibrate, and it is these new air-borne vibrations that are heard as air-borne sound.
- Impact sound: impact sources, such as footsteps, create vibrations directly in the element they strike. These vibrations then spread throughout the element and into connecting elements, forcing the air particles next to them to vibrate, these new vibrations being heard as impact sound.

Sound transmission

The interconnected materials allow the sound energy to travel through the structure; as the sound travels to other elements they vibrate and cause the air next to them to vibrate, and sound passes through the structure

Reduction of impact and air-borne sound transmission

Use of resilient materials, incorporating breaks in the structure and using dense materials, reduces sound transmission

Resilient materials can reduce impact sound; using carpet or resilient cushioning on stairs reduces the impact of foot traffic

Breaks in the structure reduce structure-borne sound. The sound energy must change back to air-borne sound to pass through structural breaks; this uses energy and reduces the sound intensity

Dense materials absorb sound energy, reducing the sound energy. Mineral packing is often used to absorb sound

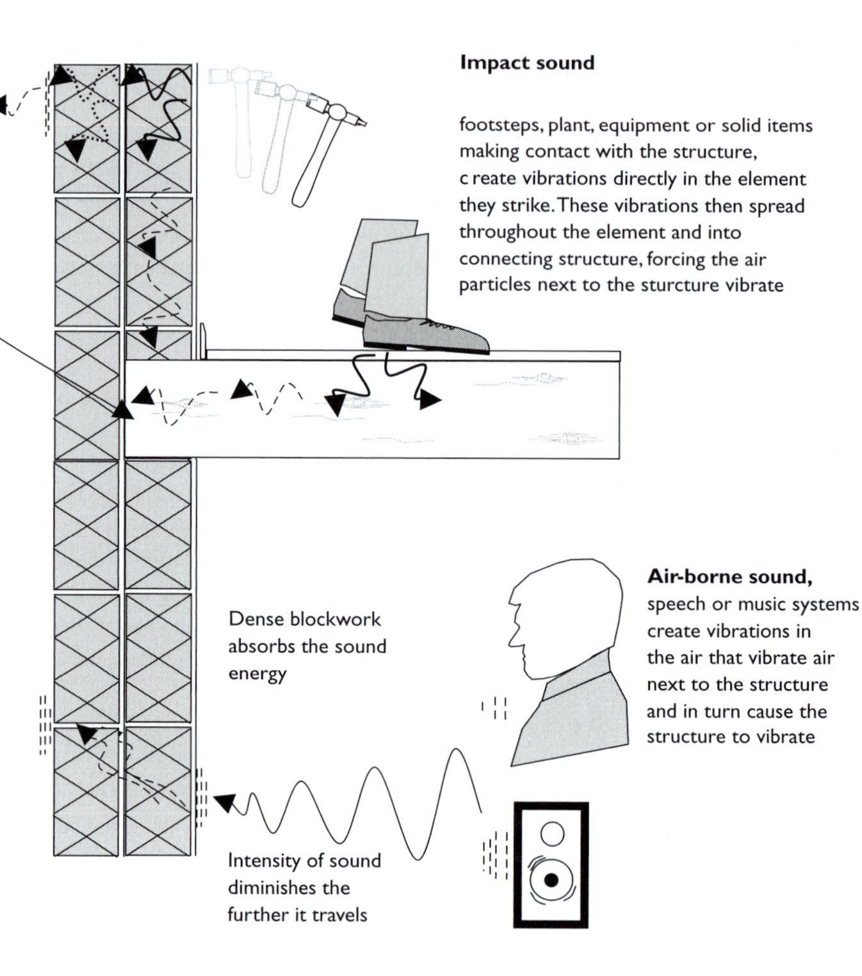

Impact sound
footsteps, plant, equipment or solid items making contact with the structure, c reate vibrations directly in the element they strike. These vibrations then spread throughout the element and into connecting structure, forcing the air particles next to the sturcture vibrate
Air-borne sound,
speech or music systems create vibrations in the air that vibrate air next to the structure and in turn cause the structure to vibrate
Dense blockwork absorbs the sound energy
Intensity of sound diminishes the further it travels

Figure 5.1
Air-borne sound and impact sound (adapted from Emmitt and Gorse, 2006)

5.3 Upgrading the acoustic performance of separating walls

Traditional brickwork or blockwork separating walls between semi-detached houses, terraced houses, flats and maisonettes will normally provide adequate sound insulation, thereby preventing noise nuisance between occupancies. However, some occupants may be less tolerant of noise, and some may generate unacceptably high levels of noise. Some old terraced houses that were built prior to the introduction of fire or sound regulations may have as little as half a brick separating dwellings and may have breaks in the walls at attic level. When renovating such properties, attention must be given to preventing the passage of fire, and consideration should be given to the introduction of sound insulation and acoustic barriers.

Building Regulation E1 requires that any wall that separates one dwelling from another dwelling or from another building, or any wall separating a habitable room in a dwelling from any other part of the same building, which is not used exclusively with that dwelling, shall resist the transmission of air-borne sound. Thus, in conversion/refurbishment work it will be necessary to ensure that such walls are constructed or upgraded to comply with the guidance contained in Approved Document E to the Regulations. The following methods are suitable where there is a need to upgrade the acoustic performance of existing brickwork or blockwork separating walls.

Timber stud frame independent leaf

One of the most effective ways of upgrading the sound insulation of an existing separating wall is to add a separate leaf on one or both sides of the wall to provide a clear air gap, together with the inclusion of a sound-absorbent material in the cavity.

If the existing masonry wall is at least 100 mm thick and plastered on both faces, a separate leaf need only be provided on one side. For any other construction, separate leaves should be built on both sides. The new independent leaf construction (see Figures 5.2 and 5.3) comprises two dense layers of 12.5 mm thick acoustic plasterboard (for example Gyproc SoundBloc; www.british-gypsum.co). Plasterboard, which is designed for sound insulation, generally makes use of a dense core to absorb sound energy. Wallboard plasterboard is fixed, with staggered joints, to a timber stud framework. The staggered joints in the plasterboard prevent air-borne sound simply passing through adjoining gaps or weaker (less dense) parts of the structure. It is essential that the studding is fixed only to the existing floor and ceiling, and not to the wall itself, since any contact with the wall will adversely affect the degree of sound reduction obtained. The best procedure is to fix the head and sole plates to the existing ceiling and floor, respectively, and insert the vertical studs between them to give a continuous gap between the stud framing and the wall. Sound-absorbent mineral wool, 25 mm thick, with a density of at least 10 kg/m^2, is hung either between the vertical timber studs or against the existing wall face by means of a timber-fixing batten along the top edge. The independent leaf should be sealed around its perimeter with mastic or tape. Adequate spacing between the new independent leaf and the existing wall is of vital importance, and Approved Document E requires a gap of at least 25 mm between the inside face of the plasterboard and the existing wall face, and a gap of at least 13 mm between the inside face of the studding and the wall face. Any other type of studding, such as proprietary metal studding, may be used as an alternative to timber.

A wider gap than the minimum required by Approved Document E will further improve the sound insulation upgrading, and the construction in Figure 5.2 shows a clear gap of 125 mm.

When addressing the use of independent composite panels, the Approved Document gives different clearance gaps depending on the fixing of the composite panel. If independent panels are not supported on a frame, they should be at least 35 mm from the masonry core (Figure 5.4). Where panels are supported on an independent frame, there should be a minimum gap of 10 mm.

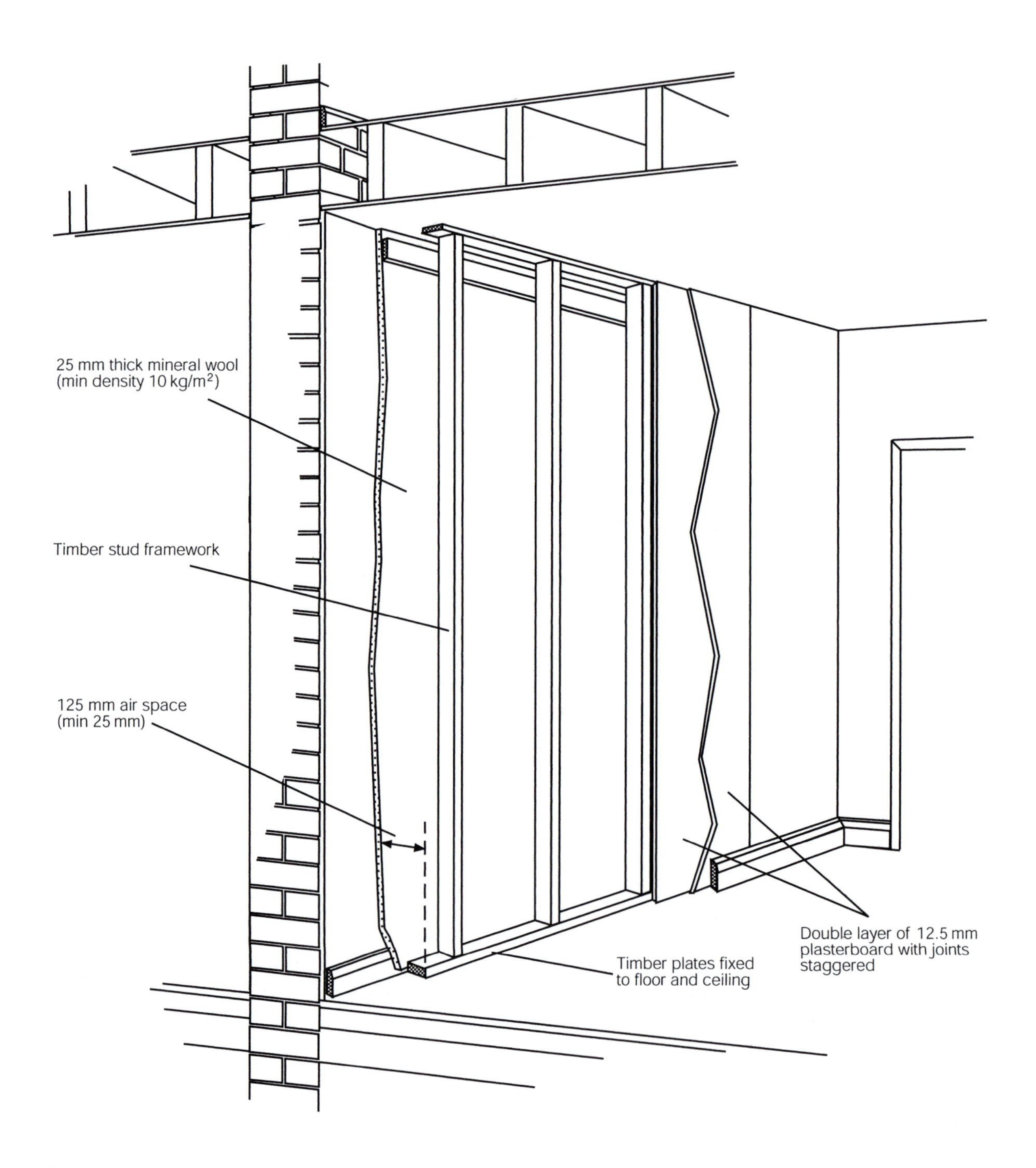

Figure 5.2 *Acoustic upgrading of separating walls: independent leaf construction*

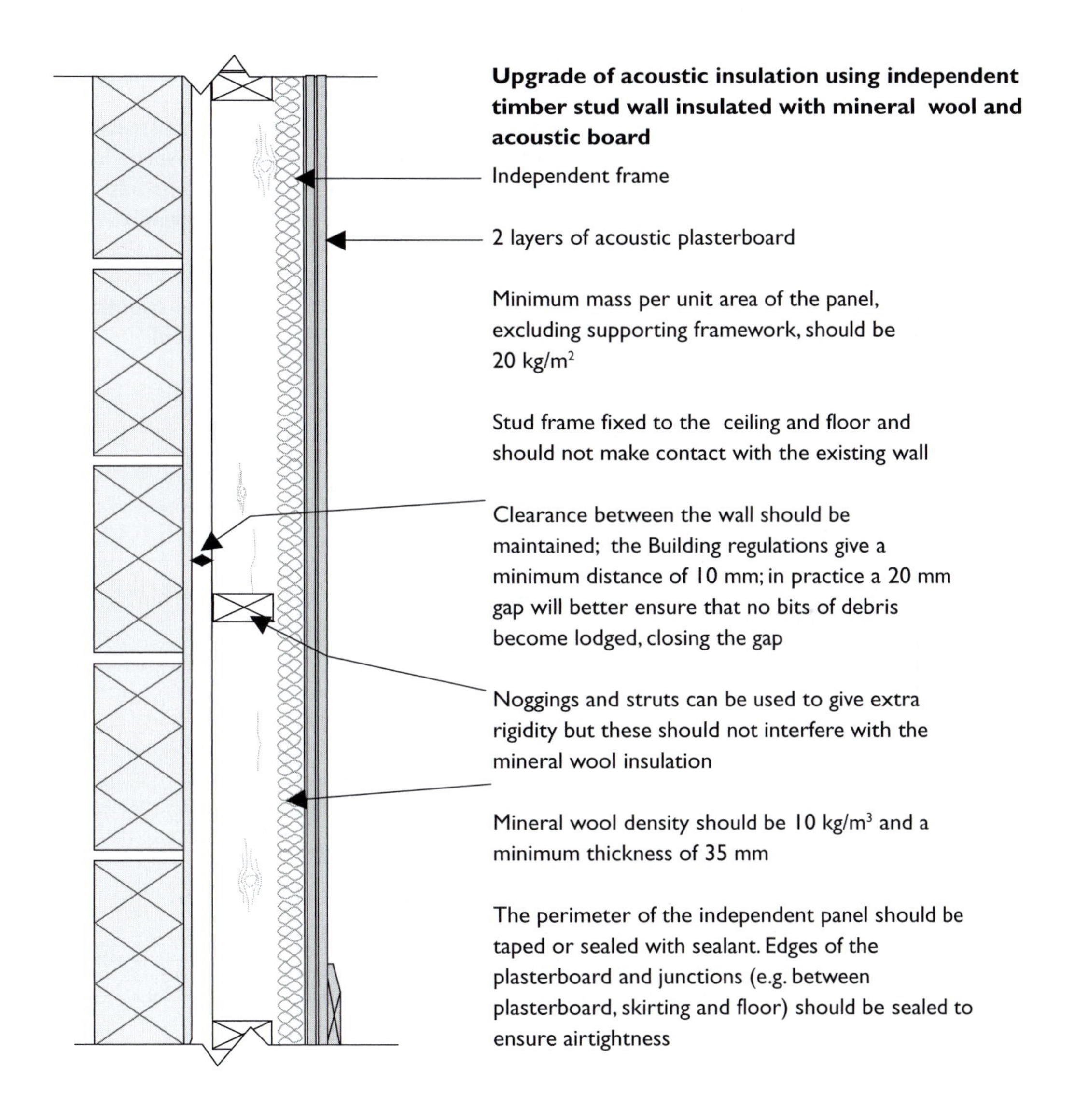

Upgrade of acoustic insulation using independent timber stud wall insulated with mineral wool and acoustic board
Independent frame
2 layers of acoustic plasterboard
Minimum mass per unit area of the panel, excluding supporting framework, should be 20 kg/m²
Stud frame fixed to the ceiling and floor and should not make contact with the existing wall
Clearance between the wall should be maintained; the Building regulations give a minimum distance of 10 mm; in practice a 20 mm gap will better ensure that no bits of debris become lodged, closing the gap
Noggings and struts can be used to give extra rigidity but these should not interfere with the mineral wool insulation
Mineral wool density should be 10 kg/m³ and a minimum thickness of 35 mm
The perimeter of the independent panel should be taped or sealed with sealant. Edges of the plasterboard and junctions (e.g. between plasterboard, skirting and floor) should be sealed to ensure airtightness

Figure 5.3 *Upgrade of acoustic insulation using independent timber stud wall insulated with mineral wool and acoustic board*

Other materials may be used to construct the independent leaf, but the principles of structural separation, high mass and airtightness must be maintained. It should also be noted that the method results in the loss of up to 150 mm of perimeter floor space, which, in smaller buildings, may necessitate the use of a narrower clear gap between the new leaf and the existing wall.

Typical improvements using this method will be from 5 to 10 dB across the audible frequency range, 100–3150 Hz.

Proprietary metal stud frame independent leaf

Figure 5.5 shows an alternative, proprietary independent leaf construction suitable for the acoustic upgrading of existing separating walls. The Gyproc Independent Wall Lining system consists of a framework of Gyproc I-metal studs to which is screwed a single or double thickness of 12.5 or 15 mm Gyproc SoundBloc WallBoard

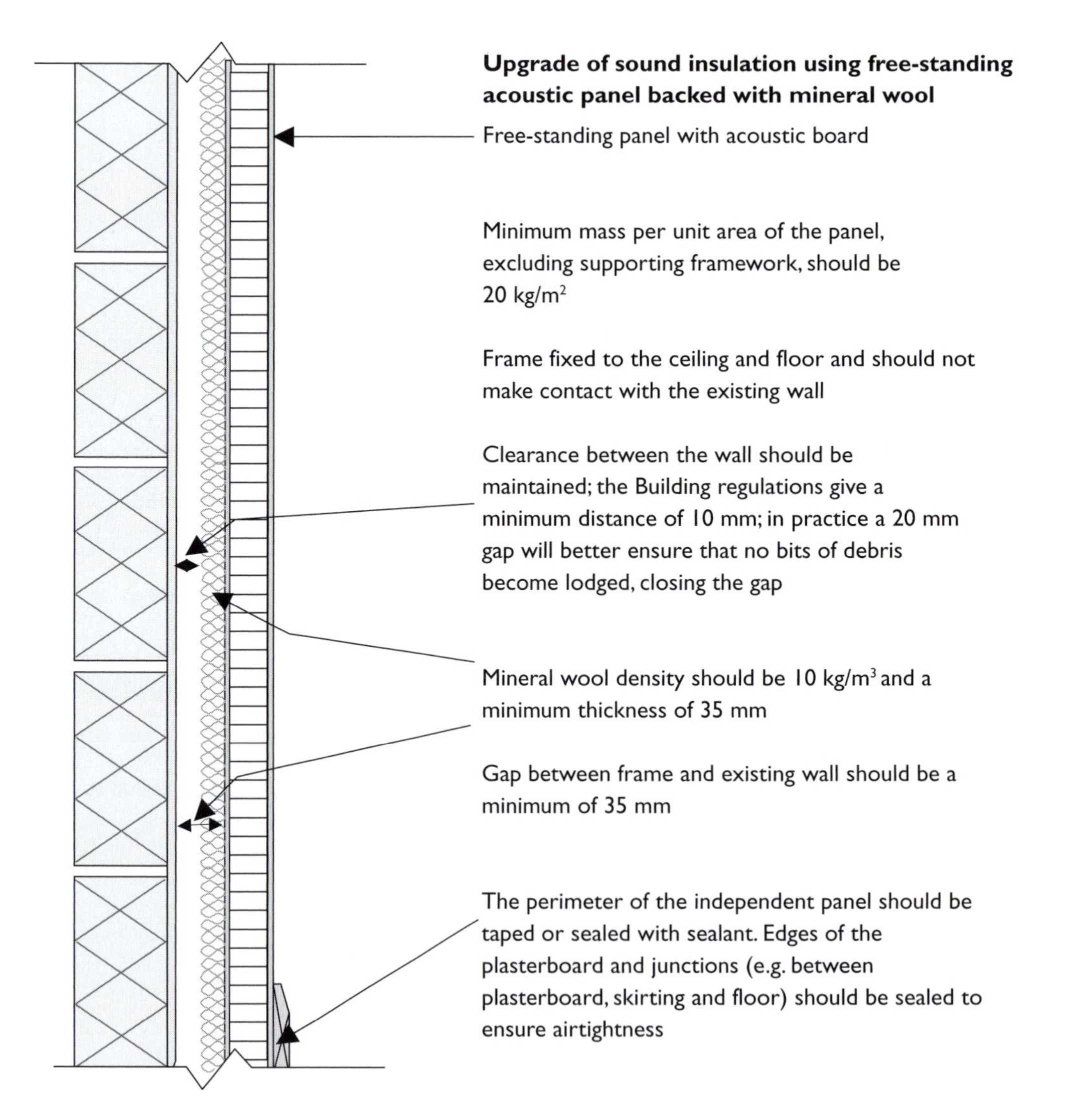
Upgrade of sound insulation using free-standing acoustic panel backed with mineral wool
Free-standing panel with acoustic board
Minimum mass per unit area of the panel, excluding supporting framework, should be 20 kg/m^2
Frame fixed to the ceiling and floor and should not make contact with the existing wall
Clearance between the wall should be maintained; the Building regulations give a minimum distance of 10 mm; in practice a 20 mm gap will better ensure that no bits of debris become lodged, closing the gap
Mineral wool density should be 10 kg/m^3 and a minimum thickness of 35 mm
Gap between frame and existing wall should be a minimum of 35 mm
The perimeter of the independent panel should be taped or sealed with sealant. Edges of the plasterboard and junctions (e.g. between plasterboard, skirting and floor) should be sealed to ensure airtightness

Figure 5.4
Upgrade of acoustic insulation using free-standing panel

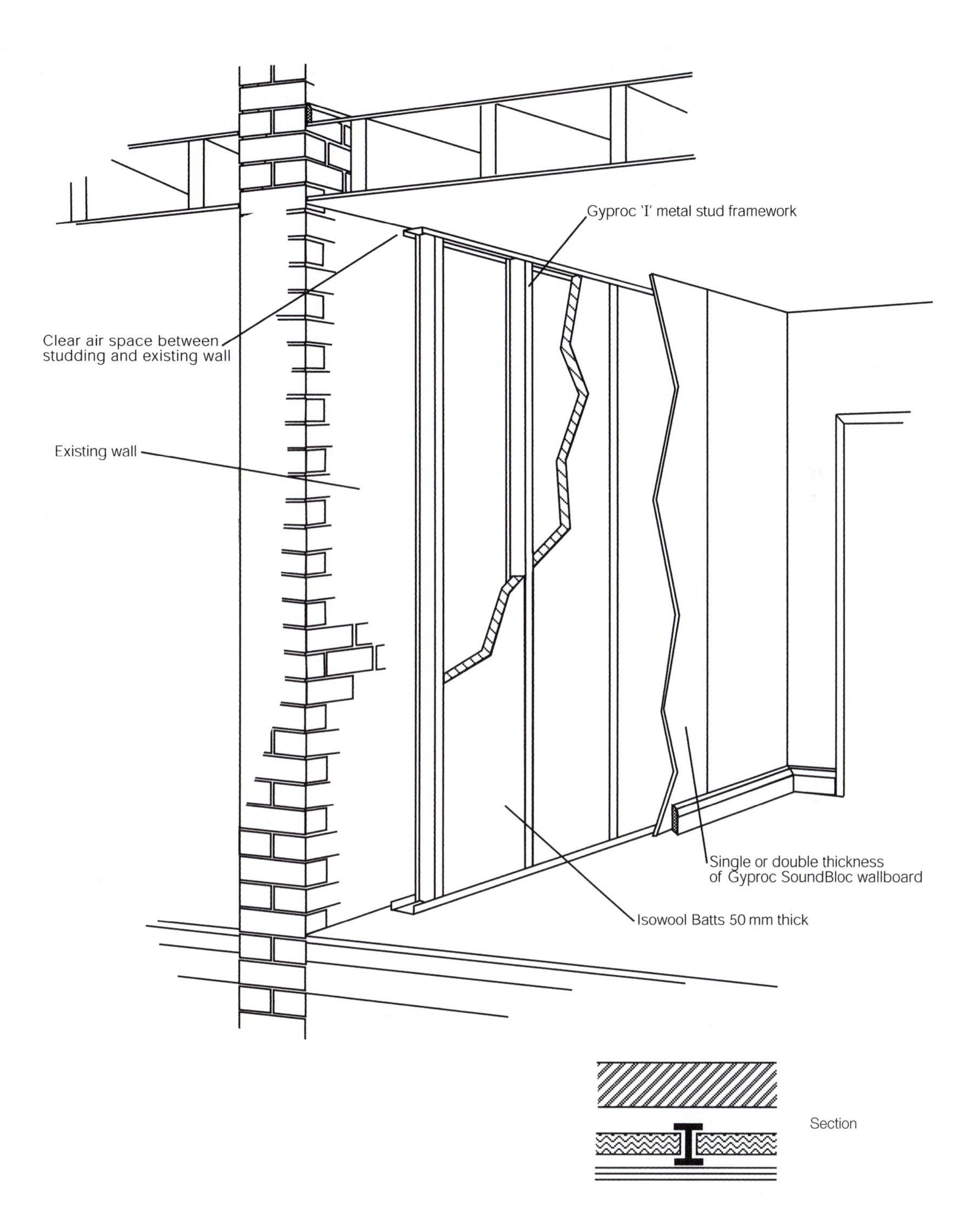

Figure 5.5 *Acoustic upgrading of separating walls: Gyproc Independent Wall Lining system*

(www.british-gypsum.com). Isowool batts, 50 mm thick, are incorporated into the studding cavity to provide a high standard of sound insulation, and the system is made airtight by the use of acoustic sealant applied between the metal framing and the existing structure.

Guidance on reducing sound transmission

The following gives brief guidance on reducing sound transmission through walls:

- Avoid, if possible, chasing walls for all cables and sockets to be recessed, as this creates weak spots in the wall (see Figure 5.6).
- If it is necessary to chase or recess walls to accommodate sockets, do not place sockets back to back; nor should voids or holes be created in the walls.
- Tape and seal all gaps, cracks or holes in the masonry.
- Do not fill the wall cavity with concrete, mortar or other masonry materials; this is likely to create a bridge for the passage of sound.

Figure 5.6
Positioning of sockets to avoid penetrating sound-resisting material

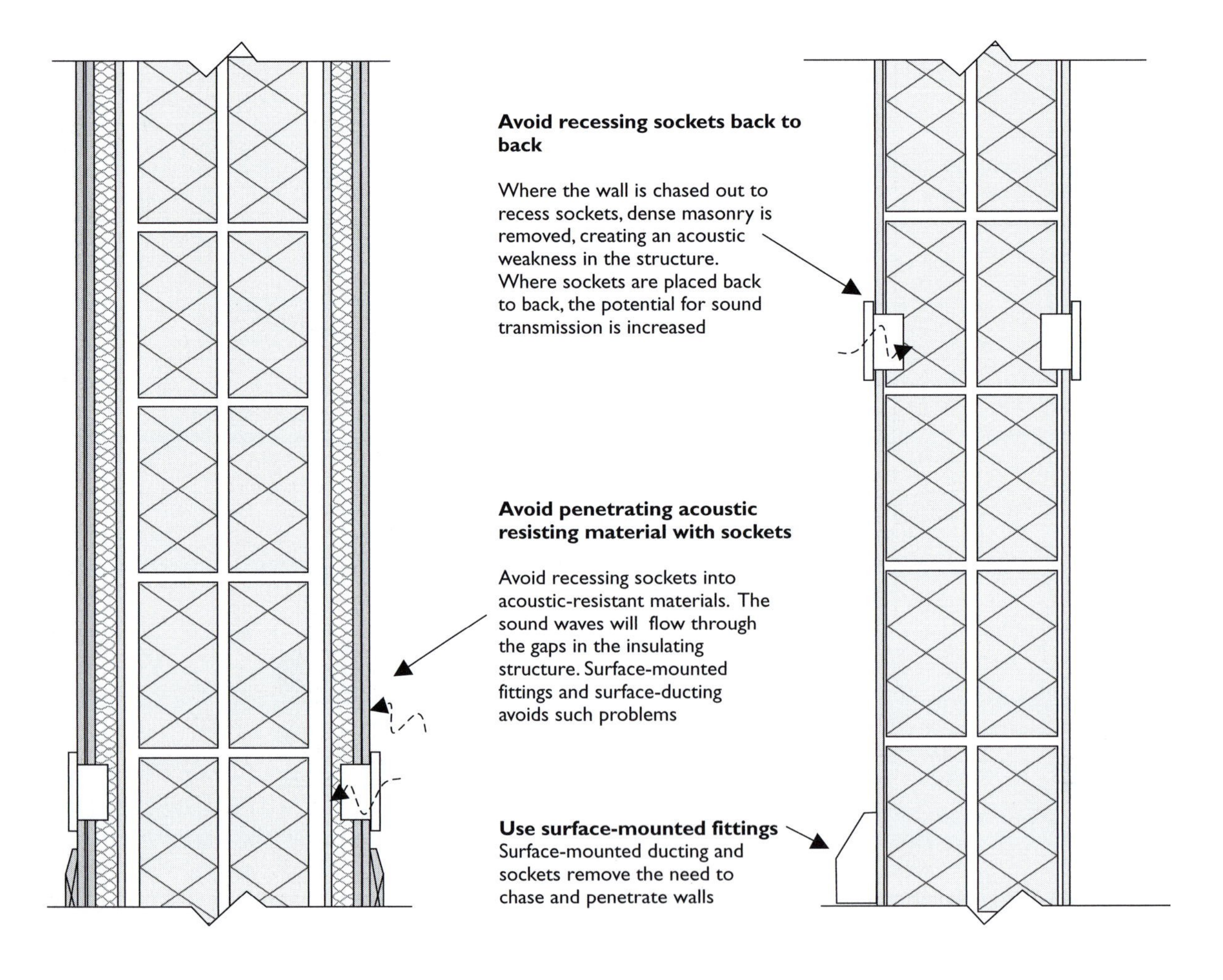

Avoid recessing sockets back to back
Where the wall is chased out to recess sockets, dense masonry is removed, creating an acoustic weakness in the structure. Where sockets are placed back to back, the potential for sound transmission is increased
Avoid penetrating acoustic resisting material with sockets
Avoid recessing sockets into acoustic-resistant materials. The sound waves will flow through the gaps in the insulating structure. Surface-mounted fittings and surface-ducting avoids such problems
Use surface-mounted fittings
Surface-mounted ducting and sockets remove the need to chase and penetrate walls

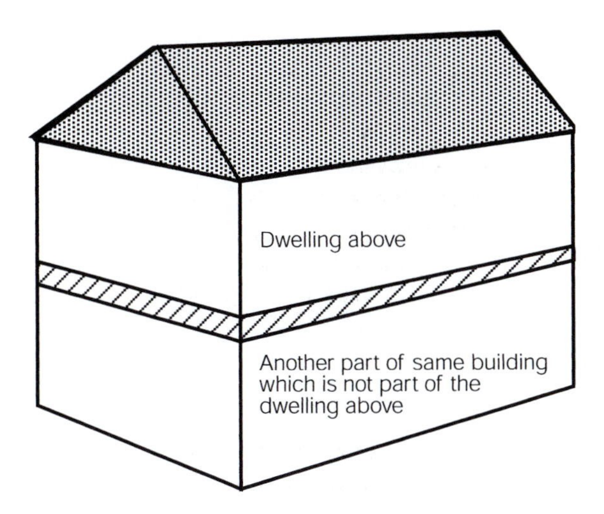

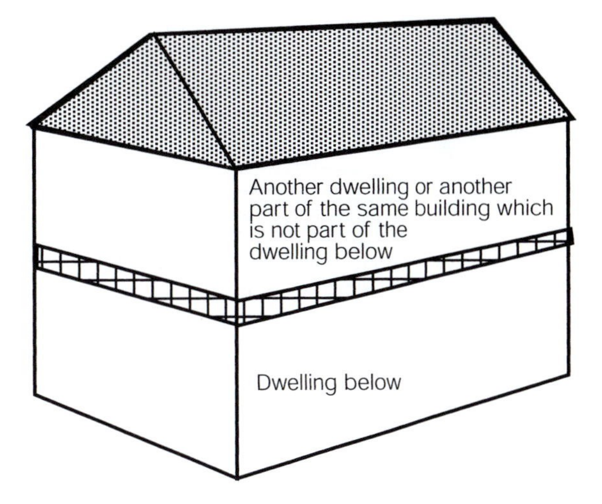

Figure 5.7
Acoustic performance of separating floors: Building Regulations Part E requirements

- Breaks in the structure, such as separate independent wall panels, reduce the passage of air-borne sound.
- Dense materials, such as dense acoustic plasterboard or dense mineral wool, absorb sound energy.
- Close cavities with flexible cavity closer or sock to prevent sound travelling along the cavity.

5.4 Upgrading the acoustic performance of separating floors

Building Regulation E2 requires that any floor separating one dwelling from another dwelling, or from another part of the same building not used exclusively as part of the dwelling, shall resist the transmission of *air-borne* sound. Building Regulation E3 requires that a floor above a dwelling separating it from another dwelling, or from another part of the same building not used exclusively as part of the dwelling, shall resist the transmission of *impact* sound. The sound insulation requirements between dwellings are explained by Figure 5.7. Five different methods of upgrading the sound insulation of existing timber separating floors, all of which meet the requirements of the Building Regulations, are described in detail below.

Floating platform floor

One of the most effective ways of upgrading the sound insulation of a separating floor is to float a new, dense, floor surface on a layer of resilient mineral wool, as shown in Figure 5.8:

- If the existing floorboarding is sound and level, it can be retained, but plain-edged floorboarding, where there are often gaps between the board edges,

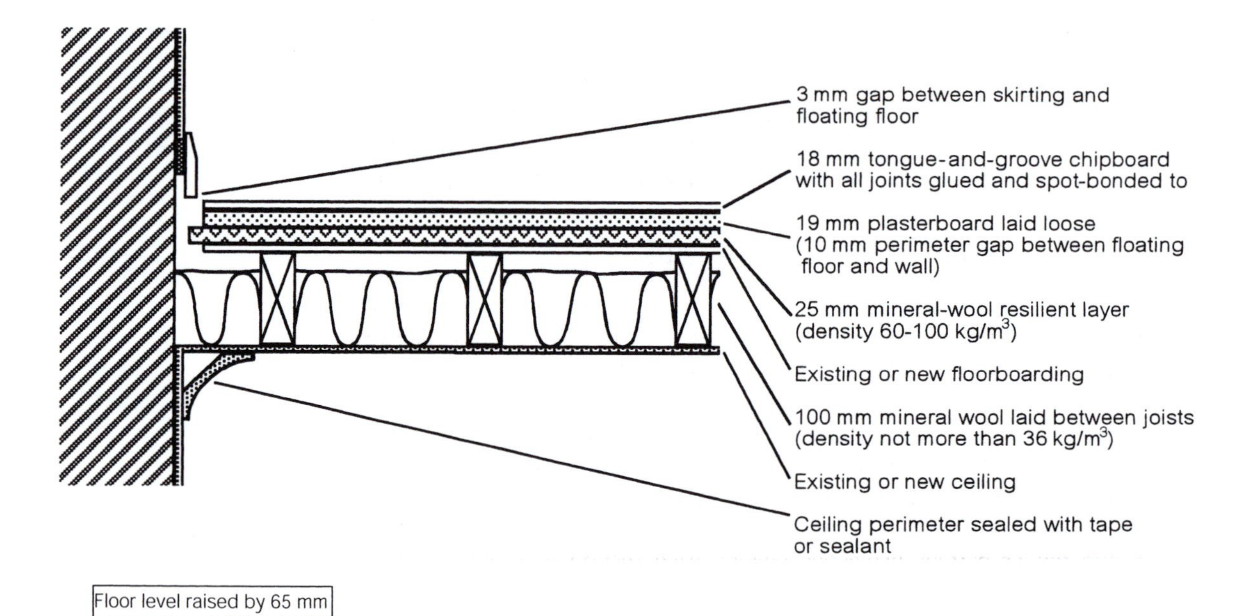

Figure 5.8
Acoustic upgrading of separating floors: floating platform floor

should be sealed with an overlay of hardboard sheeting. The existing ceiling can also be retained if it is in good condition but should be at least 30 mm thick (plasterboard and/or plaster). If it is less than 30 mm thick, it should be upgraded to this thickness using plasterboard with staggered joints.

- A layer of mineral wool, 100 mm thick, with a density of not more than 36 kg/m^3, is laid between the joists within the floor void. This entails gaining access to the void, either by lifting and re-laying the floorboarding, or, where applicable, during replacement of the existing ceiling if this proves necessary.
- 25 mm mineral wool, density 60–100 kg/m^3, is laid over the floorboarding, followed by a layer of 19 mm plasterboard laid loose.
- 18 mm tongue-and-groove chipboard is laid over, and spot-bonded to, the plasterboard, with all joints glued.
- A perimeter gap, 10 mm wide, must be left around all edges of the new floating floor to prevent sound-waves being transmitted from the floor into the walls. For the same reason, a 3 mm gap must be left between the skirting and the floating floor surface.
- All air-borne sound paths at the ceiling perimeter must be sealed with tape or acoustic sealant. All other air-borne sound paths, for example where services penetrate the floor, must also be fully sealed.
- Typical improvements in sound insulation using this method of upgrading will be from 4 to 8 dB for air-borne sound over the frequency range 100–3,150 Hz, and slightly better for impact sound.

Two important factors that must be borne in mind when considering the use of a floating platform floor, are the increase in loading and the raising of the existing floor level. The additional layers of chipboard, plasterboard and mineral wool will add to the dead loading on the existing floor joists, and their ability to carry this extra loading must be verified before installing the new floating floor. Also, the existing floor level will be raised by approximately 65 mm, affecting door thresholds, skirtings, services, sanitary fittings and so on, and the designer must ensure that the associated problems can be resolved before opting for this method of upgrading.

Floating floor on resilient strips with heavy pugging

This method, shown in Figure 5.9, entails replacing the existing floorboarding with 18 mm tongue-and-groove chipboard floated onto 25 mm mineral-wool resilient strips laid along the top edges of the existing floor joists:

- The existing ceiling can be retained if it is in good condition but should be at least 30 mm thick (plasterboard and/or plaster). If it is less than 30 mm, it should be upgraded to this thickness using plasterboard with staggered joints.
- The new chipboard surface is made up in sections nailed or screwed to 45 × 45 mm timber battens projecting beyond each section to enable adjacent sections to be screwed together when laid. The chipboard flooring is glued along all edges.
- The floor void between the existing joists receives a layer of heavy 'pugging' (or sound-deadening material) to a density of 80 kg/m^2. This may be dry sand (approximately 50 mm thick), 2–10 mm limestone chips or 2–10 mm whin aggregate (both approximately 60 mm thick). If there is doubt as to whether the ceiling is capable of supporting the pugging, it should be laid onto plywood pugging boards, supported by timber battens fixed to the sides of the joists.

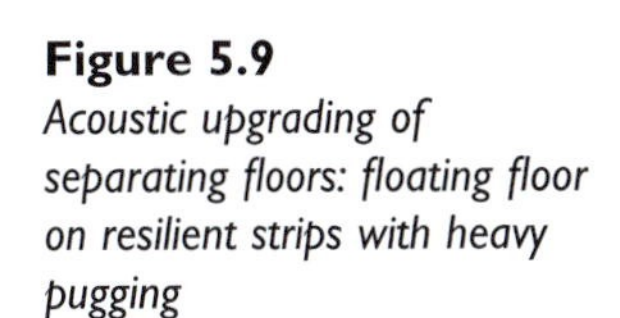
Figure 5.9
Acoustic upgrading of separating floors: floating floor on resilient strips with heavy pugging

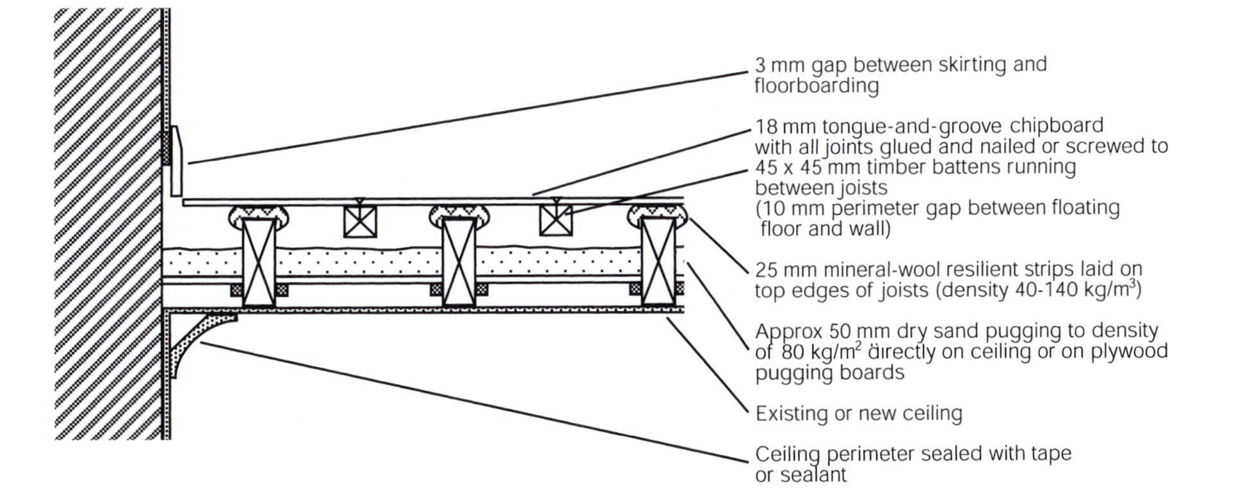

Floor level raised by 10 mm

- As with the floating platform floor, a 10 mm wide perimeter gap must be left around all edges of the new floating floor, and a 3 mm gap must be left between the skirting and the floor to prevent sound transmission into the walls. The ceiling perimeter must also be sealed, along with all other air-borne sound paths.

The principal advantage of this method, in comparison with the floating platform floor, is that the floor level is raised by only 10 mm (caused by the 25 mm mineral-wool resilient strips crushing down to 10 mm).

As with the floating platform floor, this method adds to the dead loading on the existing joists, and it is essential, therefore, that their ability to carry the extra loading is checked before any work is carried out.

Proprietary resilient flooring system

Figure 5.10 shows the Gyproc sound-insulating flooring system suitable for upgrading existing timber-joisted floors to sound-resisting floor standard:

- After removing the existing floorboarding, continuous metal channel sections, with integral resilient strips, are fitted over the tops of the existing floor joists and located on plastic clips. Gyproc Plank (plasterboard), 19 mm thick, is cut and fitted between the joists, its cut ends resting on the channel flanges.
- Isowool General Purpose Mineral Wool Roll, 100 mm thick, is laid between the joists and resting on Gyproc Resilient Bars, fixed to the underside of the joists.
- A double-layer ceiling lining of either Gyproc Plank and 12.5 mm Gyproc Sound Bloc (high-density plasterboard), or two layers of 15 mm Gyproc Sound Bloc, is screw-fixed to the resilient bars.
- The floor is completed by screw-fixing 21 mm thick floorboarding (softwood or chipboard) through the Gyproc Plank into one flange of the metal channel section.

The Gyproc floor is simple to install and raises the floor level by only 7 mm, significantly less than in the previously described methods.

As with the other upgrading methods, the loading capacity of the existing joists should be checked to ensure they are capable of sustaining the additional weight of the system.

Independent ceiling

The methods described above all affect the existing floor surface and floor void, and all raise the floor level by varying amounts. An alternative means of upgrading the acoustic performance of a separating floor, and one that does not affect the existing construction in any way, is to add a new, independent ceiling beneath it. This should be carried on its own set of joists and spaced as far below the existing ceiling as possible, with an acoustically absorbent material between the new and

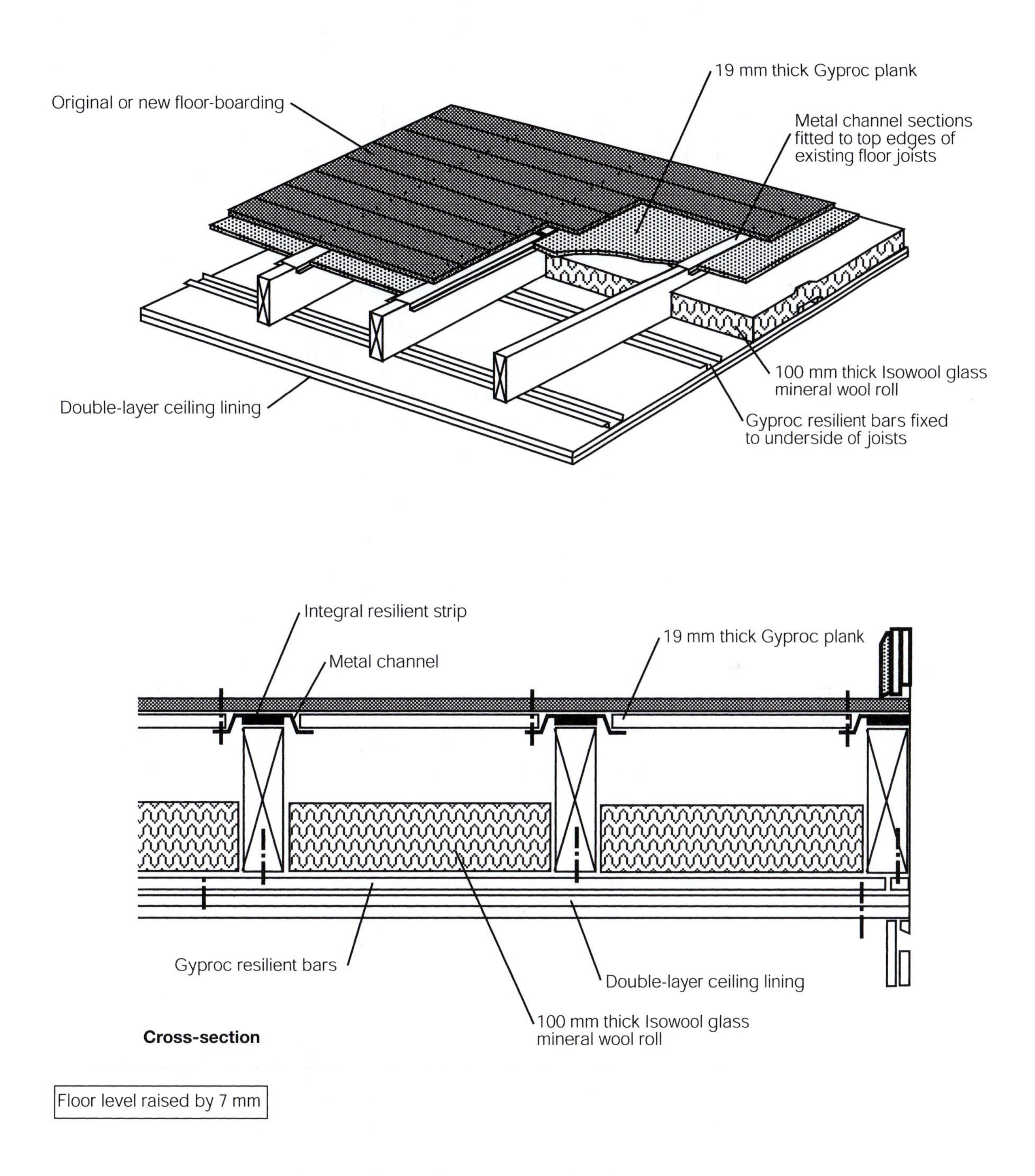

Figure 5.10 *Acoustic upgrading of separating floors: insulated with Gyproc planks*

existing ceilings. The construction method is shown in Figure 5.11 and involves the following operations:

- If the existing ceiling is less than 30 mm thick, it should be upgraded to this thickness using plasterboard with staggered joints.
- Clear air paths between the existing floorboarding, which is likely if the boards are plain-edged, should be sealed by overlaying with 3 mm hardboard sheeting.
- The new ceiling joists, which are of smaller cross-section size than the floor joists, are fixed beneath the existing ceiling. They can be supported by either notching their ends over timber bearers fixed to the walls, or by metal joist hangers.
- The new independent ceiling, comprising two layers of plasterboard having a total thickness of at least 30 mm, with their joints staggered, is fixed to the underside of the new joists. The clear gap between the new and existing ceilings should be at least 100 mm, and there should be a clear gap of at least 25 mm between the top of the new independent ceiling joists and the underside of the existing floor.
- Absorbent mineral-wool quilt, at least 100 mm thick, with a density of at least 10 kg/m^3, is laid on top of the new ceiling, between the new joists.
- The perimeter of the new independent ceiling should be sealed with tape or acoustic sealant.
- Typical improvements in sound insulation using this method will be from 5 to 10 dB for air-borne sound over the frequency range 100–3,150 Hz, and slightly better for impact sound.

It should be borne in mind that the use of an independent ceiling for upgrading sound insulation might not be feasible in all situations. If the existing headroom in the room below is limited, a further reduction of at least 130 mm, owing to the introduction of a new independent ceiling, might not be acceptable. The existing ceiling might have an ornate finish, or be an important feature in a listed building, in which case covering it with a new ceiling might not be permissible.

Proprietary laminated acoustic flooring system

Figure 5.12 shows a proprietary, laminated, sound-insulating flooring system designed for upgrading existing floors to meet Building Regulations requirements:

- The 2,400 × 600 mm laminated panels, comprising 18 mm chipboard sheet factory-bonded to a 25 mm thick resilient layer of glass fibre, density 112 kg/m^3, are laid directly onto the existing floorboards with their joints staggered. The chipboard edges of the panels are tongued and grooved, and bonded with adhesive when laid. A number of manufacturers make such panels; the construction and properties vary, but the principles are the same.
- Prior to laying the new flooring panels, the edges of the existing floorboards are sealed with an acoustic mastic, and a self-adhesive, compressible perimeter strip is fixed to the wall. When laid, the panels should compress this strip by about 2 mm to provide a seal against air-borne sound.

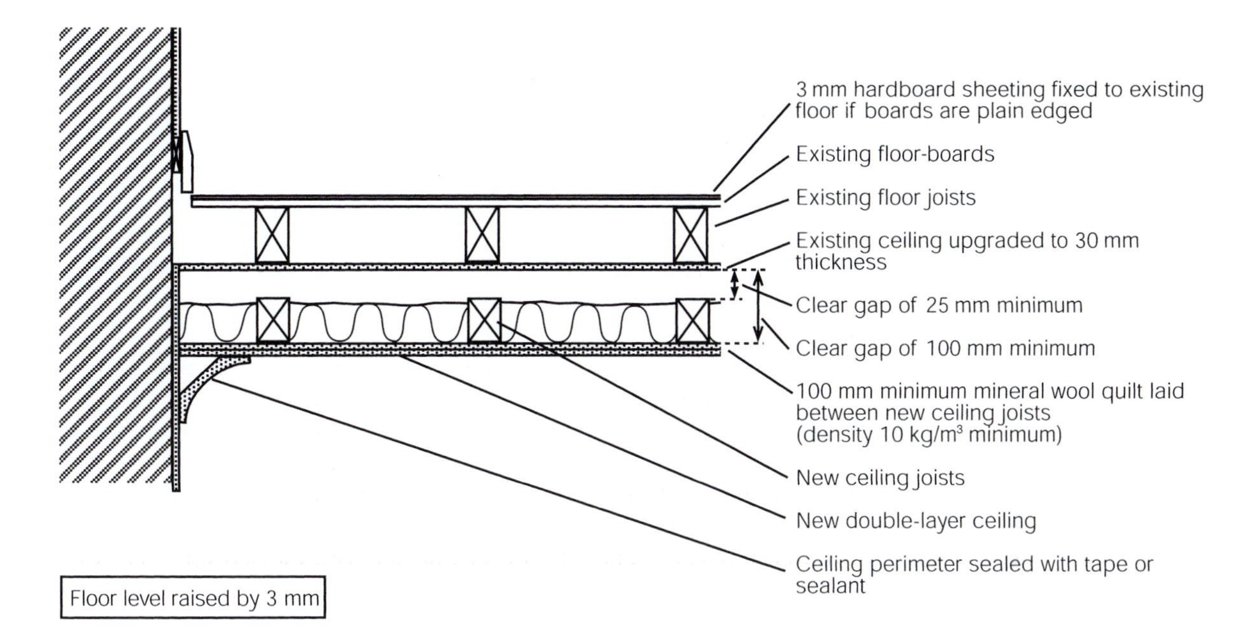

Figure 5.11
Acoustic upgrading of separating floors: independent ceiling

- When the new panels are fully installed, the skirting boards are refixed to the wall, leaving a gap of 5 mm between the bottom of the skirting and the surface of the panels to prevent transmission of impact sound from the floor into the wall.

It should be noted that this method of upgrading raises the original floor level by 43 mm, therefore requiring adjustments to doors, skirtings, services, etc.

5.5 Upgrading the acoustic performance of external walls

Upgrading the sound insulation of external walls is becoming increasingly necessary, as more dwellings are being built in busy city centres. Noisy roadway, factory or other excessive noise sources are causing increasing problems, and greater efforts are being taken to reduce the impact of external noise.

The importance of upgrading the external walls will also depend, to some extent, upon the acceptable noise levels within the building, which, in turn, depend on its proposed use after refurbishment. Houses, flats, hotels, hospitals and other buildings where people sleep (or require a minimum of intrusive noise for other reasons) may require the sound-insulation of their external walls upgraded if external noise sources cause, or are likely to cause, problems.

External noise usually enters a building via the windows, since single glazing is a very poor sound reducer. Relatively insignificant gaps around ill-fitting casements also adversely affect their acoustic performance, and, if a window is opened only slightly, its sound reduction capability will be drastically reduced.

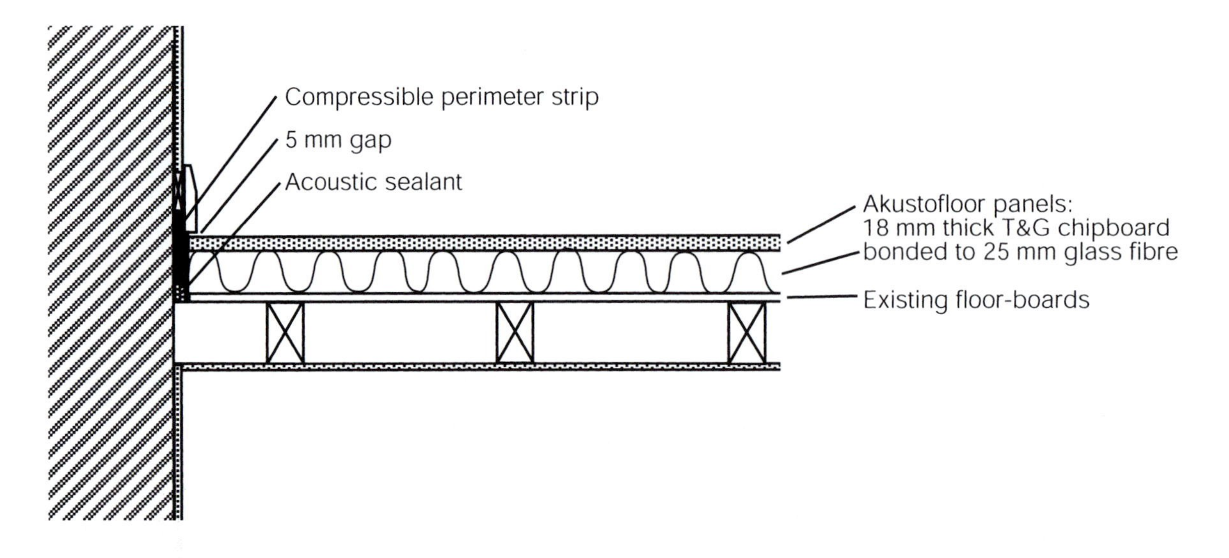

Figure. 5.12 *Acoustic upgrading of separating floors*

The most effective means of upgrading the acoustic performance of an external wall, therefore, is to deal with the windows. If the existing windows are in good condition, they can be converted into double windows by adding new glazing internally, in a separate frame with an intervening airspace of 150–200 mm. A narrower airspace will result in a poorer sound-reduction performance. If double windows are installed, it must be ensured that the casements and frames fit tightly. Flexible sealing gaskets should be incorporated between the casements and frames, and gaps between the new secondary frames and the existing window reveals should be sealed with acoustic sealant.

If the installation of double windows is impractical due, for example, to the absence of an adequate reveal in which to fit them, or if the existing windows are in poor condition, the only solution will be to fit replacement windows with double or triple glazing. As with double windows, it is essential that all clear air paths between casements and frames, and frames and reveals, are properly sealed to prevent leakage of air-borne sound. The spaces between the glazing sheets, which should be sealed around their edges, should be as wide as possible to give maximum acoustic upgrading.

Ideally, new double windows or double glazing used to upgrade acoustic performance should not normally need to be opened since, as previously stated, opening windows drastically reduces their sound-insulation capability. This may, in turn, require the installation of mechanical ventilation, which will significantly increase the overall refurbishment costs. In view of this, and the expense of providing and sealing the new windows, acoustic upgrading of external walls in this way should normally only be carried out where intrusive noise from external sources would cause severe problems or loss of amenity for the occupants.

References

BSI (1998) BS7913: 1998, *Guide to the Principles of the Conservation of Historic Buildings*, London: BSI.

Building Research Establishment (1985) *Improving the Sound Insulation of Separating Walls and Floors* (Digest 293), Watford: BRE.

Building Research Establishment (1988) *Sound Insulation of Separating Walls and Floors, Part 1: Walls* (Digest 333), Watford: BRE.

Building Research Establishment (1988) *Sound Insulation of Separating Walls and Floors, Part 2: Floors* (Digest 334), Watford: BRE.

Building Research Establishment (1993) *Double Glazing for Heat and Sound Insulation* (Digest 379), Watford: BRE.

Building Research Establishment (1999) *Improving Sound Insulation* (Good Repair Guide 22: Parts 1 & 2), Watford: BRE.

Department of the Environment and The Welsh Office (1992) *The Building Regulations 1991, Approved Document E; Resistance to the Passage of Sound*, London: HMSO.

Department of the Environment, Transport & the Regions (1999) *Manual to the Building Regulations*, London: DETR.

Emmitt, S. and Gorse, C. (2006) *Advanced Construction of Buildings*, Oxford: Blackwell Publishing.

Powell Smith, V. and Billington, M. J. (1999) *The Building Regulations Explained and Illustrated*, Oxford: Blackwell Science.

Richardson, B. A. (1991) *Defects and Deterioration in Buildings*, London: E. & F.N. Spon.

Stephenson, J. (1995) *Building Regulations Explained*, London: E. & F.N. Spon.

CHAPTER 6 Preventing moisture and dampness within buildings

6.1 General

The most common single cause of building deterioration is dampness, and it has been estimated that over 1.5 million dwellings in the UK are seriously affected by dampness problems. The principal sources of dampness are rainwater penetration through roofs and external walls, rising damp through walls and solid floors, and condensation. Because its causes and prevention are different from those of other sources of dampness, condensation is dealt with separately. Owing to the increased humidity created through modern cooking and heating devices and reductions in natural ventilation, condensation is responsible for a large proportion of dampness and mould growth. The sources of moisture can come from inside or outside the building, and it is essential that proper investigation is undertaken to determine the cause of dampness before any remedial action is taken.

6.2 Preventing moisture penetration through external walls and walls below ground level

The majority of older buildings have solid stone or brick external walls, which are inherently vulnerable to rainwater penetration, often resulting in permanent dampness and deterioration of plasterwork and internal finishes. The severity of the problem varies and, at best, might result in only a few damp patches over the internal surface of the wall. In some cases, however, the dampness may be so severe as to cause total deterioration of plasterwork and finishes over large areas of the wall. Walls below ground level enclosing basement accommodation are permanently vulnerable to groundwater penetration, and it is quite likely that such walls in older buildings will not have received any waterproofing treatment when originally constructed. In such cases their internal surfaces and finishes will be subject to dampness and deterioration.

Uninsulated solid walls are good thermal conductors of heat; they provide an easy path for heat energy to escape out of the building and create cold surfaces that are very susceptible to the formation of condensation. A solid wall provides a thermal bridge, and the cold surfaces that result from the passage of heat will lead to condensation if the moisture content in the room is high. When curing dampness, it must be determined whether the water is coming from inside or outside the building. Dampness caused by condensation is often misdiagnosed as water penetration. If it is determined that the problem is caused by condensation, there are a number of methods that can be used to reduce or cure the problem. The methods of reducing condensation are discussed at the end of this chapter; see 'Preventing condensation within buildings'.

Preventing water penetration from external sources is quite different to preventing condensation. A number of different techniques can be used to overcome the problems of moisture penetration through solid walls; these are discussed below.

Internal treatments

Dry linings

The use of dry linings for upgrading internal wall surfaces is described in detail in Chapter 3.2. In cases where the external walls are suffering only from slight dampness, a traditional timber batten dry lining, used in conjunction with an externally applied water-repellent solution, is an appropriate means of overcoming the problem. The dry lining can be fixed to provide a new, dry wall surface. Additional precautions are taken to protect it from the penetrating dampness (see Figure 6.1). The timber battens must be pressure-impregnated with preservative to prevent the risk of fungal attack, and they should be secured to the wall over strips of polythene sheeting or bitumen felt to isolate them from the damp wall. If the dampness is widespread, the whole wall surface should be lined with a polythene sheet, properly lapped and jointed, and secured to the wall by the timber battens.

Similarly, a metal channel dry-lining system, also described in Chapter 3, could be used as an alternative to a timber batten system, to prevent slight damp penetration. The metal channels should be fixed to the wall with Dri-Wall adhesive, and the existing wall surface should be treated with a waterproof EVA in the line of the adhesive dabs. The provision of ventilation in the cavity between the new lining and the existing wall is also recommended. Where it cannot be guaranteed that moisture will not penetrate or form within a structure, ventilation is advisable to allow any moisture to be carried away. Where the structure can be properly sealed and no moisture is present, then it is not necessary to ventilate voids within the structure. Wherever possible, air movement should be controlled regardless of the system used to ventilate the property. Properly controlled natural and mechanical ventilation will provide greater thermal comfort than air leaking through the structure.

If the damp penetration is severe, a dry lining fixed directly to the wall will itself be vulnerable to deterioration, and other methods of dealing with the problem must be employed.

Care should be taken to ensure that condensation does not form on the wall, timber or polythene covering. Adding a dry lining to an external wall will make the surface of the existing wall slightly colder. If not ventilated, the new cavity between the dry lining and the existing wall acts as an insulator, keeping the plasterboard warm and making the existing wall surface colder. If water vapour manages to pass behind the plasterboard or other dry-lining material, there is a good chance that

Upgrade of solid wall: dry-lined wall with vapour barrier

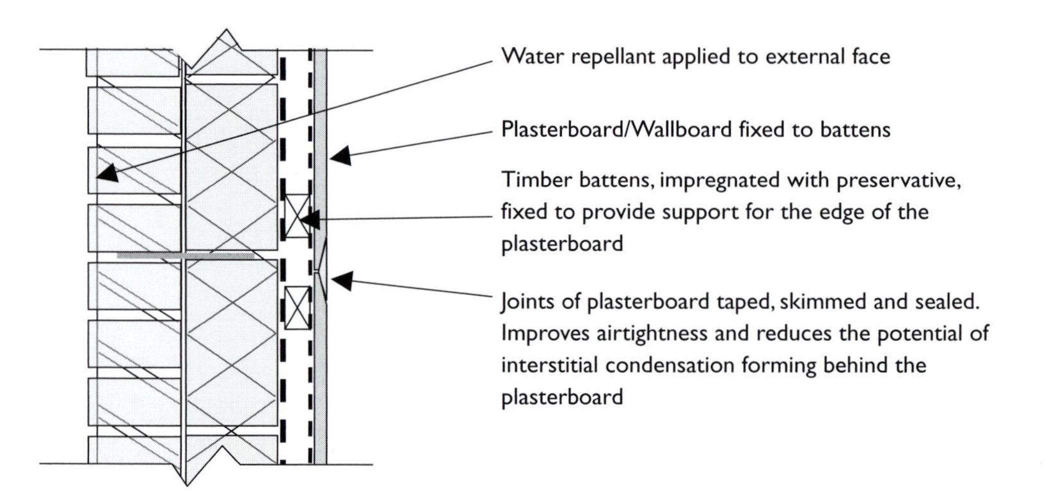

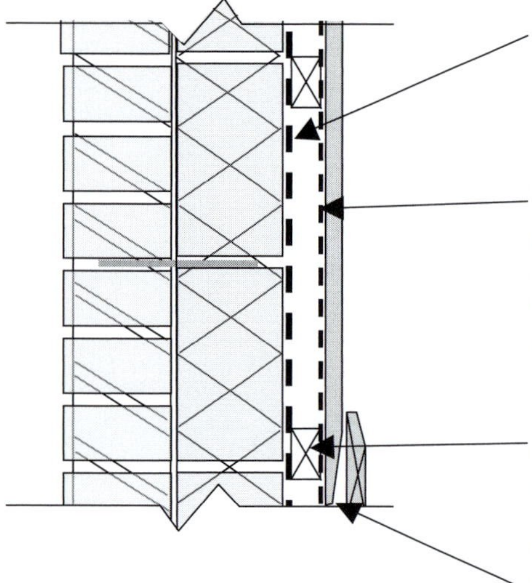

Figure 6.1 *Dry-lining a solid wall to overcome slight dampness*

it will take the form of interstitial condensation on the surface of the existing wall, timber studwork or metal studs. Interstitial condensation will lead to mould growth and bad odours. When dry linings are placed on external walls, vapour barriers should be used to prevent the warm moist air within rooms penetrating to cold wall surfaces.

Pre-formed waterproof sheeting systems: ventilated cavities and cavity drain – for slightly damp walls

A number of proprietary waterproof sheeting systems are available for application to walls, above and below ground level, that are suffering from moisture penetration.

Newlath 2000 is a damp-proof lightweight sheet material made from high-density polypropylene, 0.5 mm thick, formed into a pattern of raised studs linked by reinforcing ribs (www.newton-membranes.co.uk). The 5 mm high studs face the wall, creating channels and a cavity that allow air to circulate freely behind the polypropylene sheet. A polypropylene mesh welded onto the internal face of the waterproof sheet provides a rot-proof key for a plaster finish. The material is inert and highly resistant to water, alkalis, saline solutions and organic acids, and is not affected by minerals. It is also resistant to bacteria, fungi and other small organisms. Polypropylene cavity drains are particularly suited to refurbishment and improvement work where rising and penetrating damp is a serious problem in the existing walls. It is supplied in 1.5 m wide rolls, 10 m in length, and works on the principle of providing a ventilated, moisture-proof barrier between the existing wall surface and the newly applied finish, as shown in Figure 6.2 and in Photograph 6.2.

The system in its ventilated cavity form should only be used where slightly damp conditions exist. The ventilated systems are not suitable where water is likely to penetrate and trickle down the face of a wall.

All damp or crumbling plaster, where it exists, should be removed prior to fixing the polypropylene sheet to the wall surface. The material can then be fixed to the background, using 50 mm long polypropylene plugs at not more than 300 mm vertical and horizontal centres. Closer centres should be adopted on uneven or curved surfaces. For fixing to wood or other nailable surfaces, galvanised clout nails can be used. All joints between adjacent sheets should be lapped by not less than 100 mm. Newlath 2000 is available in brown or clear polypropylene. The clear version provides the added capability of visually identifying adequate fixing positions in the substrate, especially useful when the latter is of variable materials.

The provision of an adequate gap and through ventilation between the polypropylene studded sheet and the damp wall is essential, and this is achieved by using Newlath Profile Strips along the top and bottom of the wall. The profile strip prevents the new finish from making contact with the original surface and stimulates the flow of air to carry away any residual dampness into the atmosphere. The small quantity of moisture involved is easily absorbed into the large volume of air in the room. Use of the profile bead also ensures a constant depth of plaster. If it is not possible to use the profile, ventilation gaps should be provided at both bottom and top. The bottom edge of the studded polypropylene must be raised 20–25 mm above the floor, and a 2–3 mm gap must be left at ceiling level. Once the plaster finish has been applied and has dried, the ventilation gaps can be

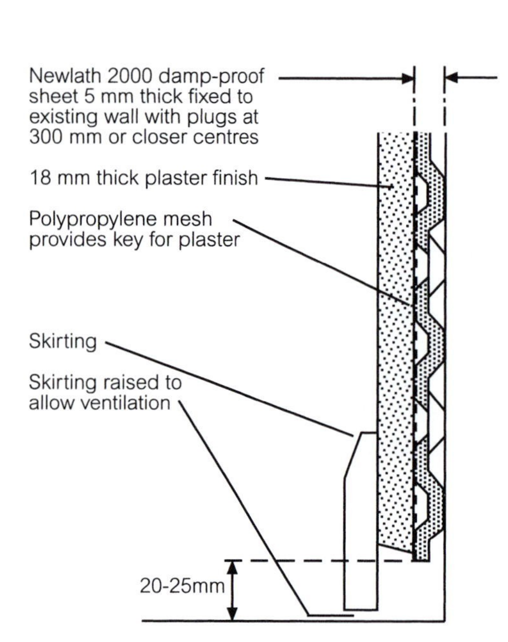

Newlath 2000 damp-proof sheet 5 mm thick fixed to existing wall with plugs at 300 mm or closer centres
18 mm thick plaster finish
Polypropylene mesh provides key for plaster
Skirting
Skirting raised to allow ventilation
20-25mm

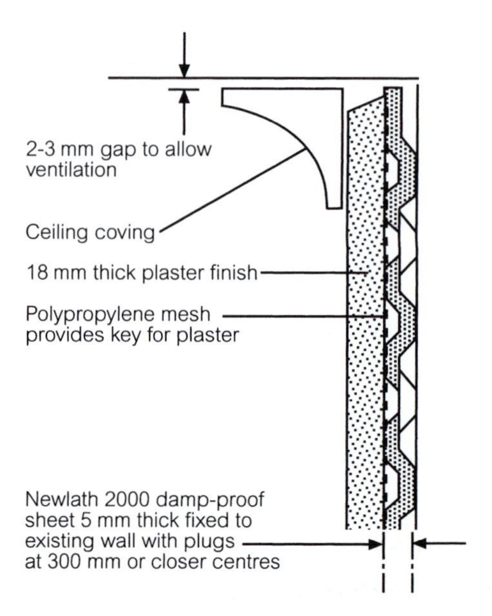

2-3 mm gap to allow ventilation
Ceiling coving
18 mm thick plaster finish
Polypropylene mesh provides key for plaster
Newlath 2000 damp-proof sheet 5 mm thick fixed to existing wall with plugs at 300 mm or closer centres

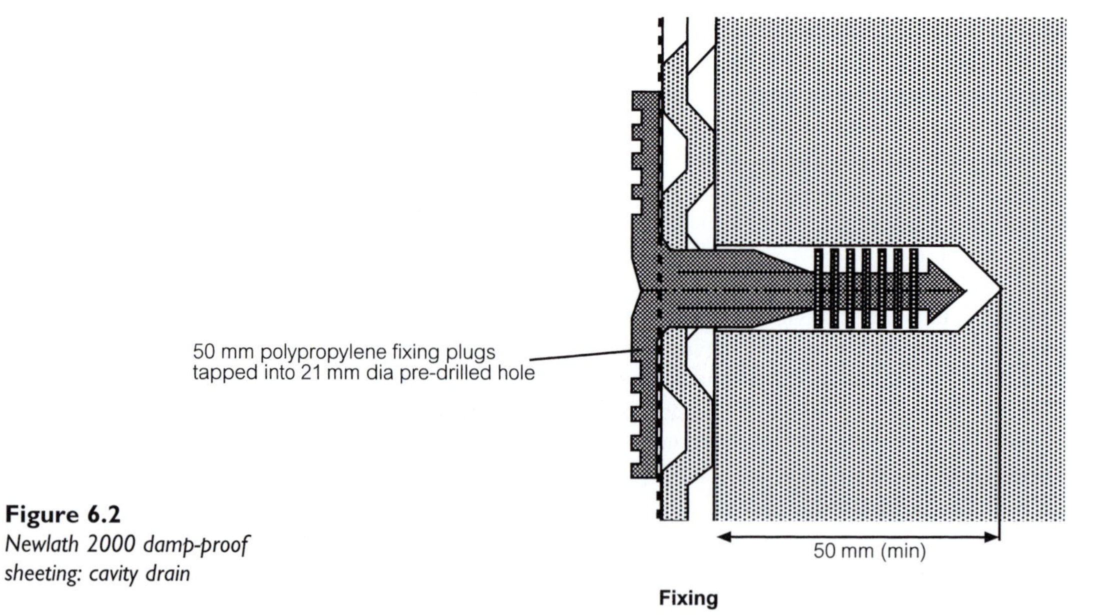

50 mm polypropylene fixing plugs tapped into 21 mm dia pre-drilled hole
50 mm (min)
Fixing

Figure 6.2
Newlath 2000 damp-proof sheeting: cavity drain

Photograph 6.1
Polypropylene cavity drain ready for fixing to the internal surface of a basement wall

concealed by a wooden skirting and coving. (Figure 6.2 shows a detail of the ventilation gap, cavity drain system and skirting board.)

Newlath 2000 will accept plasters specified by the manufacturer; they should be applied to a thickness of 13–15 mm to provide the new, damp-proofed wall finish. In addition to providing a damp-proof base for internal wall finishes, Newlath 2000 can be used for arches and vaults, and also as a base for external render finishes (Photograph 6.2).

Photograph 6.2
Polypropylene cavity drain fixed with polypropylene studs ready to receive plaster finish

If water penetration is a greater problem, the Newton 500 system can be used (www.newton-membranes.co.uk). This is a high-density extruded membrane, 0.6 mm thick, with a raised stud formation 8 mm high, similar in principle to the system described above but capable of withstanding more aggressive groundwater conditions. The material is also suitable for damp-proofing floors and is described in detail in Section 6.4. The heavier version of the cavity drain is designed for tanking rather than just preventing damp penetration. One difference is that the cavity is sealed in the waterproofing system. It is used where water penetration into the structure is expected, rather than just dampness. The 8 mm wall cavity is used to channel the water down the wall to the floor. At the floor, a polypropylene studded sheet, with 20 mm studs, is laid to create a large cavity that allows the water to flow to a sump where it is then pumped out of the building. RIW also produces similar cavity drain tanking systems (www.riw.co.uk).

Waterproof coatings (cement-based)

A number of proprietary materials are available specifically for brush or spray application to walls suffering from damp penetration both above and below ground level.

Thoroseal, supplied by Master Builders Technologies, is suitable for the interior and exterior waterproofing of brickwork, stonework and concrete, above and below ground level, including basements (www.thoroproducts.com). It comprises a blend of Portland cements, well-graded sands and chemical modifiers supplied in powder form. The material is site-mixed with Acryl 60, an acrylic polymer in the proportions of one part Acryl 60 to three parts water. The wall surface being treated should be completely clean, structurally sound and mechanically keyed.

The wall/floor joint, which, in basements, is usually the point of greatest water ingress, should be cut out, cleaned and filled with Waterplug. Waterplug is a special fast-setting mortar designed to stop water seepage at joints, cracks and holes in the substrate. When the Thoroseal has been mixed to a thick, batter-like consistency, in the proportions 5.2 litres of Acryl 60/water, to 25 kg of powder, a first coat is brush-applied to the pre-dampened wall surface. This coat should be well brushed into the surface, a typical application being 1.5 mm thick. This first coat should be left at least overnight to cure before the second coat is applied to approximately 1 mm thickness. Thoroseal is available in grey and white, and, to ensure proper coverage, the second coat should be white over a grey first coat.

The Thoroseal itself may act as the final finish, and can either be textured or given a smooth finish. Alternatively, a cement-based renovating plaster may be applied, in which case the final coat of Thoroseal should be applied with horizontal brush strokes to provide a better key for the finish.

Waterproof coatings (bitumen-based)

RIW Liquid Asphaltic Composition is a solution of natural and petroleum bitumens in white spirit, applied cold in two coats to form a damp-proof membrane that

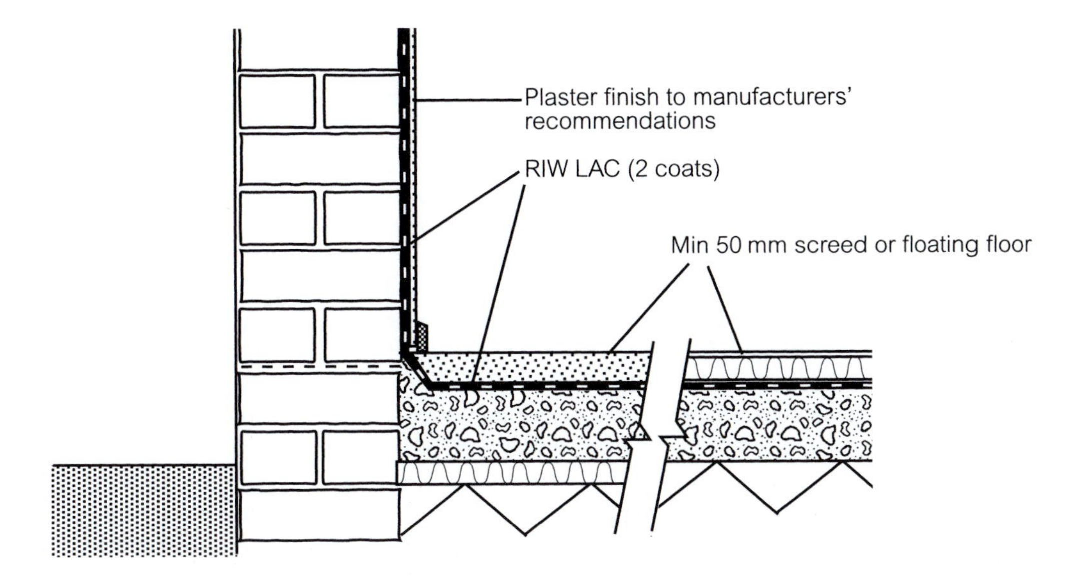

Walls above ground level

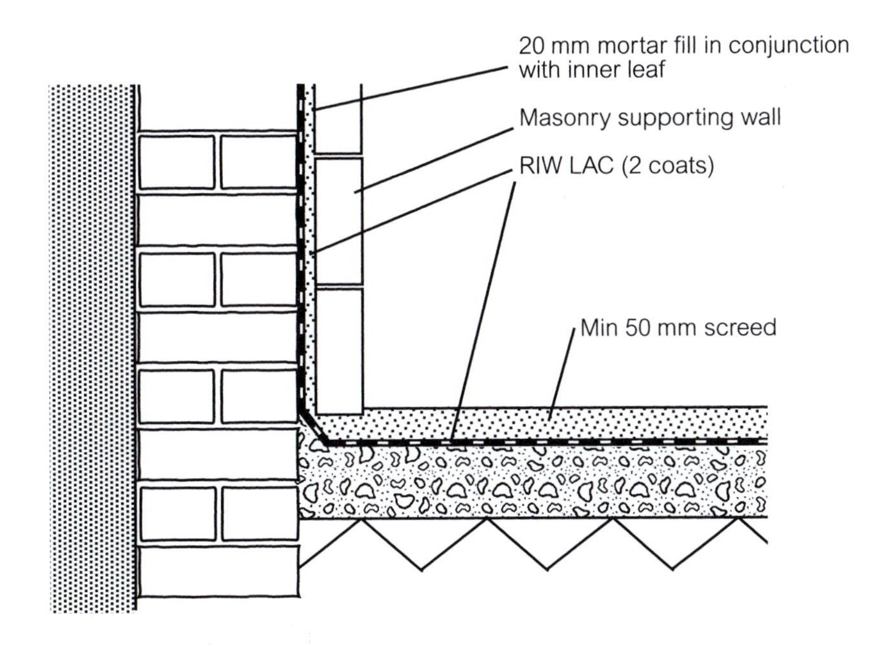

Walls below ground level

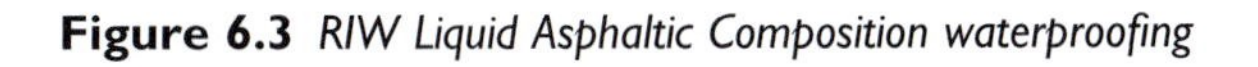

Figure 6.3 *RIW Liquid Asphaltic Composition waterproofing*

dries to a uniform gloss black finish (www.riw.co.uk). The material can be applied to the internal face of external walls above ground level to prevent moisture penetration, and is also used as a damp-proof membrane to ground floors and for the waterproofing, or tanking, of basements.

All surfaces to be treated must be perfectly clean and dry, to a depth of 1–2 mm, with any voids or hollows made good to a flush finish with Portland cement mortar. Brickwork or stonework should be sound, with joints flush-pointed before the membrane is applied.

The RIW Liquid Asphaltic Composition should be applied by brush, roller or spray in two coats at a minimum application rate of 1.7 litre/m^2 for the first coat and 2.5 litre/m^2 for the second. A minimum of 24 hours should elapse before application of the second coat.

Where Liquid Asphaltic Composition is used above ground level to combat penetrating moisture, it can be finished with a direct application of two-coat plaster to the manufacturer's recommendations. Below ground level, where penetrating moisture is subject to hydrostatic pressure, a masonry supporting wall is required. A wall of brick, block or concrete should be constructed immediately after the membrane has cured. If brick or block is used, a cavity should be left between the membrane and the loading skin, and the cavity should be filled with mortar as the work proceeds. Figure 6.3 shows the application of RIW Liquid Asphaltic Composition both above and below ground level.

Waterproof coatings (urethane-based)

An alternative to RIW Liquid Asphaltic Composition, produced by the same manufacturer and suitable where the situation demands a higher-performance waterproofer, is RIW Flexiseal (www.riw.co.uk/flexiseal.htm). This is a solvent-free, thixotropic liquid, based on urethane prepolymers, that, on contact with atmospheric moisture, cures to give a tough, rubber-like coating. The material is applied in the form of a blue basecoat and a black topcoat by brush, roller or spray. Finishing and protection is the same as for RIW Liquid Asphaltic Composition, described above.

Traditional dense polythene membrane wall linings

As an alternative to the use of the dry-lining and proprietary waterproofing systems described above, moisture penetration can be successfully prevented using the more traditional solution of applying impermeable dense polythene sheeting to the inner surface of the wall. The sheeting, which must be lapped and taped at all joints, is held in place against the wall surface while a brick, block or concrete loading wall is erected to protect and support it. If brick or block is used, a cavity should be left between the sheeting and the new loading skin, and the cavity should be filled with mortar as the work proceeds. This technique is generally only used in basements and is shown in Photographs 6.3 and 6.4. Note the loss of floor space inherent in using this technique. The photographs also show the use of a thermal board dry lining (see Section 4.3) applied to the new loading skin to upgrade the thermal performance of the existing basement walls.

Photograph 6.3
Traditional dense polythene wall-lining with block-work loading wall and thermal-board dry lining. (Note: the untreated section of wall is above ground level)

Photograph 6.4
Traditional dense polythene wall-lining with block-work loading wall and thermal-board dry lining. (Note: the untreated section of wall is above ground level)

External treatments

Water-repellent solutions

A number of proprietary water-repellent solutions are available, and these provide a relatively simple and inexpensive means of preventing rainwater penetration through external walls when the problem is not too severe. Liquid Plastics K501 Masonry Waterproofing Solution is a water-based silane siloxane compound specifically designed to provide waterproofing to external building surfaces above ground level (www.liquidplastics.co.uk). After application, the solution is virtually invisible, therefore preserving the natural appearance and texture of the substrate being waterproofed.

K501 is a surface-impregnating solution for porous substrates such as brick and stone, and penetrates from 1 to 4 mm into the surface pores, rather than coating the surface, giving it good protection from abrasion, weathering and ultra-violet deterioration. This water-repellent lining prevents absorption of water, providing excellent protection against rainwater penetration for at least ten years.

K501 is applied in two flooding-brush or airless-spray coats, the second applied as soon as the first coat is dry. The substrate should be dry, clean and free from surface material such as dirt, dust, grease and organic growth prior to application of the solution.

Cementone Water Seal is a water repellent designed to prevent penetration of rainwater through brickwork, masonry and render above ground level. Prior to treatment, the wall surface should be clean and dry and free from efflorescence, moss, grease, oil and loose dust. Defective guttering and pointing should be made good before application. Cementone Water Seal is applied in one generous flooding coat using a brush or spray, aiming for a 'run-down' of at least 300 mm on surfaces of average porosity. The waterproofer does not darken, stain or alter the appearance or texture of the existing wall surface and penetrates up to 4 mm into the surface, lining the pores with a water-repellent coating and preserving the original porosity. Full waterproofing effectiveness is achieved 24 hours after application.

External renders

The application of an external render finish to a building's elevations is a more expensive and time-consuming means of overcoming the problem of rainwater penetration. However, a render finish can also be used to fulfil other upgrading requirements, such as improving the building's appearance, or upgrading the thermal performance of the external walls (see Section 4.3). Thus, where upgrading of appearance and/or thermal performance is necessary, in addition to a rainwater penetration problem, it may be convenient and economical to apply a render finish. All of the proprietary exterior thermal insulation systems described in Section 4.3, in addition to upgrading the thermal properties and appearance of external walls, will also provide total resistance to rainwater penetration.

Waterproof masonry paints

The majority of conventional masonry paints are not capable of providing total resistance to rainwater penetration and are therefore unsuitable as a means of solving this problem. However, recent developments in paint technology have led to the introduction of exterior coatings that are capable of fully waterproofing external wall surfaces.

Monolastex Smooth, produced by Liquid Plastics Ltd., is a water-borne styrene acrylic co-polymer coating system that, when applied in two coats, will provide full waterproofing to existing walls (www.liquidplastics.co.uk). The paint has good bonding properties, is permeable to water vapour, allowing underlying moisture to escape without causing flaking or blistering, is resistant to mould, fungal and algae growth and is elastic, allowing it to accommodate movement. The coating is available in a wide range of matt colours and can be applied by brush, roller or airless spray.

The existing surface should be thoroughly cleaned and free from surface material such as dirt, dust, grease and organic growth prior to application. Porous, absorbent backgrounds, such as brickwork or stone, should initially receive an application of Liquid Plastics Bonding Primer.

If a textured waterproof external wall finish is required, Monolastex Textured, produced by the same manufacturer, may be used. This is also a water-borne styrene acrylic co-polymer system and is available in either matt white or magnolia. It is particularly well suited, because of its texture, for hiding surface defects.

Tough-Cote Superflex RW2 (www.carrspaints.com), is a water-borne, textured, wholly waterproof decorative wall coating based on an elastomeric resin blended with fine aggregate, titanium dioxide, light-fast earth pigments and special preservative agents. The coating, available in four colours and white, is applied by roller or spray, in one or two coats, to the substrate, which should be thoroughly clean and dry. Porous friable substrates, such as brickwork or stone, should first receive a one-coat application of Glixtone Weatherproof Stabilising Solution SSOI. Tough-Cote Superflex RW2 accommodates substantial substrate movement without cracking or loss of bond, is effective at hiding surface defects and, if applied in two coats, provides a decorative, protective finish for a minimum of 20 years.

6.3 Preventing rising damp in walls

The penetration of ground moisture, in the form of rising dampness in walls, is a common problem in many old buildings, and it can result from any of the following:

- the lack of a damp-proof course (DPC) in the original construction;
- deterioration and failure of the existing DPC because of age;
- bridging of the existing DPC.

Bridging of the existing DPC in an external wall by the building-up of soil, or by the addition of new paving, to a level above that of the DPC, is a common cause of rising dampness. However, this 'short-circuiting' of the existing DPC can easily be alleviated simply by lowering the adjacent soil or paving level back to 150 mm below that of the DPC. Provided the DPC is in good condition, the problem will not recur once the residual moisture has dried out from the wall. Other common causes of bridging are where a new external render has been applied and taken over the DPC, or where new mortar pointing has been carried out over the outer edge of the DPC. These also have the effect of 'short-circuiting' the DPC by providing a path for rising moisture to pass around it, but the problems can easily be overcome by cutting back the rendering to above the DPC level, or by raking out the offending mortar pointing.

The installation of DPCs was made mandatory by the Public Health Act 1875, but in practice their use was not universal immediately. The majority of pre-1900 buildings are, therefore, without DPCs and, as a result, many of them are found to be suffering from severe rising dampness and its associated problems. In such cases, the only means of overcoming the problem is to install a new DPC to cut off further rising ground moisture from entering the building. The installation of a new DPC will also be essential in those older buildings where a DPC was incorporated initially, but where this has deteriorated and failed with age. DPC failure is quite common in pre-1920 buildings, where the felts and slates used for DPCs were often of poor durability, although DPC failure is also not uncommon in more recent buildings.

The installation of a new DPC is both time-consuming and expensive, but it will be imperative in any building suffering from rising dampness caused by the lack of, or failure of, a DPC.

The installation of new DPCs

New DPCs can be installed using a number of different methods:

- Remove two courses of bricks, a short length at a time, and replace with two courses of dense engineering bricks. Alternatively, the same bricks can be replaced and a new DPC incorporated during the process.
- Physically insert a new DPC by cutting a slot in a suitably located horizontal mortar joint and inserting metal sheet, bitumen felt, dense polythene or other suitable material.
- Pressure-inject a chemical, water-repellent fluid into the wall at a suitable position to provide a 'band' of masonry that will resist rising damp.

The first two methods described above are applicable to certain types of wall only: the first method can be used only where the existing walls are of brickwork, and the second method can be used only for walls with continuous horizontal mortar joints and is not, therefore, suited to uncoursed masonry.

Until the early 1980s, the physical insertion of a new DPC into a slot cut into an existing mortar joint was the most widely used means of installing new DPCs. However, the chemical injection method has now overtaken physical insertion techniques and is almost universally used as a means of overcoming rising dampness in walls.

Pressure-injected chemical DPCs

Pressure-injected DPCs involve the use of silicone-resin-based water repellents or aluminium stearate polymeric water repellents. A large number of proprietary injection systems are available from specialist companies, which normally provide a full diagnosis and treatment service. Water repellents work on the principle of lining, rather than blocking, the pores within the material being treated. This allows the passage of some water vapour, but prevents the rise of liquid moisture. The procedures used for the injection of a typical chemical DPC are described below and illustrated in Figures 6.4 and 6.5.

Figure 6.4
(facing page) Pressure-injected chemical DPCs

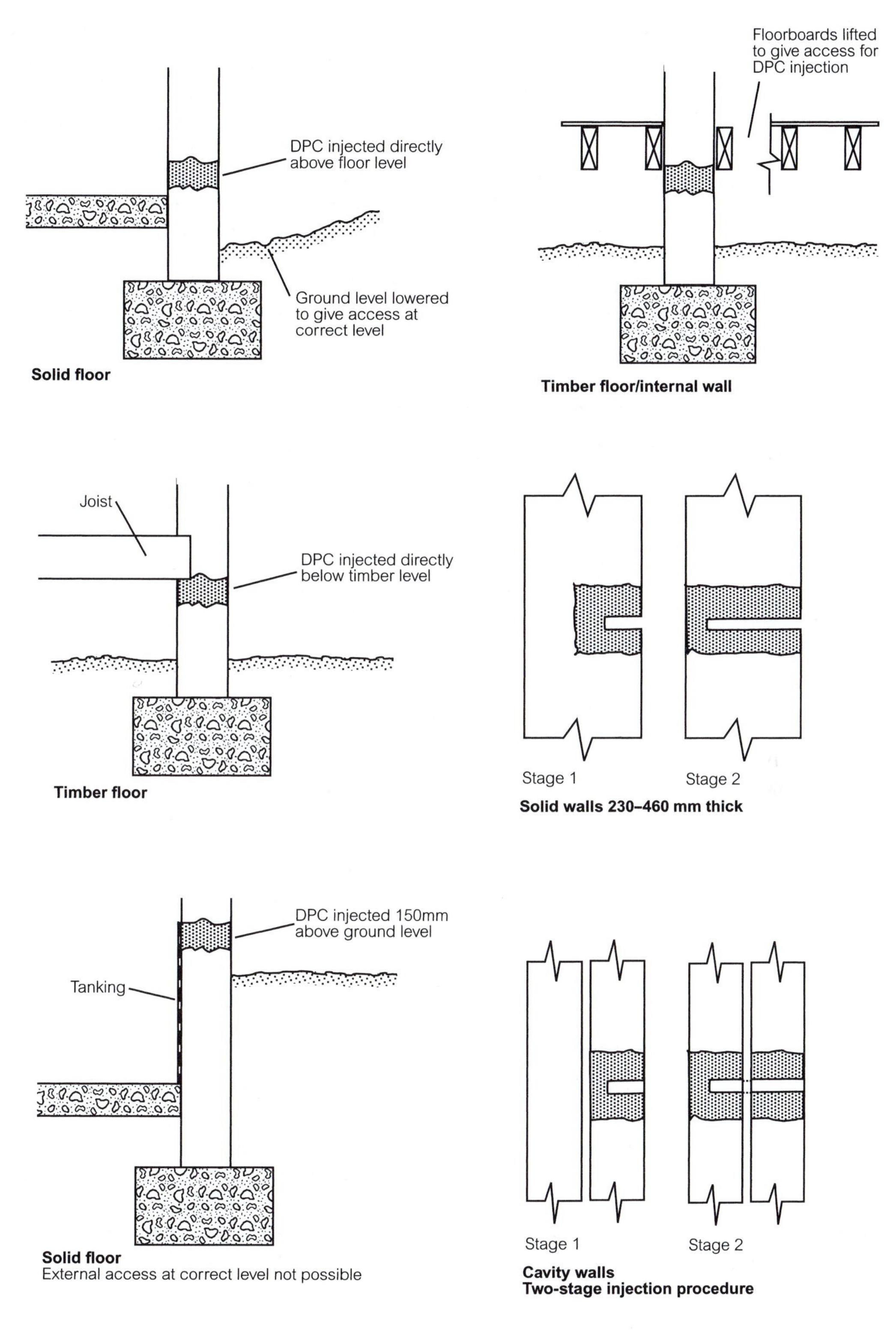
DPC injected directly above floor level
Ground level lowered to give access at correct level
Solid floor
Floorboards lifted to give access for DPC injection
Timber floor/internal wall
Joist
DPC injected directly below timber level
Timber floor
Stage 1
Stage 2
Solid walls 230–460 mm thick
DPC injected 150mm above ground level
Tanking
Solid floor
External access at correct level not possible
Stage 1
Stage 2
Cavity walls
Two-stage injection procedure

Preparation

- Expose the walls externally to at least 150 mm below the proposed new DPC level. This may involve digging a trench (see Figure 6.4).
- Remove all skirtings and floorboards adjacent to the walls suffering from rising dampness and check for timber decay.
- If timber decay is evident, this should be treated and additional underfloor ventilation provided if necessary.
- Cut away the plaster, or other surface rendering, to 450 mm above the last visible signs of dampness. If dampness is not evident, expose 230 mm of wall along the proposed DPC line.

Treatment

- Select the course of masonry to be treated. With a timber floor this should ideally be below the timber level, and, with a solid floor, the course immediately

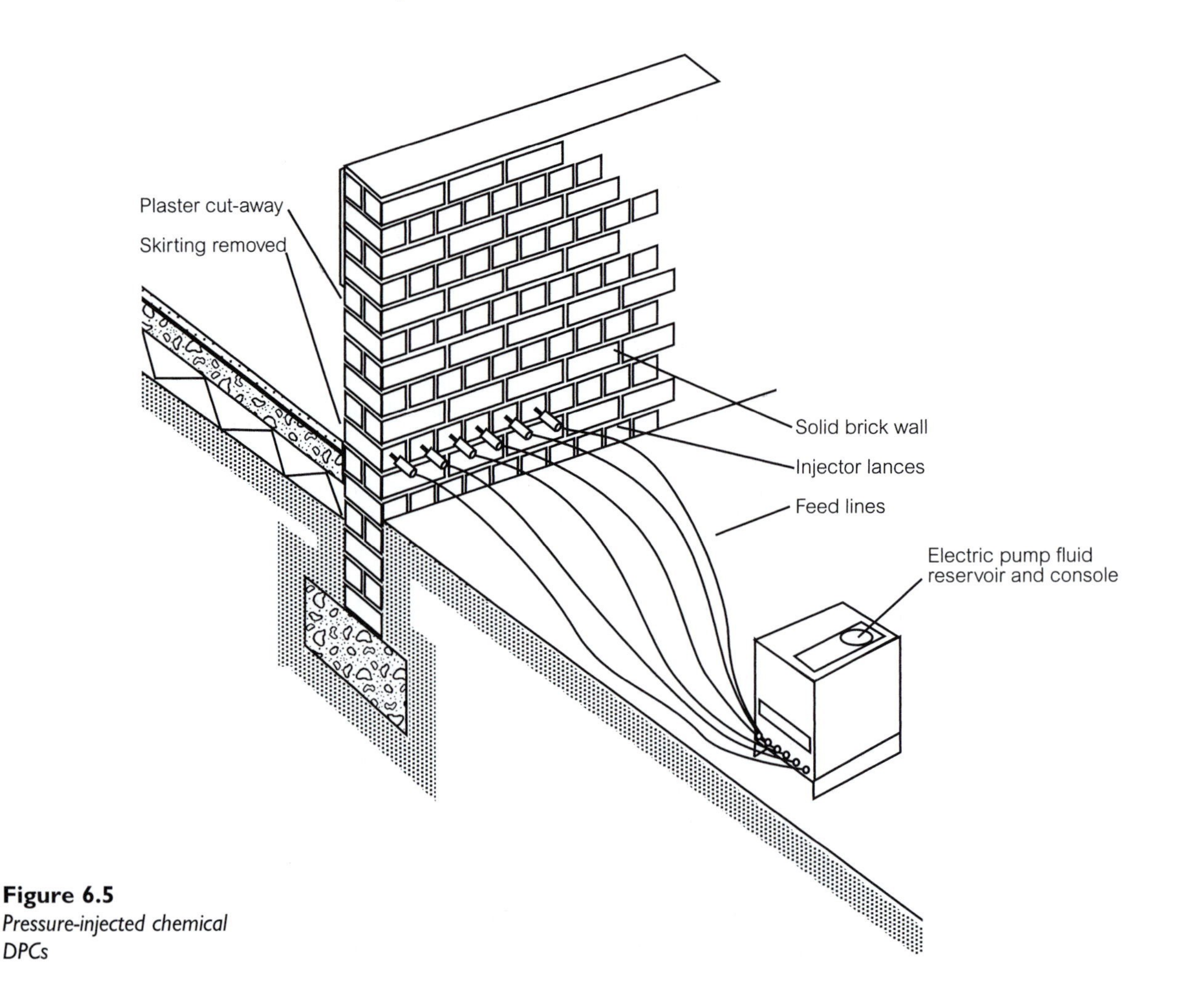

Figure 6.5
Pressure-injected chemical DPCs

above the floor level (see Figure 6.4). Avoid engineering bricks or similar dense masonry.

- Drill the holes for injection of the damp-proofing solution. If, owing to ground conditions, access to this course externally is not possible, the injection should take place 150 mm above ground level, and the section of wall below should be tanked internally (see Figure 6.4).
- Connect the electric pump, feed lines and injector lances to the pre-drilled holes and seal the mouth of each hole (up to six injector lances can be used at once – see Figure 6.5).
- Open the control valves and commence the pressure injection process. When the section of wall being treated is saturated, move to the next set of drillings.

The 9.5 or 12.7 mm diameter injection holes should be drilled horizontally into the wall at the rate of two per stretcher, to a depth of 75 mm, and one per header, to a depth of 190 mm. For solid brick walls between 230 and 460 mm thick and for cavity walls, the injection is carried out in two stages, as illustrated in Figure 6.4. After the outer zone/skin of the wall has been treated, further drilling takes place, and the injector lances are passed through the original holes to treat the inner zone/skin. For stone walls, the drilling and injection procedures are similar to those for brickwork, with injection holes drilled at 120–150 mm centres, to a depth of two-thirds of the thickness of the wall. Thicker walls should be treated in two stages, as for brickwork.

The walls of many older buildings consist of dry, loose rubble fill between an outer facing and an inner skin, and in order to achieve an effective DPC this infill must be treated separately. After injecting the outer facing and inner skin, separate drillings are made directly through into the rubble, which must be flooded with injection fluid to extend the moisture-resistant band across the full thickness of the wall.

6.4 Preventing rising damp in solid ground floors

In pre-1939 construction, solid ground floors were not normally provided with damp-proof membranes. Floors of concrete, stone flags or quarry tiles in older buildings are therefore often found to be suffering from rising dampness. Originally these floors were not usually covered, and the rising dampness was therefore allowed to evaporate and did not cause problems. However, if, in a refurbishment scheme, a new covering is applied to such a floor, evaporation of the rising dampness will be prevented, and the covering will rapidly become damp and may ultimately lift and/or deteriorate.

It is therefore essential, first, to check for rising dampness and, if it is present, to provide a new damp-proof membrane before laying the new floor covering. An effective method of checking for rising dampness is to lay a piece of impervious material, such as polythene, on the floor. The underside of the polythene will become wet within a few days if rising dampness is present.

Mastic asphalt and pitch-mastic are suitable materials for damp-proofing solid ground floors. Two 10 mm coats should be applied, and the coating should be set into a 25 × 25 mm chase in the wall, bonding with the DPC. Provision of the new

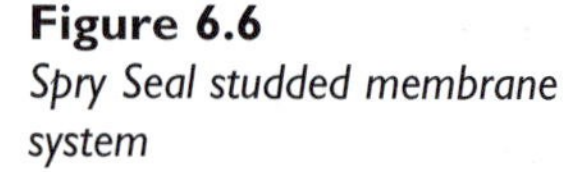

Figure 6.6
Spry Seal studded membrane system

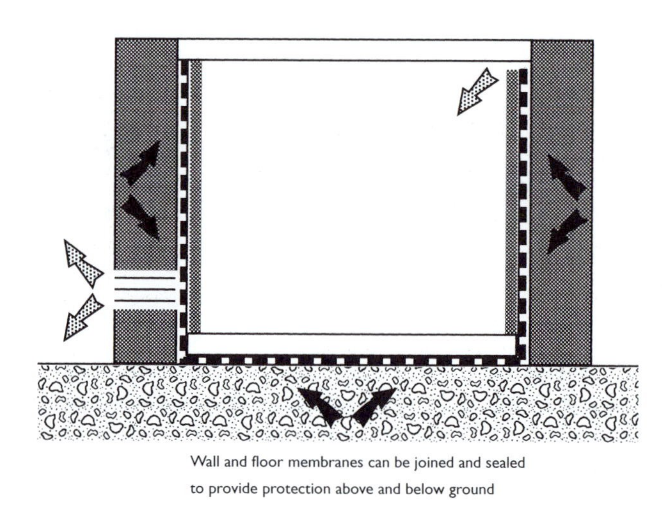

damp-proof membrane will usually involve removing existing skirting boards together with the wall coverings and plaster and, possibly, some of the floor edge itself, in order to expose the DPC in the wall (where one exists) and bond the new membrane to it to seal the walls and floors completely against rising dampness.

Wall and floor membranes can be joined and sealed
to provide protection above and below ground

As an alternative to asphalt or pitch-mastic, a proprietary damp-proofing system may be used. The Spry Seal studded membrane system (www.spryproducts.co.uk) comprises a black, studded sheet of high-density polyethylene, 0.6 mm thick, supplied in 20 m long rolls, 0.5–2.4 m wide. The membrane is laid on the floor with its 8 mm high studs against the damp surface, no fixings being required. (Where Spry Seal is applied to walls, sealed nylon fixings are required.) All joints are made watertight with sealing tape prior to the laying of the floor surface, which may be cement/sand screed, a minimum of 50 mm thick, chipboard sheet, floorboards or other suitable surfacing.

The raised studs provide a ventilated airspace beneath the new floor surface which is vented into the room by gaps left at the edges, or to the outside air through airbricks (see Figure 6.6). If the moisture problem is severe, or running water is present, a completely sealed, unvented, system may be used, with drainage gullies or pumps below the membrane.

The Newton System 500 (www.newton-membranes.co.uk) comprises a high-density extruded membrane, 0.6 mm thick and with 8 mm high studs, supplied in rolls. The membrane is laid with its studs against the damp floor surface and fixed using special water-stop plugs inserted into pre-drilled, 11 mm diameter holes. All joints in the membrane are made watertight with sealing tape before laying of the floor surface, which may be a cement/sand screed, a minimum of 50 mm thick, chipboard sheet, floorboards or other suitable surfacing (see Figure 6.7).

The air gap of 5.5 litres/m^2 created by the 8 mm high studs enables moisture and water vapour to move unhindered in all directions over the whole area of the floor being treated, achieving an equalisation of damp-pressure. This enables the whole area being treated to take the damp loading, rather than only the weakest areas, as occurs with conventional damp-proofing systems.

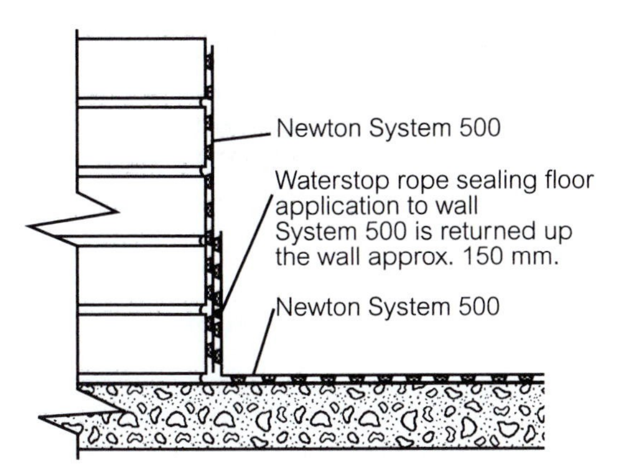

Figure 6.7
Newton System 500 (www.newton-membranes.co.uk)

Wall-floor details on sealed system

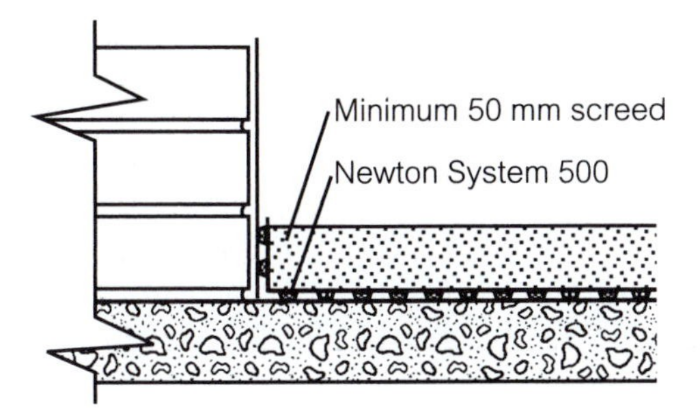

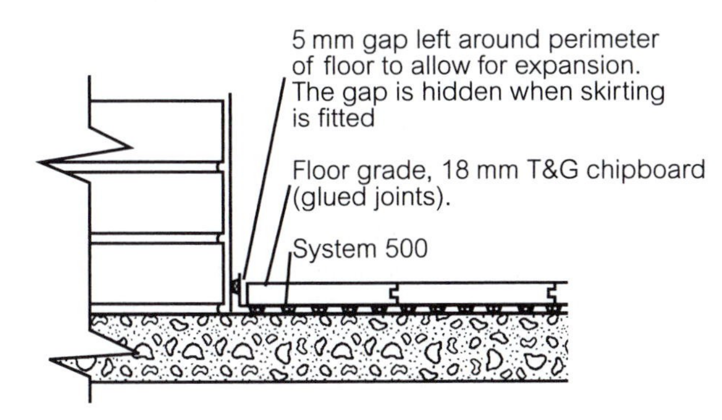

For basement floors, a fully sealed Newton System 500 is recommended, with no ventilation to the outside air. Where severe dampness or running water is present or likely, provision for drainage (or sump and pump equipment) will be required. This fully sealed system will also involve applying the membrane to the basement walls. In above ground situations, where the intrusive dampness is not severe, a ventilated system is suitable, with venting around the edges at the junction of the floor and walls.

6.5 Preventing rainwater penetration through roofs

Rainwater penetration through the roof of a building, in addition to causing inconvenience to the occupants, often results in deterioration and decay of the roof structure, thermal insulation, ceilings and internal finishes. It is, therefore, essential in the refurbishment of buildings that existing roofs are thoroughly examined and, in cases where rainwater leakage is evident, that proper remedial action is taken.

Preventing rainwater penetration through pitched roofs

A vital barrier against rainwater penetration into pitched roofs is the lining of impervious sarking felt provided immediately beneath the tiles or slates. Any rainwater penetrating the roof covering, because of wind pressure or damaged tiles or slates, is effectively drained away by the sarking felt and prevented from entering the roof space. However, sarking felt has been in use only since around 1938, and, therefore, the roofs of many older buildings, particularly those whose coverings have deteriorated or been damaged, are often found to be suffering from rainwater penetration and its effects.

The only effective solution, where there is evidence of rainwater penetration through a roof without sarking felt, is to take off the existing covering, add a new lining of sarking felt, and re-cover the roof. Whether or not the original tiles or slates are re-used will depend on their condition. In many older buildings, the tiles or slates will have deteriorated because of frost, chemical attack or mechanical damage, and, generally, if more than 10 per cent of the total are in bad condition, an entirely new roof covering will be advisable.

When roof tiles are laid properly and are in good condition, no rain should pass through to the felt. Exceptionally, wind pressure may blow the rain back up the roof and under the tiles, causing the rain to penetrate the tiles. As well as discharging any water that manages to pass through the tiles, the sarking felt prevents drafts and wind blowing through the roof and into the roof space.

Preventing rainwater penetration through flat roofs

Flat roofs are inherently more susceptible to rainwater penetration because of the much slower rate of run-off. The points most vulnerable to leakage are at flashings, upstands and parapets, at points where openings are formed for soil pipes, flues and so on, and at joints between separate sheets of the roof covering.

Lead was widely used as a flat roof covering until the late nineteenth century, when it was replaced by zinc, which is lighter and cheaper. The relative lifespans of these materials are 80–100 years and 40 years, respectively, and, if they are encountered in refurbishment work, it is likely that they will be near the end of their useful life and in poor condition. It is therefore recommended, if an old lead or zinc flat roof is found to be leaking, to strip and entirely replace the covering, since localised repair is likely to give only temporary relief before other faults develop in different parts of the roof.

The flat roofs of more recent buildings are likely to be covered with either asphalt or built-up felt, these materials having lifespans of 20–40 years and 10–15 years, respectively. In the event of rainwater leakage, it is possible to carry out localised patch repairs, but these tend to be unsatisfactory and provide only temporary relief. In view of the potentially serious effects of rainwater penetration, it will therefore be advisable, in the majority of cases, to strip the existing roof covering and replace it with a new covering. Alternatively, a new covering can be laid over the existing covering.

Where a roof has suffered from rainwater penetration over a lengthy period, it is likely that associated decay and deterioration of the roof structure and ceilings below will have occurred. It is therefore essential, in addition to repairing or replacing the defective roof covering, that the presence and nature of associated defects are established and that proper remedial work is carried out.

6.6 Preventing condensation within buildings

The causes and effects of condensation

Before considering how to prevent condensation, it is necessary to have some understanding of its causes and effects. Water vapour in varying quantities is always present in the air, and the quantity of water vapour that the air can carry depends upon its temperature; the warmer the air, the greater its vapour-carrying capacity. Condensation occurs in a building when warm air, containing water vapour, comes into contact with a cold surface, which reduces its temperature and, therefore, its vapour-carrying capacity. Any excess water vapour that the air is incapable of carrying because of its reduced temperature is deposited as condensation on the cold surface.

For any given air condition (temperature/water vapour content) there is a corresponding 'dew-point temperature'. If air containing water vapour comes into contact with surfaces that are at, or below, this dew-point temperature, it will deposit some of its water vapour as condensation. This usually shows itself as mist, beads of condensation, or damp patches on windows, walls and other surfaces, including fabrics, but it will be most obvious on the harder, more impervious surfaces.

Condensation can also occur within a permeable building element, where the dew-point temperature occurs at some point within its thickness, rather than on its surface. This *interstitial* condensation is potentially more harmful than surface condensation, since it is not visible, and the resulting dampness may remain undetected until substantial decay and damage have been caused.

The adverse effects that are possible as a result of condensation occurring within a building are:

- misting up of windows;
- moisture deposited on window frames and sills, leading to mould growth and rot;
- moisture deposited on wall, floor and ceiling surfaces (most evident on hard, impervious surfaces);
- moisture on and within the surface layers of absorbent surfaces (may not be evident until the surface becomes saturated);
- mould growth on all surfaces affected by condensation;
- internal breakdown of materials and elements where interstitial condensation has occurred;
- a general deterioration in the internal environment owing to the occurrence of associated dampness, smells caused by mould growth and rot, and deterioration of the appearance of surface finishes.

Preventive measures

Condensation is a widespread problem, and its adverse effects are capable of leaving a building uninhabitable if no preventive measures are in place. It is therefore essential that the following steps are taken when carrying out refurbishment work in order to ensure that condensation does not occur:

- Check for the presence of condensation in the existing building and, where applicable, eliminate it.
- Ensure that the upgrading or replacement of any existing elements does not increase the risk of condensation.
- Where a proposed change of use is likely to introduce new conditions that are conducive to condensation, ensure that appropriate measures are taken to prevent its occurrence.

Condensation is caused by a combination of different factors, but is most likely to occur in those buildings where large quantities of water vapour are produced, such as buildings housing certain industrial processes and, particularly, housing where modern living habits – for example, drying washing indoors and not opening windows – create conditions that are conducive to condensation. The risk of condensation occurring in buildings can be significantly reduced by paying attention to three specific factors, that is, ventilation, heating, and thermal insulation and vapour barriers.

Ventilation

Ventilation helps to remove air containing water vapour from the building and is particularly important where large quantities of vapour are released, such as areas housing certain industrial processes, kitchens, bathrooms, showers, etc. Good ventilation is best achieved using powered extractor fans, and their installation is recommended in areas where large quantities of water vapour are produced. In areas where water vapour levels are lower, natural, rather than mechanical, ventilation will usually be satisfactory, and this can be achieved by means of openable windows or ventilators.

Heating

Adequate heating reduces the risk of condensation in two ways: first, it warms room surfaces, keeping them above dew-point and preventing surface condensation; and second, it increases the moisture-carrying capacity of the ventilated air. Different heating methods vary considerably in their efficiency at reducing condensation. Short periods of high-level heating with no heating in between are conducive to condensation. While the heat is off, the interior surfaces become cold and may fall below dew-point. The heating periods usually correspond with a significant increase in water vapour production while the building is occupied, and this, in conjunction with the cold interior surfaces, aggravates the problem of condensation.

Leaving some rooms unheated while the remainder of the building is heated increases the possibility of condensation. The cold interior surfaces of the unheated rooms will be susceptible to condensation as warmer air, with a higher vapour content, migrates to them from heated rooms.

Flueless oil and gas heaters release large quantities of water vapour into the atmosphere, and their use significantly increases the possibility of condensation.

Continuous background heating of the whole building, used in conjunction with the main heating periods, is the most effective means of preventing condensation, and, where possible, a system that is capable of economically providing this should be installed.

Thermal insulation and vapour barriers

Thermal insulation of walls, floors and roofs helps to reduce the risk of surface condensation by ensuring that their internal surfaces are kept above dew-point temperature. However, the provision of thermal insulation, while preventing surface condensation, can increase the risk of interstitial condensation within the building element. This is because the insulation has the effect of moving the position of the dew-point temperature from the element's surface to a point within its thickness. Water vapour will then diffuse into the insulation and the element until it reaches the position of the dew-point temperature, where interstitial condensation will occur. To prevent interstitial condensation, the water vapour must be prevented from diffusing into the insulation and the element, and this is achieved by providing a vapour barrier on the 'warm side' of the insulation.

Several materials are resistant to the passage of water vapour, including certain paints, wallpapers, polythene sheeting, plastic and aluminium foils. In refurbishment work, vapour barriers are often provided pre-bonded to other materials, one of the most common proprietary examples being vapour-check grade Gyproc Thermal Board (www.*british-gypsum.com*). This is similar to the thermal boards described in Section 4.3 and comprises gypsum wallboard, factory-bonded to a backing of expanded polystyrene insulation, with a vapour-resistant membrane incorporated between the two. These composite boards are widely used to upgrade the thermal performance of existing walls and roofs (see Chapter 4) and, when fixed, the vapour-resistant membrane is on the warm side of the polystyrene insulation, therefore preventing the passage of water vapour into the insulation and the existing element.

Where existing elements are insulated by other methods (see Chapter 4) but still require a new internal lining, normal plasterboard with a pre-bonded vapour-resistant membrane can be used. Gyproc Duplex WallBoard (www.british-gypsum.com) comprises gypsum wallboard, capable of direct decoration, with a metallised polyester film on the inner face that provides resistance to the passage of water vapour.

Ideally, the vapour barrier should be continuous over the whole area of the element, but this is difficult to achieve in practice, especially where the above types of material are used. Joints between the boards and holes around pipes etc. allow some 'leakage' of water vapour, but, in normal circumstances, this should not cause problems. However, in pitched-roof spaces with insulated ceilings, and in 'cold' flat roofs with insulated ceilings (see Section 4.4), the risks of condensation are much greater, and additional precautions are essential. In such cases, the roof void above the insulated ceiling should be properly ventilated in order to remove any water vapour that succeeds in penetrating joints in the vapour barrier or 'leaking' through holes around pipes, badly fitting loft trapdoors, etc.

Part F of the Building Regulations, which deals with ventilation in buildings and condensation in roofs, requires that 'Adequate provision shall be made to prevent excessive condensation in a roof void above an insulated ceiling', and it is essential that this is complied with if condensation within cold roof voids is to be avoided.

Preventing condensation in pitched-roof spaces

The addition of thermal insulation to the ceiling beneath a pitched roof significantly increases the risk of condensation within the roof space. The additional insulation has the effect of lowering the temperature within the roof space and, consequently, increasing the condensation risk, as warmer, moist air from the building migrates upwards into the roof. The incorporation of a suitable vapour barrier at ceiling level will considerably reduce the amount of water vapour migrating into the roof space, but, in practice, large quantities of vapour will still enter the roof at 'weak points', such as around loft access trapdoors, at ceiling roses and where service pipework penetrates the ceiling. The water vapour entering the roof will condense on any cold surface, such as the sarking felt, or will be absorbed by the timber components of the roof. Over a period of time, this can lead to serious defects, including:

- fungal decay of roof timbers;
- deterioration of the roof insulation owing to water dripping off the sarking felt and saturating the insulation material;
- short-circuiting of electrical wiring.

The thermal upgrading of existing elements is now a major feature of building refurbishment, with roofs receiving particular attention because of the excessive heat loss that can take place through them. With the majority of existing pitched roofs, the additional insulation is provided at ceiling level, and it is therefore essential that special provision is made to reduce the associated increased risk of condensation and its damaging effects.

The most effective means of reducing the condensation risk within pitched roofs is to provide adequate ventilation, which will remove the water vapour from the roof space before it has the opportunity to condense. It is now mandatory for all new buildings to incorporate roof space ventilators, and, in order to alleviate the condensation problem in existing buildings, a number of proprietary ventilators have been developed for installation into existing roofs.

Soffit ventilators

Glidevale soffit ventilators (www.glidevale.com) have been specifically designed for incorporation into existing roofs during refurbishment work, and two ventilators from the range are described below and illustrated in Figures 6.8 and 6.9. Note that both types of ventilator should be installed in conjunction with Glidevale Universal Rafter Ventilators to ensure that the ventilation path does not become blocked by the roof insulation material (see Figure 6.8).

Twist and lock soffit ventilators (see Figure 6.8) are manufactured from ultraviolet-resistant injection-moulded polypropylene with an integral insect grille. The 70 mm diameter ventilators, available in black, white or brown, are inserted by a simple 'twist and lock' action into 70 mm diameter holes formed through the existing soffit board using a special hole saw. The soffit ventilators should be inserted at 200 mm centres, and, to ensure fully effective and permanent ventilation, they should be used in conjunction with universal rafter ventilators, described below.

Glidevale spring wing soffit ventilators (see Figure 6.9) are manufactured in polypropylene and comprise an integral insect grille and spring clips, which enable them to be easily installed into 270 mm long × 92 mm wide holes sawn through the existing soffit board. The ventilators should be inserted at 1,200 mm centres, and, to ensure fully effective permanent ventilation, they should be used in conjunction with Glidevale Universal Rafter Ventilators described below.

Glidevale Universal Rafter Ventilators (see Figure 6.8) are designed for pushing into the eaves to prevent spillage of roof insulation material into the soffit and consequent blocking of the soffit ventilators (described above). The ventilators, which are effective with both quilt and granular-fill insulation, are available in two sizes to suit 600 mm and 400 mm rafter spacings. They are manufactured from rigid PVC sheet, formed to provide a series of channels through which air may pass. When pushed into the eaves, the ventilators automatically adjust to the correct roof pitch. The quilt or granular-fill insulation is then laid into the ventilators to complete their installation.

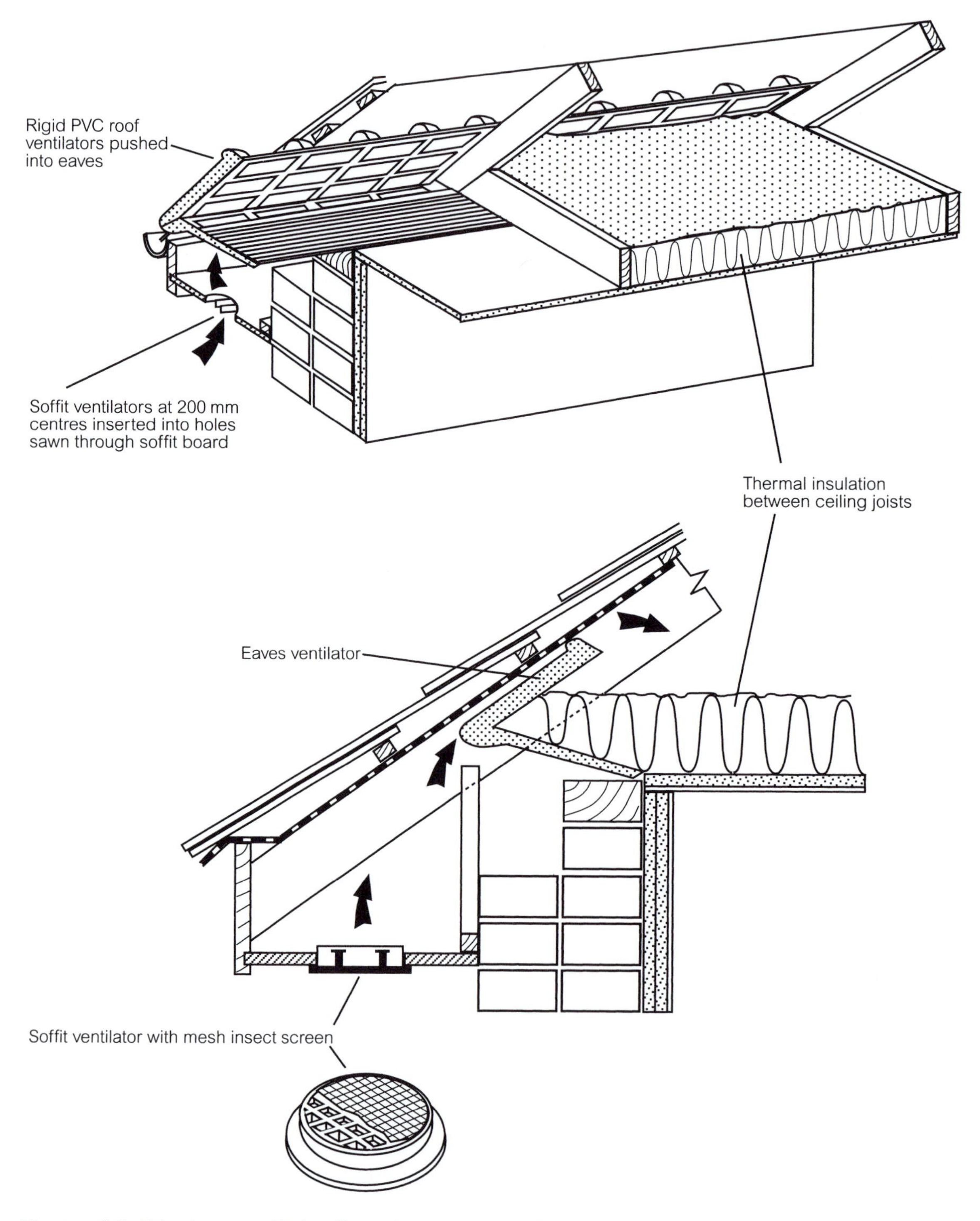

Figures 6.8 *Glidevale twist and lock soffit ventilator and roof ventilator*

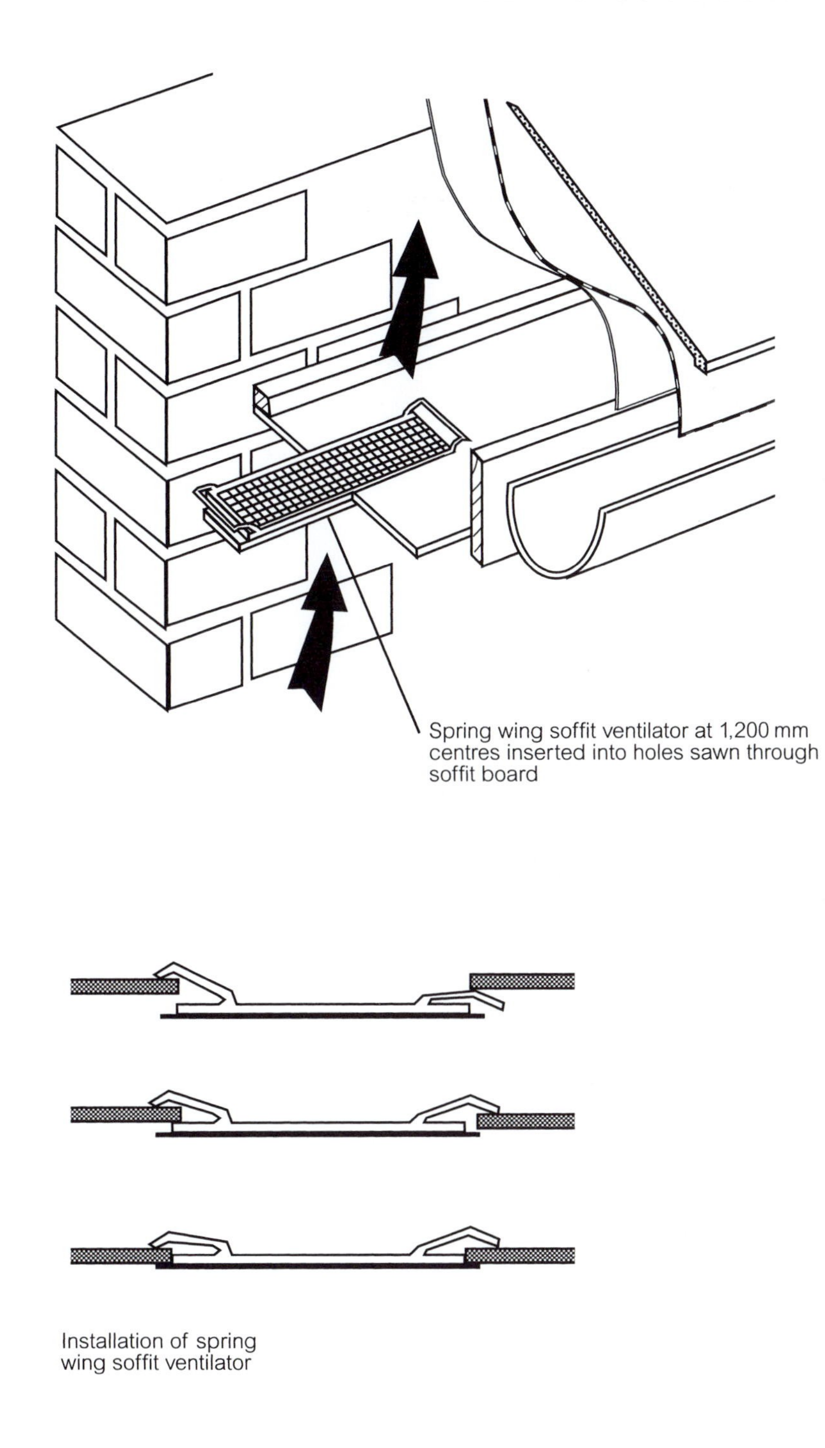

Figures 6.9 *Spring wing soffit ventilator (www.glidevale.com)*

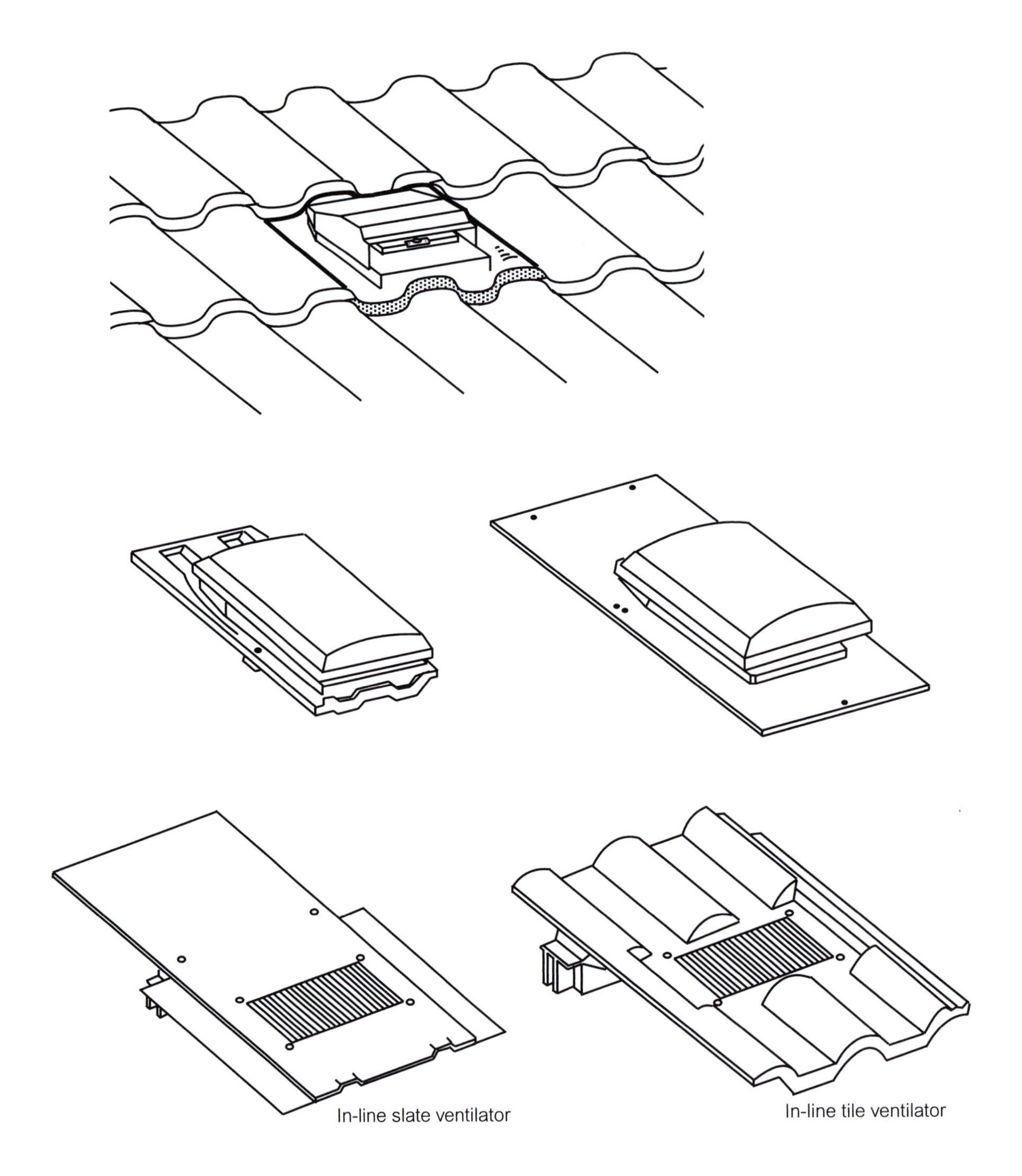

Figure 6.10 *Tile and slate vents (www.glidevale.com)*

Tile and slate vents

Eaves-to-eaves ventilation of roof spaces is the preferable method and should be employed wherever possible. However, certain types of roof construction do not lend themselves to this solution, and it is therefore essential to use some other means of ventilation. A very effective means of providing adequate roof space ventilation, where eaves-to-eaves ventilation is not practicable, is to employ purpose-made tile and slate vents, fixed at strategic positions over the roof area in place of the normal tiles or slates. Glidevale tile and slate vents are manufactured from an ABS/PVCu blend, treated with ultraviolet-resistant polymeric resin, and are available in a wide range of sizes, colours, textures and profiles, allowing them to be installed in conjunction with the main tile and slate manufacturer's products. Typical tile and slate vents are shown in Figure 6.10.

The majority of proprietary tile and slate vents are designed for installation where an entirely new roof covering is being provided, which is not uncommon in refurbishment work. However, if the building in question does not require re-roofing, it will be necessary to insert tile or slate vents individually into the existing roof.

To install a ventilating tile, one existing tile is removed, the sarking felt is cut and folded in a prescribed way, and the new ventilating tile is positioned and secured. To install a ventilating slate, it is necessary to remove a number of slates from the area where the vent is to be positioned.

Airbricks

A relatively simple, inexpensive means of improving roof space ventilation in buildings where the roof has gable ends, and where eaves-to-eaves ventilation or the installation of tile or slate vents is not practicable, is to insert new airbricks into the gable walls. Airbricks of any standard size can be inserted into the roof's gable ends after carefully cutting out sections of the existing brickwork or masonry, and these will provide a fairly effective means of ventilation. However, this will not be as effective as eaves-to-eaves ventilation.

References

Building Research Establishment (1993) *Damp Proofing Existing Basements* (Good Building Guide 3), Watford: BRE.

Building Research Establishment (1997) *Diagnosing the Causes of Dampness* (Good Repair Guide 5), Watford: BRE.

Building Research Establishment (1997) *Treating Rising Damp in Houses* (Good Repair Guide 6), Watford: BRE.

Building Research Establishment (1997) *Treating Condensation in Houses* (Good Repair Guide 7), Watford: BRE.

Building Research Establishment (1997) *Rain Penetration* (Good Repair Guide 8), Watford: BRE.

Building Research Establishment (1999) *Treating Dampness in Basements* (Good Repair Guide 18), Watford: BRE.

Garratt, J. and Nowak, F. (1991) *Tackling Condensation*, Watford: BRE.

Hutton, T. (1998) 'Rising Damp', in J.Taylor (ed.), *The Building Conservation Directory 1998,* Tisbury: Cathedral Communications Ltd: pp. 38–40.

Newman, A. J. (1988) *Rain Penetration through Masonry Walls: Diagnosis and Remedial Treatment,* Watford: BRE.

Richardson, B. A. (1991) *Defects and Deterioration in Buildings,* London: E. & F.N. Spon.

Introduction of new floors and access between levels

CHAPTER 7

7.1 General

It is nearly always prudent for developers to introduce new floors into buildings, thus increasing the rentable or saleable floor area of the property. Conversion of attic or roof space, making it a habitable room, is often the most economic method of doing this. Clearly, the floors will need to be strengthened, dormer windows or roof lights (such as Velux windows) can be introduced, and party wall issues will need to be addressed. It will be necessary to ensure there are necessary precautions to prevent passage of fire, and even with party walls it is prudent to upgrade thermal insulation. Recent research has shown considerable heat is lost through party walls due to the movement of air within the cavity. Walls in attics are often just half a brick thick; the addition of acoustic barriers to reduce the passage of sound out of and into new rooms is often desirable.

The pitch of the roof, existing beams and unique windows can make the roof space one of the most interesting areas of a building. The inclusion of such features can make the idea of the conversion of the roof space attractive, even if it proves to be costly (Photographs 7.1–7.3).

7.2 Roof spaces: accommodating stairs – access and clearance

To ensure that the conversion is possible, attention will also need to be paid to the access into the loft space. In a recent decision by the Secretary of State (reported in *Building Engineer* 2007), while regulations for vertical access were not relaxed, a level of tolerance within the scope of the Building Regulations Part K was identified. The main problem associated with introduction of a new floor in an existing property is where to locate the stairs and whether there is sufficient headroom and width to comply with the Regulations. Accommodating a new casement of stairs into a listed property often needs considerable thought.

The current Building Regulations require a clear head height of 2 m (Figure 7.1), and where the stairway is used for access to a single room housed in a loft space the requirement is reduced, allowing for a 1.8 m headroom at the external edge of the stairs, increasing to 1.9 m at the centre (Figure 7.2).

The case considered by the Secretary of State involved the conversion of listed barns to three two-storey dwelling units, each with four bedrooms. The properties

Photograph 7.1
Room introduced into roof space of existing building: façade retention project

Photograph 7.2
Steel frame introduced into a terraced house to add additional room space making use of the attic

Photograph 7.3
Timber gluelam beams used to complement existing trusses in a barn conversion

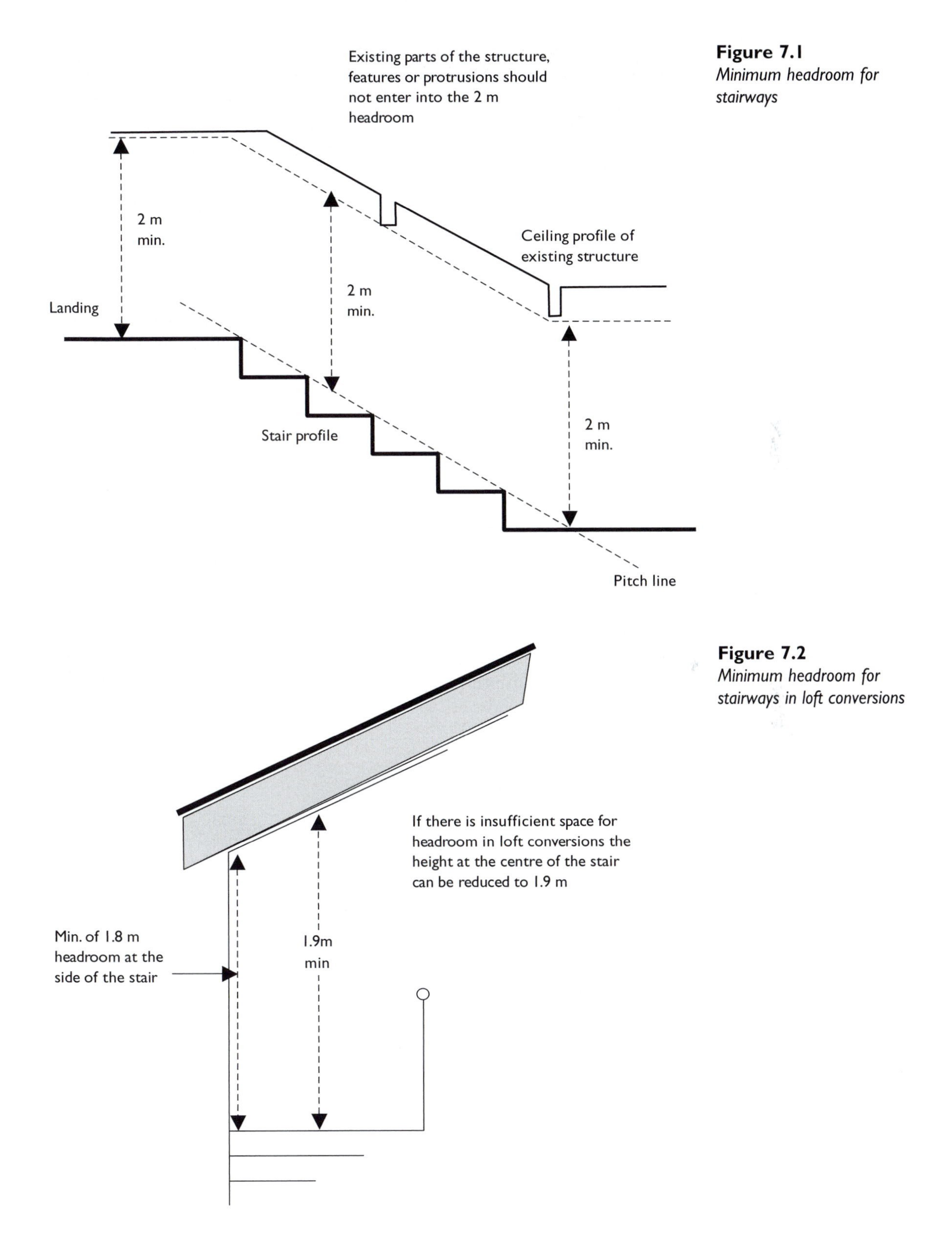
Existing parts of the structure, features or protrusions should not enter into the 2 m headroom
2 m min.
Landing
2 m min.
Ceiling profile of existing structure
2 m min.
Stair profile
Pitch line
If there is insufficient space for headroom in loft conversions the height at the centre of the stair can be reduced to 1.9 m
Min. of 1.8 m headroom at the side of the stair
1.9m min

Figure 7.1
Minimum headroom for stairways

Figure 7.2
Minimum headroom for stairways in loft conversions

were completed; however, the developer had altered the positioning of the stairs from that shown on the original drawings. It was considered necessary to alter the position of the stairs due to two existing oak beams, which were obstructing the route of the stairway. To reduce the impact of the beams on the headroom, the landing had been lowered, giving more headroom under the beams. The alteration still left the headroom at the centre of the upper landing at 1.95 m and 1.6 m at the side of the landing. The developer had noted that considerable thought had gone into the location of the stairs, and consultation with stair manufacturers had not produced alternative layouts that gave the additional clearance required. The council had considered that the stair could have been located to provide greater headroom and could comply with Requirement K.

Requirement K states that 'stairs, ladders and ramps shall be so designed, constructed and installed as to be safe for people moving between different levels in or about the building'.

The Secretary of Stated recognised the difficulty of undertaking work in listed buildings and noted that in this case the headroom was above the 1.9 m threshold used in conversions, but was less than the prescribed 1.8 m at the lower edge, being only 1.6 m high. The landing, which was 720 mm wide, did allow persons to pass unobstructed on the higher side. With the greater clearance offered on the inside edge of the landing, and the 2 m clearance that was achieved on the main stair, the Secretary of State considered it unlikely that any person would fall down the stair as a result of a collision with the ceiling on the lower side of the landing.

Although the design fell short of the ideal situation, the Secretary of State concluded that the stair and landing offered a reasonable level of safety for the purpose of complying with Requirement K. While the Secretary of State did not relax the regulation, a level of tolerance was indicated in the advice.

Clearly it is easier to address such issues with the council during the planning and approval stages. However, with refurbishment there are often unknowns;

Photograph 7.4
Mezzanine floor introduced into a barn conversion making use of the large roof space

as the structure is exposed situations may be encountered that prevent the initial design being adhered to. In such situations it may be useful to know what tolerance exists in the guidance offered by the Approved Documents.

Photograph 7.4 shows an existing column protruding into the stairwell; in the situation shown the column is included in this open area as a feature; the stair was designed with the existing structure in mind, and clear headroom and stair widths are properly accommodated within the design.

Further information on appeals and determinations by the Secretary of State can be found at www.planningportal.gov.

References

Building Engineer (2007) 'Building Act 1984 – Section 39', *Building Engineer,* July: 32–33.

Emmitt, S. and Gorse, C. (2005) *Barry's Introduction to Construction of Buildings*, Oxford: Blackwell Publishing.

Emmitt, S. and Gorse, C. (2006) *Barry's Advanced Construction of Buildings*, Oxford: Blackwell Publishing.

CHAPTER 8 Timber decay and remedial treatments

8.1 General

The majority of older buildings contain many more timber components than modern buildings, and it is not unusual to encounter timber decay in refurbishment work, particularly where the building has suffered from neglect and lack of maintenance. The latter often results in the ingress into the building of moisture and dampness, which represents one of the principal causes of decay in timber components. Timber decay can result from fungal attack or insect attack, both of which cause a gradual weakening of the timber, and, if remedial action is not taken, its eventual disintegration.

Where timber decay is evident in a building that is about to be refurbished, it is advisable to call in one of the many companies that specialise in its diagnosis and treatment, such as Rentokil (www.rentokil.co.uk/propertycare) or Peter Cox (www.petercox.com). A specialist company will carry out a detailed survey and diagnosis of the timber decay, and provide a treatment and eradication package backed by an extensive guarantee, usually for a minimum of 30 years after treatment.

The nature and treatment of the main forms of timber decay are outlined below.

8.2 Fungal attack

Fungal attack occurs only where sufficient moisture is present in the timber, and it is usually caused by one of two wood-destroying fungi, *Serpula lachrymans* or *Coniophora puteana.*

Dry rot

The best-known wood-destroying fungus is *Serpula lachrymans* or 'dry rot', and this typically grows on timber remaining moist, rather than very wet, over long periods, resulting from moisture penetration into the building, leaking plumbing or condensation, often combined with bad ventilation. Dry rot has the unique property of being itself able to produce the moisture it needs for further growth, even when the moisture content of the timber has been reduced to below the level needed to sustain fungal growth (around 20 per cent).

The most vulnerable parts of the building include timber ground floors, joists built into solid external walls, and roof timbers. However, all timber within the

Photograph 8.1
Deep cuboidal cracking resulting from dry rot attack (Rentokil Property Care: www.rentokil.co.uk/propertycare)

building will be vulnerable, particularly in the event of a severe attack. Once established in moist timber, the fungus will readily extend into adjoining dry timbers and will penetrate masonry and plasterwork in order to reach other sound timber, which then, in turn, becomes infected. The fungus grows from very fine air-borne spores, rusty red in colour, that have alighted on damp timber, and it gradually spreads, either as a silky white sheet or a greyish, felted skin. The fungus feeds on the timber, causing it to lose strength, and reduces it to a dry, brittle state. The rotted timber has a pale brown colour, and the surface splits into cuboidal or brick-shaped sections formed by a combination of deep, longitudinal and transverse cracks (see Photograph 8.1). The early symptoms may be a characteristic musty, mouldy smell with, later on, the presence of a rusty red powder (the spore dust) and the appearance of flat, mushroom-like growths (sporophores) through joints in the timber.

Treatment of dry rot fungus

Because dry rot is so prolific and capable of spreading through other materials and attacking dry timber, its eradication treatment involves not only the replacement of all affected timber with pretreated timber, but also the treatment of adjoining timbers, brickwork, plasterwork and adjacent areas well away from the point of decay. All of the following operations must be carried out in order to ensure that the fungus is completely eradicated:

- Locate and eliminate the dampness responsible for the original dry rot attack. This may involve providing a new DPC, preventing moisture penetration through walls, repairing leaking roofs and plumbing, eliminating condensation, clearing the bridging of DPCs and improving ventilation, including the provision of additional airbricks, etc.

- Cut out, remove from the site and burn all defective timbers showing cuboidal cracking, pale-brown coloration, white fungal mycelium or soft areas when probed. All apparently sound timber within 300 mm (HSE 2001), or, to provide a safety margin, 600 mm (PCA 2008), of defective timber should also be cut out, removed and burned, together with all debris and loose material within roof voids, sub-floor voids and other areas in the vicinity of the attack.
- Surface spraying of timbers will not stop hyphae growing across the surface to get from A to B.
- Carry out a thorough check of all other timbers within the building to which the fungus might have spread. This may involve removing skirtings, architraves, wall panels or ceilings, lifting floorboards, etc.
- Strip off all wall plaster that may contain fungal strands to 300 mm beyond the observed limit of growth (HSE 2001; PCA 2008), and clean down all exposed masonry by wire brushing. Remove all stripped plaster, debris and dust from the site.
- Surrounding masonry should be cleaned and treated by surface application of masonry biocide; preservative plugs or pastes inserted into holes drilled into localized problem areas; and/or irrigation of fungal solution inserted via holes drilled into the wall (PCA 2008). The use of controlled heat has been suggested as a means of sterilizing masonry adjacent to the affected area; however, there is little reported evidence of the use and effectiveness of this. Fungicide fluids for treatment against dry rot in masonry are water-based and normally require total saturation; owing to the health and safety risks caused by the preservative, this should be avoided wherever possible (HSE 2001). Preference is for the use of preservative plugs, paste or, if used, organic, solvent-based fungicidal products.
- Replace all timber that has been cut out with new, well-seasoned timber that has been pretreated with fungicidal preservative.
- Make good all work that has been disturbed during investigation, treatment and eradication of the dry rot attack, including renewing stripped plasterwork, decorative finishes, etc.

Wet rot

Coniophora puteana or 'wet rot' is the other common wood-destroying fungus, and this requires much wetter timber than dry rot in order to develop. It is therefore more common in exterior joinery exposed to rain, such as windows, fascia boards, timber cladding, etc., although interior timbers that have become wet because of excessive moisture penetration or leaks in plumbing are also vulnerable.

Unlike dry rot, the wet rot fungus is incapable of spreading to, and infecting, dry timber, and outbreaks are therefore confined to the affected wet timber and its immediate vicinity. In an advanced stage, the wet rot fungus produces slender, thread-like, dark brown or black strands, and there are seldom any signs of a fruiting body or of the olive-brown spores. The decayed timber is dark in colour, and any cracking is less deep than in timber affected by dry rot (see Photograph 8.2).

Photograph 8.2
Shallow cracking resulting from wet rot attack (Rentokil Property Care: www.rentokil.co.uk/propertycare)

Treatment of wet rot fungus

Because wet rot is not as prolific as dry rot, its outbreaks usually being much more localised, treatment and eradication are simpler, and usually only involve replacement of the affected timber with new pretreated timber, together with the elimination of the original cause of the wet rot attack.

A wet rot attack can be remedied by cutting out the affected wood well back from the point of decay and splicing-in new, preservative-treated timber. However, where the attack is widespread, as is often the case with neglected window frames, localised repair is often not justified, and complete replacement with new units will be preferable. Where the existing windows are badly rotted, their replacement with 'maintenance-free' metal or PVCu units is advisable, since these will neither rot, nor require the expense of regular painting throughout the life of the refurbished building.

Where exterior woodwork has suffered an attack of wet rot, the usual cause is breakdown of the protective paint finish, allowing rainwater to penetrate the timber and create suitable conditions for attack. In the case of internal elements, the cause is usually water penetration through the building envelope, or leaking plumbing, and, in addition to replacing the rotted timber, it is essential that the source of the water or moisture is ascertained and eliminated in order to rule out the risk of further outbreaks.

8.3 Insect attack

The symptoms of insect attack are disfigurement of the timber's surface by small circular- or oval-shaped holes, accompanied by deposits of bore dust, and a gradual reduction in strength. Although insects will attack dry timber, it will be more vulnerable if it is damp or weakened by fungal decay.

Wood-boring insects lay their eggs on the surface or in crevices in timber and, after the larvae hatch out, they bore into the timber, feeding on it and growing in the process. After a period of between one and several years, the larvae pupate near the surface of the timber, and the beetles emerge, leaving a hole that is characteristic of the particular species of insect. The exit holes are usually clear, sharp edged and accompanied by bore dust and are thus not easily confused with man-made nail or pin holes.

The most prevalent wood-boring insect is the Common Furniture Beetle or *Anobium punctatum* (see Photographs 8.3–8.5), which is responsible for around three-quarters of all cases of insect attack in buildings. The insect's name is misleading, since it attacks all structural timbers, both hardwood and softwood. Its exit holes are circular and about 1.5 mm in diameter.

Wood-Boring Weevils (*Euophryum confine and Pentarthrum huttoni*) are responsible for around 5–6 per cent of all cases of insect attack and usually attack very damp or decayed hardwoods and softwoods, often where fungal decay is also present. As the attack by the wood-boring weevils only occurs in timber that is already damp or where fungal decay is present, the damage is secondary to that of the already decaying wood. No treatment for the wood-boring weevil is necessary as, once the timber dries out, the beetles die and will not inhabit the timber. Infestation of sound, dry timber is not possible. The exit holes are oval- or slit-shaped, with ragged edges, and are only 0.5–1 mm wide.

The Death-Watch Beetle or *Xestobium rufovillosum* (see Photographs 8.6 and 8.7), which accounts for about 5 per cent of all cases of insect attack, leaves larger exit holes, approximately 3 mm in diameter; the adult is approximately 7 mm long. The Death-Watch Beetle attacks mainly hardwoods, and particularly oak. Death-Watch Beetle has been found in softwoods, but this is not common. It is therefore usually found only in much older buildings where the use of oak was common (see Photograph 8.8).

The *Lyctus* Powder Post Beetle or *Lyctus brunneus* also attacks only hardwoods, including oak, ash and particularly elm. The beetle will eat the sapwood of hardwoods, especially those hardwoods with high starch contents and large pores. It is quite common for block-board and plywood to be attacked. Its exit holes are circular and 1–2 mm in diameter, and its bore dust is a fine, talcum-like powder, distinguishing it from that of the Death Watch Beetle, which produces a much coarser dust. Beetles need moist timber to lay their eggs into, as the larvae need the water and sapwood, which contains minerals and nutrients to survive. Dry timber and kiln-dried timber are the best cures against infestation.

The House Longhorn Beetle or *Hylotrupes bajulus* attacks only softwoods, and its area of activity is restricted to the south of England, mainly around Surrey. The insect causes rapid deterioration of the timber it attacks, leaving very large oval exit holes, approximately 10 mm in length and 6 mm wide. This insect and the *Lyctus* Powder Post Beetle are together responsible for only about 1 per cent of all cases of insect attack. When the weather is warm, and in areas of large infestation, a scraping noise may be heard, which is the larvae feeding on the timber.

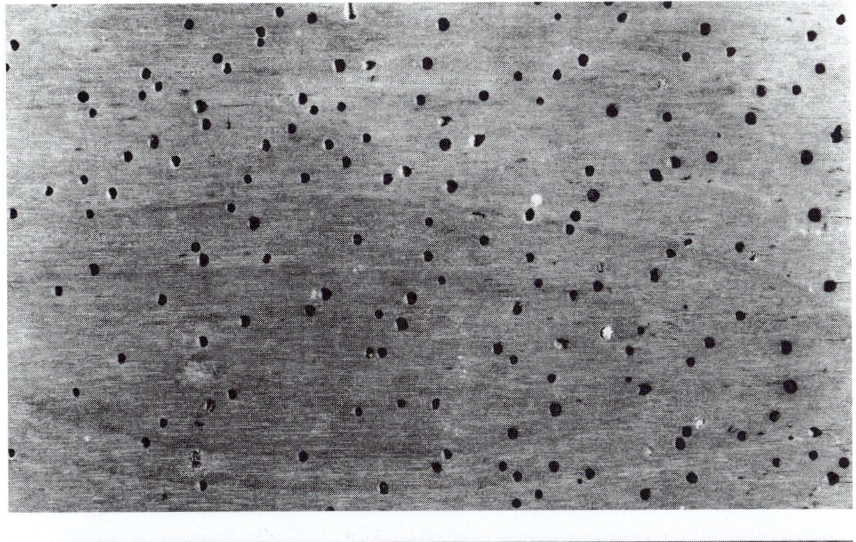

Photograph 8.3
Furniture Beetle exit holes (Rentokil Property Care: www.rentokil.co.uk/propertycare)

Photograph 8.4
Furniture Beetle exit holes in a roof truss; exit holes 1–1.5 mm

Photograph 8.5
Furniture Beetle damage (Rentokil Property Care: www.rentokil.co.uk/propertycare)

Photograph 8.6
Death-Watch Beetle larva (Rentokil Property Care: www.rentokil.co.uk/propertycare)

Photograph 8.7
Death-Watch Beetle damage (Rentokil Property Care: www.rentokil.co.uk/propertycare)

Photograph 8.8
Oak roof truss with exit holes from Furniture (small exit holes, 1.5 mm) and Death-Watch Beetles (large exit holes, 3–4 mm)

Treatment of insect attack

Where the extent of the insect attack is not advanced, usually indicated by few and scattered exit holes, causing negligible reduction in the strength of the timber, it is recommended that the timber be treated with water-based insecticide and applied to minimise the exposure to people, pets and the environment. Where the attack goes to the heart of the timber, the insect infestation must be assessed so that the amount of organic solvent-based treatment matches the degree of penetration. If the attack is widespread, paste or injected solvent-based formulations should be applied to all timbers (HSE 2001). This will kill any insects that are still active and leave the wood toxic to wood-boring insects, therefore preventing further attack. If the attack is localised, treatment need only be applied 300 mm beyond the area of the affected timber. Modern proprietary insecticides are capable of penetrating timbers deeply and being retained by the timber for very long periods, irrespective of other causes of deterioration. However, the cure for both insect attack and mould growth is to remove the source of the damp conditions and environment.

Timbers that have been structurally weakened by insect attack must be either treated with insecticide and strengthened, or cut out, burned and completely replaced with new, pretreated timber. The primary control method to prevent insect infestation and fungal attack is to keep timber dry. Excessive use of even modern, water-based treatments is not recommended. If the treated timber ever becomes wet, the chemicals may leach out. Although many older buildings are likely to suffer from both insect and fungal attack, it is advisable, when carrying out a programme of treatment and eradication, to treat the affected and immediately adjacent areas only. Use of chemicals 'just in case' is not advised (www.askjeff.co.uk).

8.4 *In situ* injection techniques for the preservation of timber components

The *in situ* injection of preservatives into timbers that have been affected by, or are susceptible to, fungal and insect attack is a relatively recent development. Where fungal or insect attack in timber components is not sufficiently advanced to have caused a significant loss in strength, specially designed plastic nozzles, inserted into the timber, can be used to pressure-inject preservatives much more deeply than can be achieved by surface application. The technique can also be used on components that have not suffered attack, but that are considered vulnerable.

The minimum size of timber that can be treated by *in situ* injection is 50 × 25 mm, and typical applications include external softwood joinery, such as window- and door-frames, floor joists, roof timbers, beams and lintels.

In the Wykamol timber injection system (see Figure 8.1), hollow polypropylene injectors (36 mm long × 9.5 mm diameter, or 24 mm long × 6.5 mm diameter) are inserted into pre-drilled holes in the timber (www.wykamol.com). The injector holes must penetrate to within 12 mm of the component's far face and be at least 40 mm deep for the larger injector, and 30 mm for the smaller. The outer end of each injector has an injector nipple, containing a non-return valve, that is left protruding from the face of the timber, and to which the injection line and pump

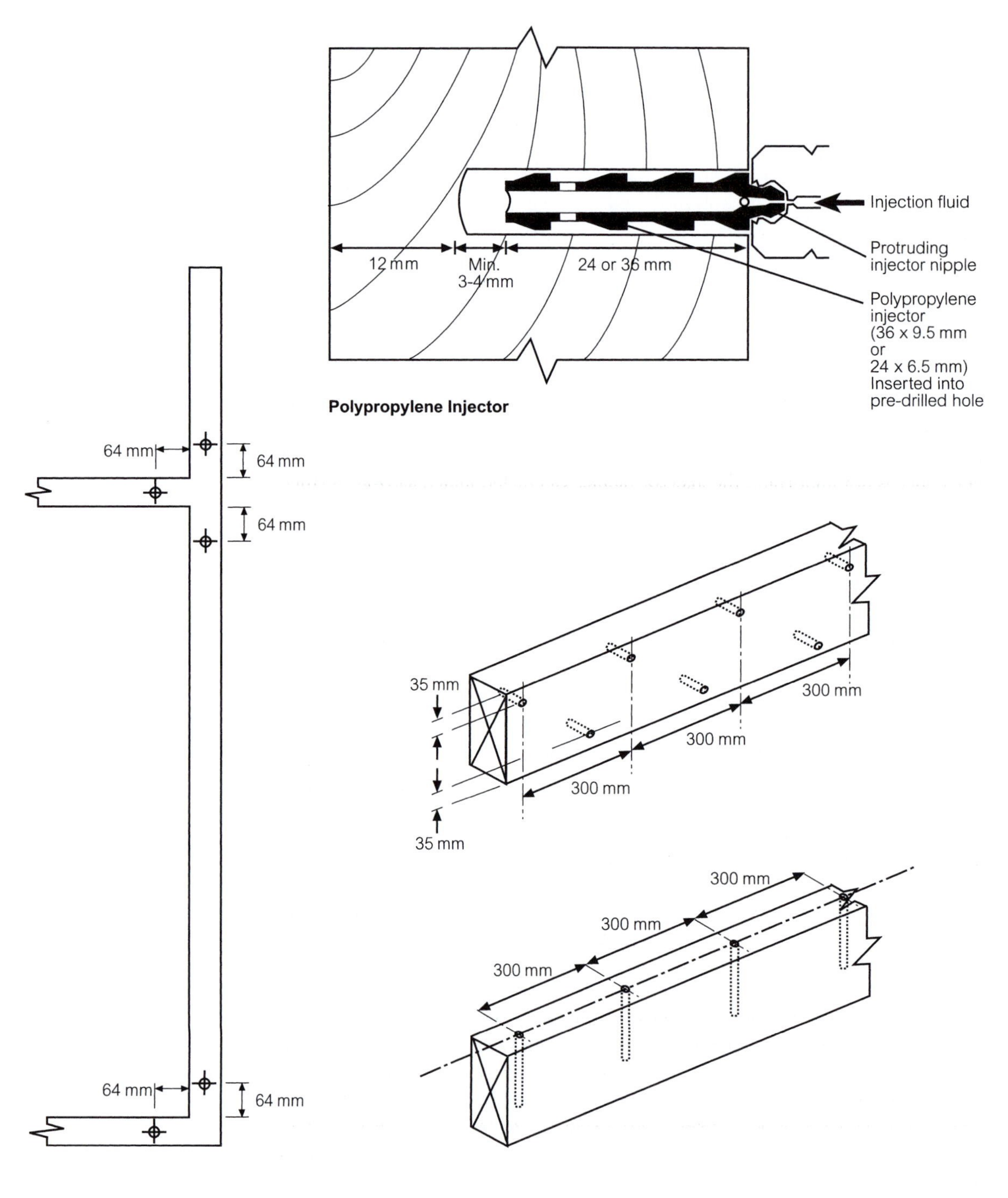

Figure 8.1 In situ *injection technique for the preservation of timber components*

are attached. The organic preservative fluid is then injected under pressure for at least 2 minutes, or until the timber is seen to be saturated.

The positions and number of polypropylene injectors required will vary according to the type and size of component being treated. For example, a 200 × 75 mm beam on edge would require injectors inserted at 300 mm staggered centres, in two rows, 35 mm from the top and bottom edges of the wide face; or, where the upper narrow face is accessible, in one row at 300 mm centres (see Figure 8.1). Small-section external softwood joinery, such as window and door frames, can be treated using the smaller, 24 mm long injectors, which should be inserted from the exterior in areas susceptible to decay, such as the lower and intermediate joints of frames, including sills. They should be inserted into each member at a distance of approximately 64 mm from each side of the joints and, if required, at regular centres between joints.

After the injection procedure has been completed, the injector nipple can be cut off flush with the surface, and the holes can be sealed with putty and matching paint. Alternatively, the injectors and nipples can be driven fully into the timber, and plastic sealing caps can be inserted. In particularly high-risk areas, where appearance is not important, the injector nipples can be left intact, allowing further injections of preservative fluid in the future where this is considered necessary. However, under normal circumstances, this technique of timber preservation should ensure that sections are free from fungal and insect attack for at least 10 years after treatment. Ideally, the injection treatment should be combined with normal painting maintenance and repairs to stop further decay in affected areas and protect sound joints from attack. Injection of the organic preservative does not impart strength to rotted wood nor render unnecessary the application of normal standards in deciding the extent of repairs and/or replacements, which would always precede the protective injection where necessary.

An alternative timber injection treatment to the above employs preservative in the form of a gel. Wykamol Boron Gel 40 is a glycol-based wood preservative containing disodium octaborate, designed for injection into larger-section timbers, such as beams or joists, to protect against fungal and insect attack. The gel can be used to treat sound (new or existing) timbers or timbers where decay is already present, provided sources of moisture are removed and ventilation is improved. It is not suitable for timbers exposed as a decorative feature, as the gel causes staining to the timber and to adjoining porous materials.

The 10 mm diameter injection holes are drilled to within 15 mm of the full depth of the timber, at intervals and spacings according to the size of the section. After injection of the gel, the holes are sealed with plastic plugs or timber dowelling.

An alternative to the above fluid- and gel-injection systems involves the use of boron rods, which are implanted into the timber sections where there are signs of decay, or earlier as a preventative measure. Boron rods are cylindrical rods composed of anhydrous boric oxide. Often referred to by its trade name Borax (also known as DOT, disodium octaborate tetrahydrate), the material is toxic to fungi and insects.

The boron rods contain the maximum level of borate preservative available in rod form. The preservative, which controls both insect and fungal attack, is mobilised when exposed to a moisture content in excess of 25 per cent within the timber. Typical applications are for timbers that may be exposed to wetting in service, such as windows and external doors. The rods are inserted into pre-drilled holes, approximately 1–2 mm greater in diameter, and at least 10 mm longer, than the rods used. The holes should extend to not more than 10 mm from the rear face of each timber section being treated. Once inserted, the rods are sealed in position with a suitable filler, plastic cap or timber dowelling. Standard rods are 8 × 65 mm long; other sizes are available to order. The number and configuration of the rods, and their installation, must be in accordance with the manufacturer's instructions. Further information can be found on the 'diy doctor' website (www.diydoctor.org.uk).

8.5 Localised repair techniques for decayed timber window frames and other joinery

Timber window frames are highly susceptible to wet rot attack and subsequent localised decay, and it is not unusual to find extensive damage to window frames in older buildings, particularly where maintenance has been neglected.

Provided the areas of decay are not too extensive, and the larger proportion of timber remains unaffected, it is possible to carry out localised repair to good effect. A polyester resin-based filler system, such as Cuprinol Ultra Tough Wood Filler system, can be used for localised repair of decayed timber windows and other joinery (www.cuprinol.co.uk). The polyester resin-based filler hardens when mixed with catalyst paste. It hardens by chemical action and is fast-setting, non-shrink and weather-resistant, with excellent adhesion to wood. It can be sanded, filed, drilled, screwed and nailed and is sufficiently flexible to accommodate small movements in the surrounding wood without loss of bond at the filler–timber interface. Ultra Tough Wood Filler is available in natural and white, and its application involves the following operations:

- Identify the full extent of any decay by probing the wood with a sharp instrument. Cut out all badly decayed and very soft wood.
- Remove paint, varnish, dirt or loose material from the area for repair and immediately around it. If the wood is wet, allow it to dry thoroughly.
- Apply Rapid Drying Wood Hardener, an organic, solvent-based liquid containing wood-hardening components, to the repair area. The hardener should be applied liberally, by brush, in two or three coats, allowing each to be absorbed before the next is applied. To maximise long-term preservation of the timber, the repair area should be pre-treated with Wood Preserver Clear, an all-purpose preserver for the protection of sound wood against rot and woodworm.
- To obtain improved anchorage of the filler in holes deeper than 25 mm, screws should be inserted into the base of the holes, leaving the heads and stems exposed inside the holes.
- Mix Wood Filler with catalyst in the proportions indicated on the pack. The mixed filler remains usable for approximately 10 min, the setting time reducing in warm weather and increasing in cold weather. Apply a thin coat

of mixed filler to the repair, pressing well in to obtain good adhesion. Immediately fill in the remainder of the hole, leaving the filler slightly proud for sanding down.

- The filler will set hard in approximately 30 minutes under normal conditions, after which it can be smoothed to the required profile using sandpaper or a file. The repair can then be finished with either paint or wood stain.

Where maintenance has been neglected over an extensive period, window frames will not be the only timber elements that are vulnerable to decay, and this technique is equally applicable to the localised repair of other components such as doors and frames, claddings, fascia boards etc.

8.6 Decay of structural timbers

As discussed, intrusive moisture is one of the principal factors that can lead to timber decay, the most vulnerable structural elements being roofs and floors.

Sarking, the universally accepted second line of defence against wind and rainwater penetration through roofs, has only been in general use since around 1938, and thus the roof structures of many older buildings are highly vulnerable. Any damage to, or lifting of, tiles or slate coverings permits direct rainwater penetration into the roof timbers, often resulting in the onset of fungal attack. Particularly vulnerable are the ends of roof trusses and rafters, which will need repairing or strengthening if ultimate collapse of the roof is to be avoided. Roof truss and rafter ends are also vulnerable if rainwater gutters become blocked, since the overflow is invariably absorbed into the top of the wall in the vicinity of where these timbers are built in.

Ground floors in older, solid-walled buildings are vulnerable to rising dampness into the brickwork or masonry surrounding the ends of built-in floor joists. Although the use of DPCs was made mandatory by the Public Health Act of 1875, their use did not become universal until around 1900 and, even where they were installed, they may well have deteriorated. It is also not unusual to encounter failed DPCs in more recent buildings. Built-in floor joists or beams in buildings where DPCs are non-existent or have failed are, therefore, often in need of replacement or repair owing to moisture-related decay. The ends of upper-floor joists and beams are also vulnerable to rainwater penetration directly through solid walls, and to other causes of intrusive dampness because of, for example, broken downpipes or leaking gutters, which can result in walls becoming drenched with large volumes of water.

8.7 Mechanical repair of decayed structural timbers

Where structural timber members, such as floor beams, joists and roof trusses, have suffered decay to the extent that they are structurally weakened, it will be essential either to replace or repair them. Unless the building has been seriously neglected and exposed to the elements for a prolonged period, any timber decay will tend to be only localised. For example, beams, joists and roofing members will generally be decayed only at their ends where they are built into supporting walls, since it is here that they are most vulnerable. The bulk of the member is likely to be perfectly sound and unaffected by decay, and therefore it will make economic sense to repair only the decayed parts, rather than undertaking a complete replacement

The traditional, and still most widely used, techniques for repairing decayed structural timbers involve introducing steel angles, channels or plates, often combined with the splicing-in of new timber in order to replace or strengthen those parts that can no longer fulfil their structural function.

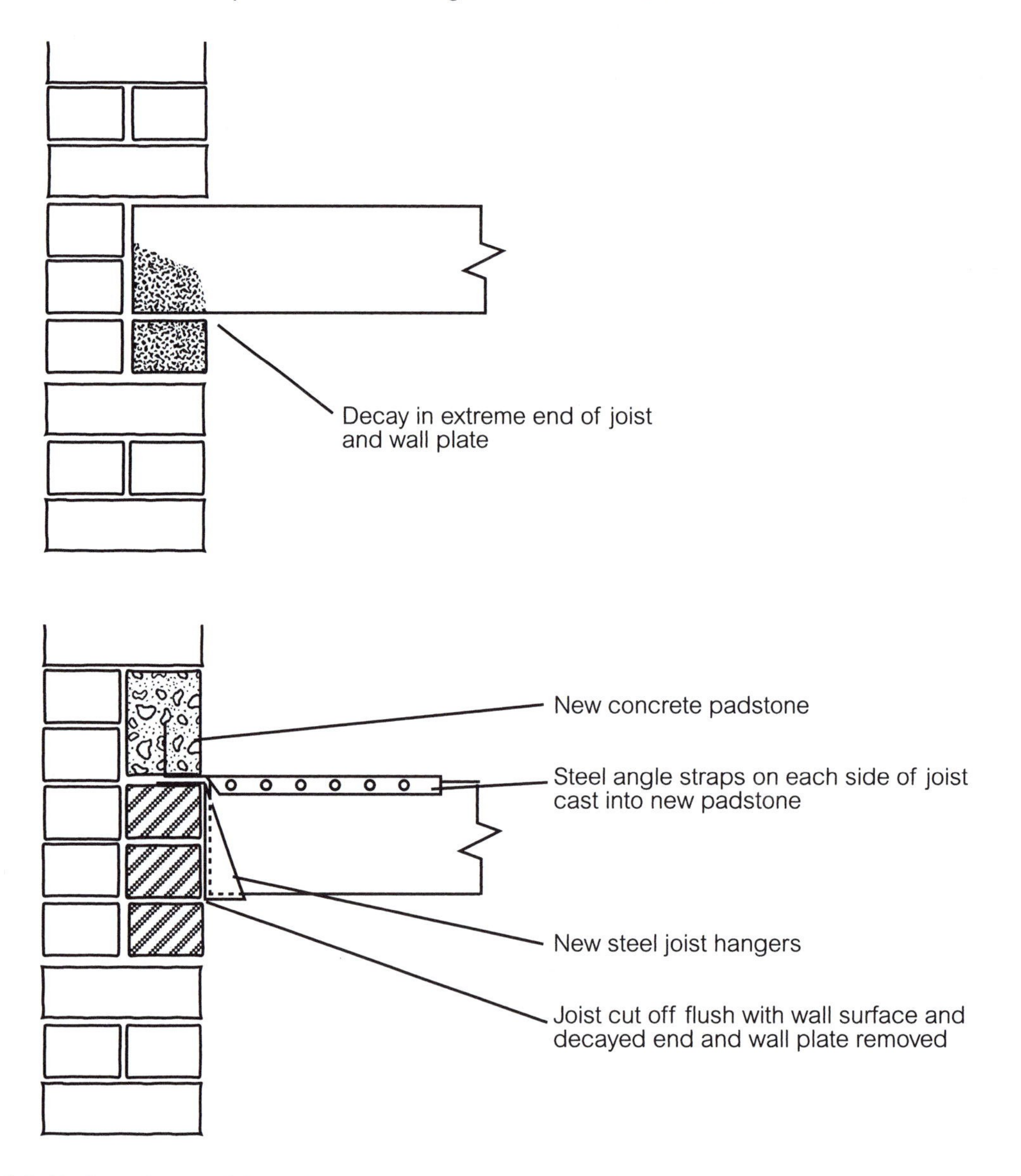

Figure 8.2 *Mechanical repair of decayed joint ends using straps and joint hangers*

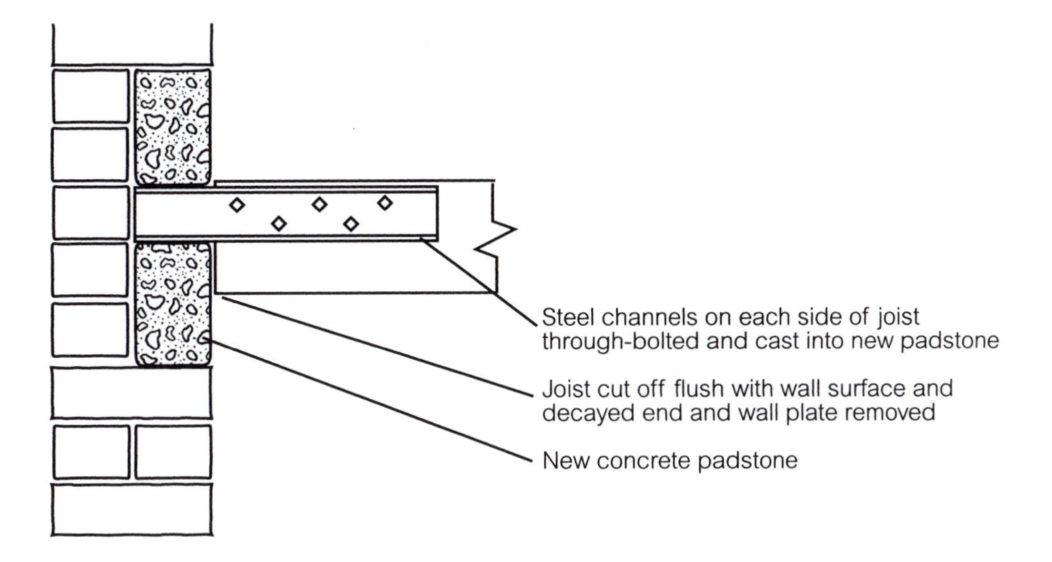

Figure 8.3 *Mechanical repair of decayed joist ends using steel channels*

Repair of decayed joist and beam ends

Where the decay is restricted to the extreme end of the joist or beam, that is, only that part which is built into the supporting wall, and no decay has spread into its exposed section, the member can be cut off flush with the wall surface, and steel supports can be used to reconnect it to the wall. In such cases, it is highly likely that the timber wall plate will also have suffered decay, and this will therefore need to be removed and replaced. The steel supports are mechanically connected to the end of the joist or beam using bolts and normally cast into a new concrete padstone built into the wall.

Figure 8.2 shows the use of a steel joist hanger and steel straps connected to the end of the joist or beam and cast into a new concrete padstone.

Figure 8.3 shows the use of through-bolted steel channels on each side of the joist or beam cast into a new concrete padstone. In both cases the remaining, unaffected parts of the joists or beams should receive brush or spray preservative treatment for a minimum of 2 m from the decayed areas.

Repair of decayed roof truss ends

Where decay has occurred in the ends of timber roof trusses, and if it has resulted in serious loss of strength, there will be no choice but to replace the decayed parts with new, preservative-treated timber. This involves cutting away the decay and using splice joints and through-bolts to mechanically connect the new timber to the sound, unaffected timber. The connection of the new truss end to the existing is completed by fixing steel through-bolted splice plates to each side of the splice

Figure 8.4
Repair to decayed roof truss ends using replacement timber and steel splice plates

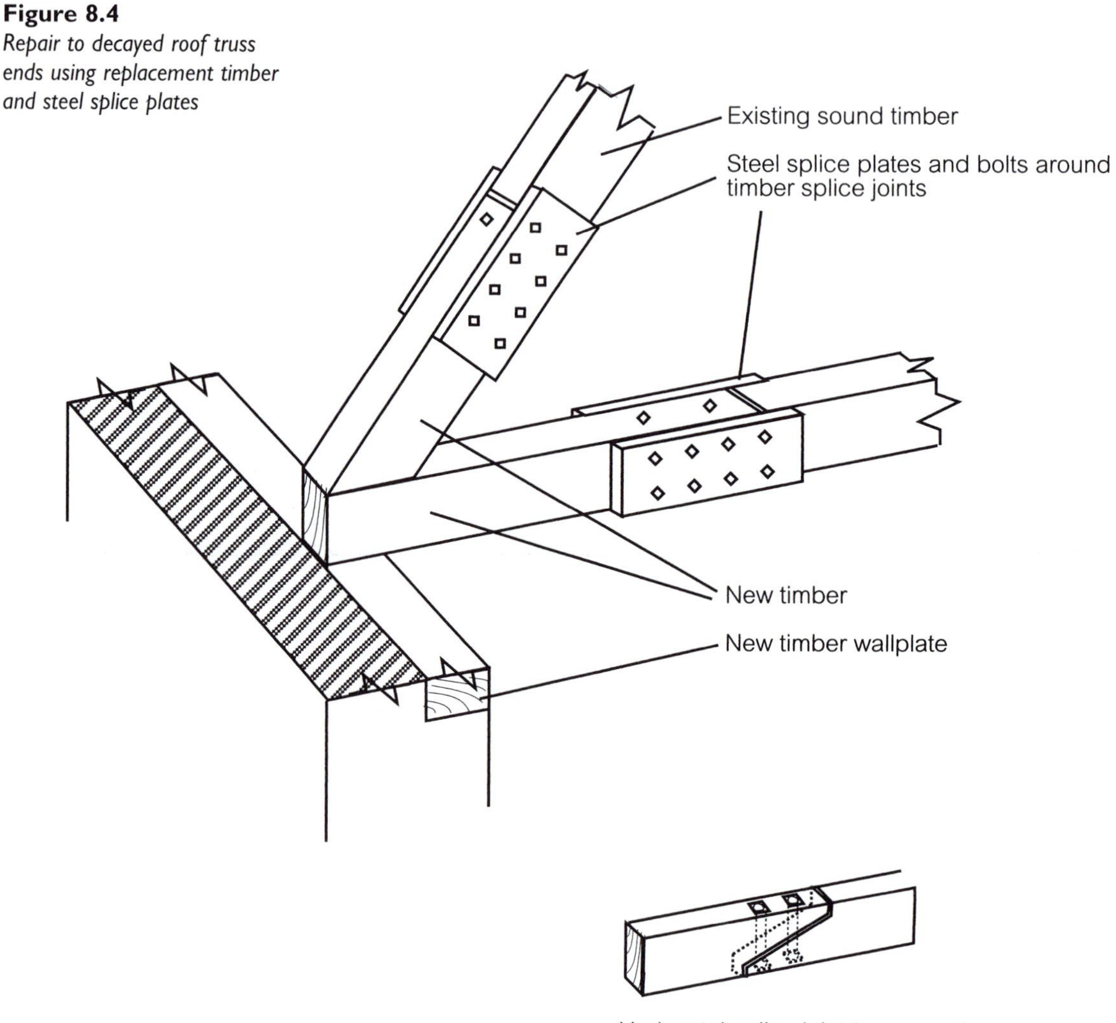

Photograph 8.9
(facing page, above)
Splice joint repair to decayed roof truss end

Photograph 8.10
(facing page, below)
Splice joint repair to decayed roof truss end showing through-bolted steel splice plates

joints. This technique is illustrated in Figure 8.4 and shown in Photographs 8.9 and 8.10.

Providing there is some residual strength in the roof truss end, despite the occurrence of decay, a simpler alternative to the above method is to treat the decayed timber with preservative and fix steel gusset plates on each side. The preservative-treated, weakened end of the truss is retained as a base for fixing through-bolted triangular steel gusset plates to each side. The gusset plates should extend beyond the decayed area, overlapping by at least 150 mm with the sound, unaffected timber.

Figure 8.5 and Photograph 8.11 illustrate the technique using steel gusset plates, and Photograph 8.12 shows the use of nailed plywood gussets as an alternative to steel plate.

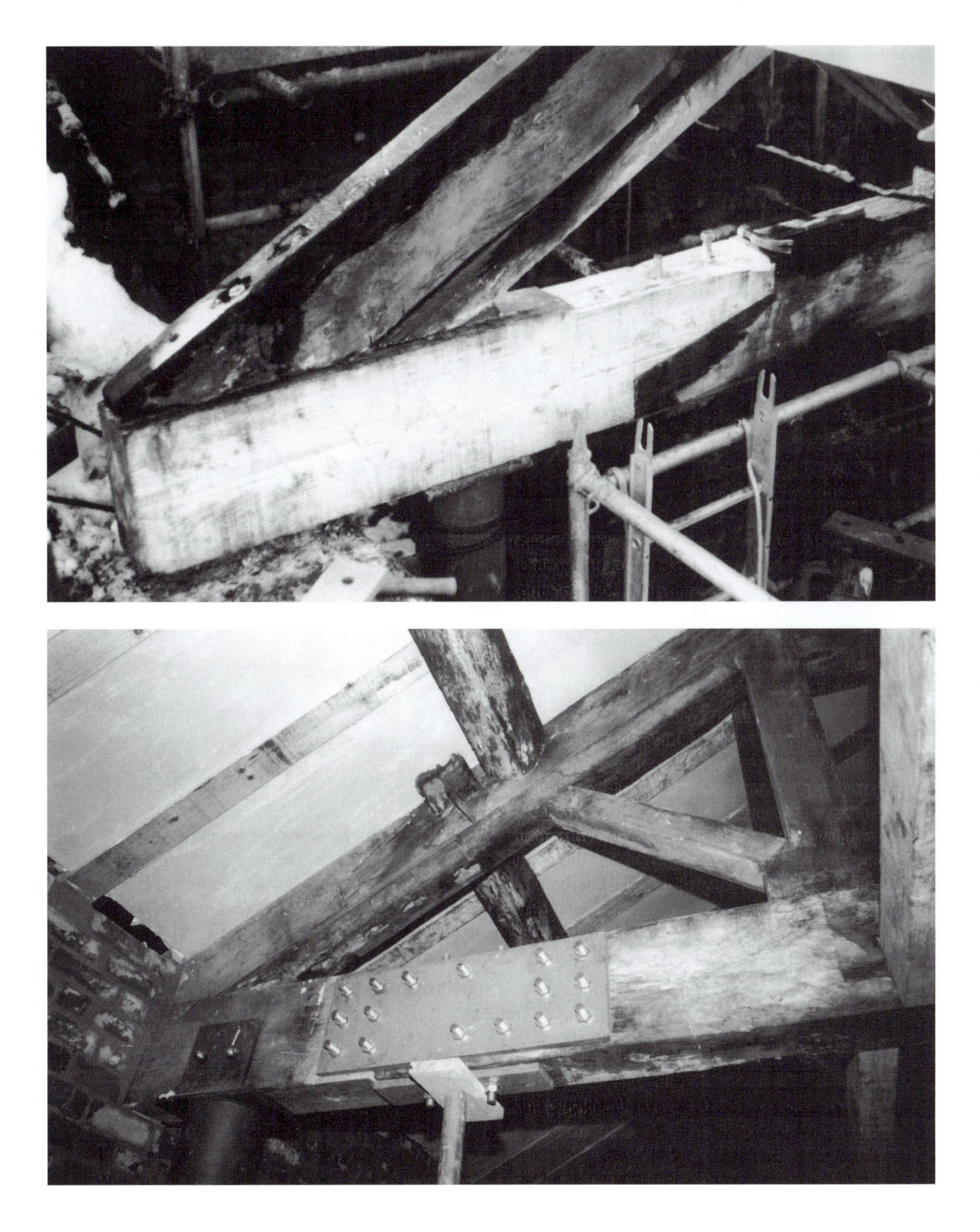

Figure 8.5
Repair to decayed roof truss ends using steel gusset plates

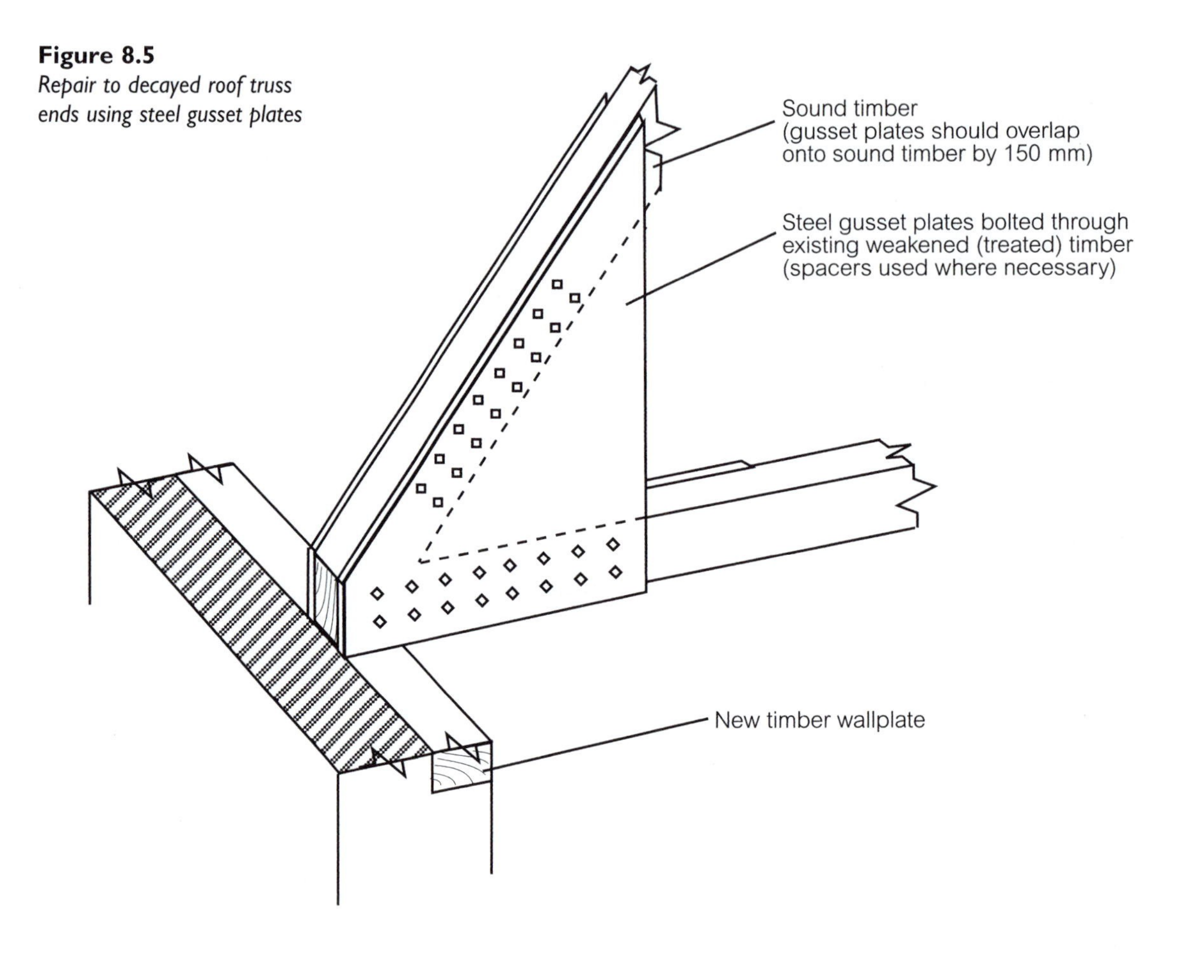

8.8 Epoxy resin-based repair and restoration of decayed structural timbers

An alternative method to mechanical techniques for the repair of decayed structural timbers, and one that has increased in popularity during the last 20 years, involves the use of epoxy resin-based systems. The use of synthetic resins for the repair of damaged or defective concrete members had been common for many years prior to their more recent successful application to timber repair.

Photograph 8.11
(facing page, above)
Steel gusset plate repair to decayed roof truss end

Photograph 8.12
(facing page, below)
Plywood gusset repair to decayed roof truss end

With timber restoration, the bonding capability of the synthetic resin is not the only criterion in effecting a successful repair: reinforcing rods of glass-fibre or polyester and steel reinforcing bars and plates also play a major role in the techniques employed. It is essential, where reinforcement is used, that it is set deep into the sound portion of the timber to ensure an effective connection and load transfer, and the epoxy resins used must have good fluidity in order to obtain maximum penetration and impregnation of the wood fibres to obtain a good bond. Epoxy resin-based repair involves highly specialised skills, and the selection of resin mixes, reinforcement types and solutions to individual problems requires a thorough understanding of the techniques available. It is therefore essential, where

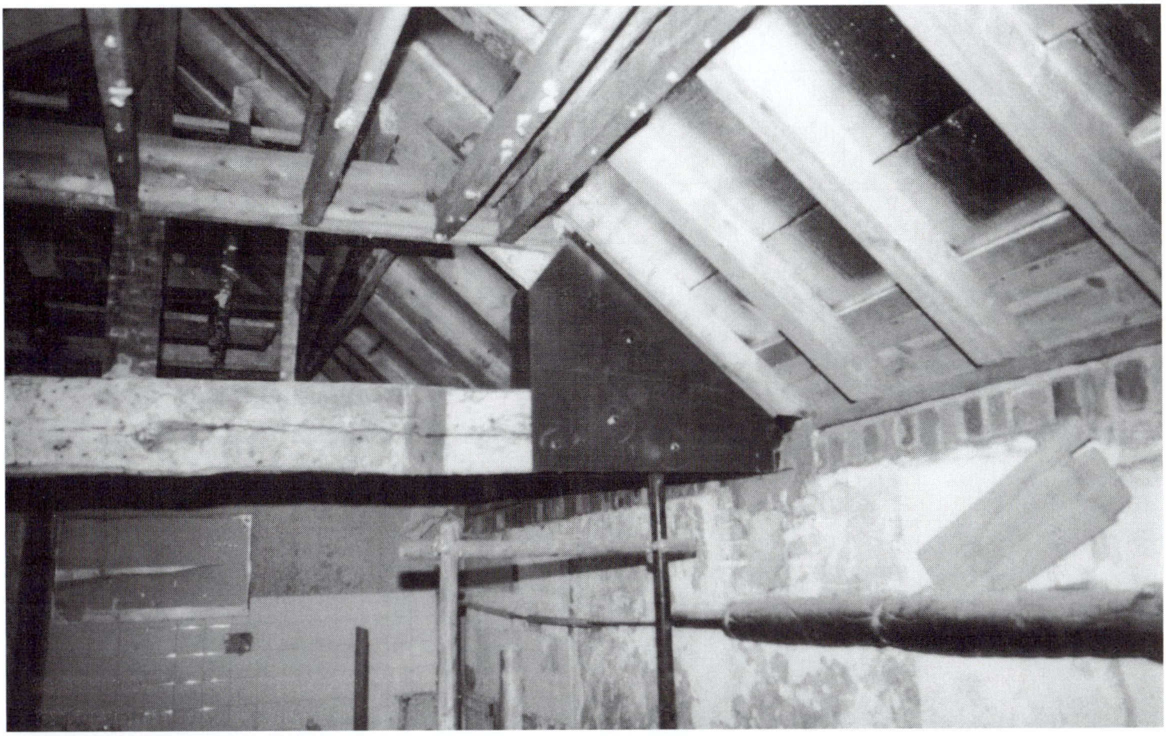

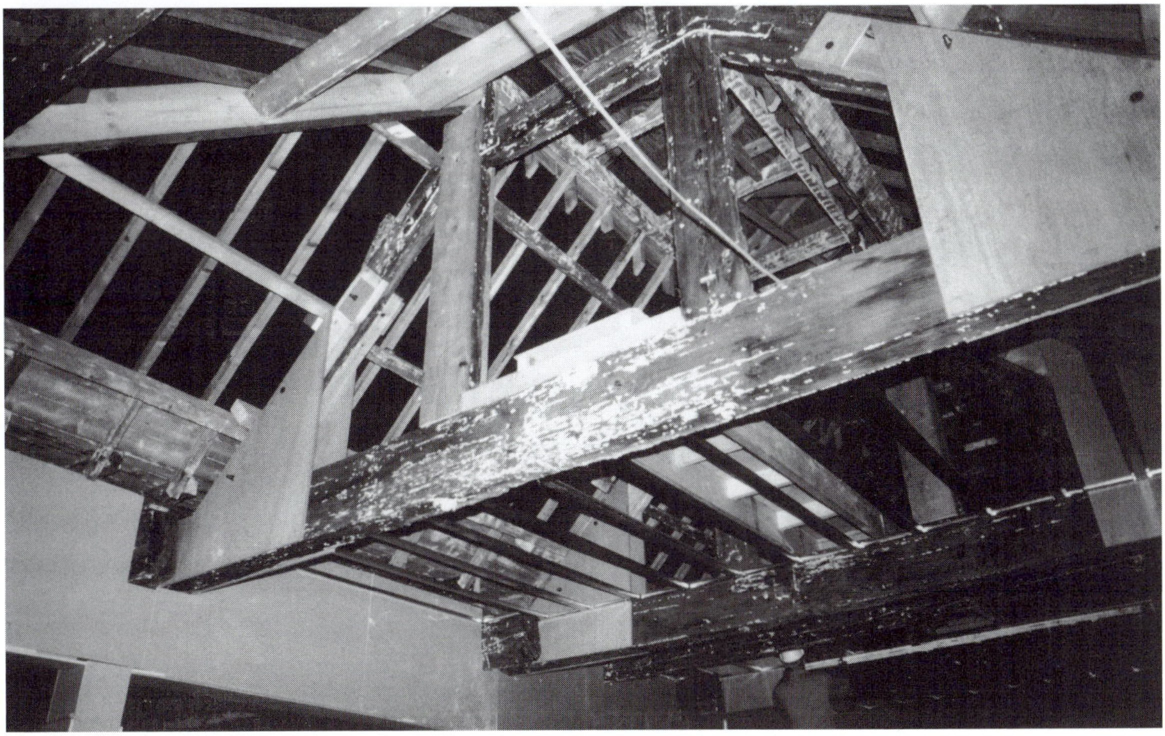

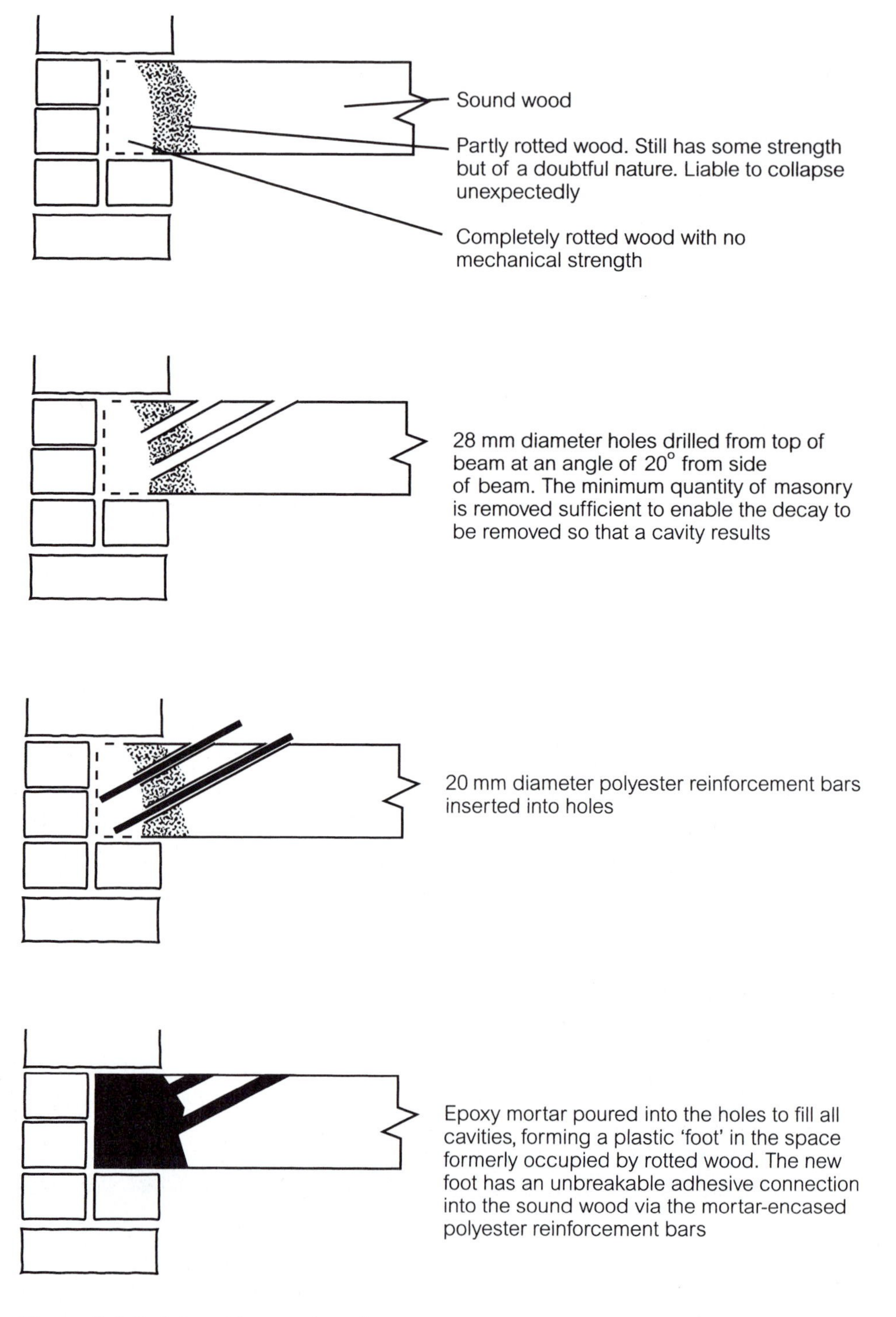

Sound wood
Partly rotted wood. Still has some strength but of a doubtful nature. Liable to collapse unexpectedly
Completely rotted wood with no mechanical strength
28 mm diameter holes drilled from top of beam at an angle of 20° from side of beam. The minimum quantity of masonry is removed sufficient to enable the decay to be removed so that a cavity results
20 mm diameter polyester reinforcement bars inserted into holes
Epoxy mortar poured into the holes to fill all cavities, forming a plastic 'foot' in the space formerly occupied by rotted wood. The new foot has an unbreakable adhesive connection into the sound wood via the mortar-encased polyester reinforcement bars

Figure 8.6 *Resin-based beam end repair system*

it is considered that this type of repair and restoration might be appropriate in a particular refurbishment scheme, to call in one of the specialist contractors experienced in such work.

Repair of decayed joist and beam ends

In situ, resin-based repair techniques are now widely used as an alternative to the traditional repair methods, which involve splicing-in new timber and using steel plates, angles and bolts. A typical resin-based beam end repair system is illustrated and described in Figure 8.6, and a completed repair is shown in Photograph 8.13. The work involves removing the decayed wood at the beam end and replacing it with a new epoxy-resin 'foot', which is bonded to the sound timber using polyester reinforcement rods embedded in epoxy resin.

Repair of decayed roof truss ends

An effective technique for carrying out structural restoration of decayed roof truss ends is described below and illustrated in Figure 8.7. Where a roof truss has decayed at the wall plate, a reinforcing plate is introduced into the main tie beam, and the decayed end timber is cut out and replaced with epoxy-resin mortar. Additional corner reinforcement in the form of reinforcing bars can be inserted into holes drilled through the end of the rafter to pass each side of the reinforcing plate in

Photograph 8.13
Epoxy resin-based repair to decayed beam end

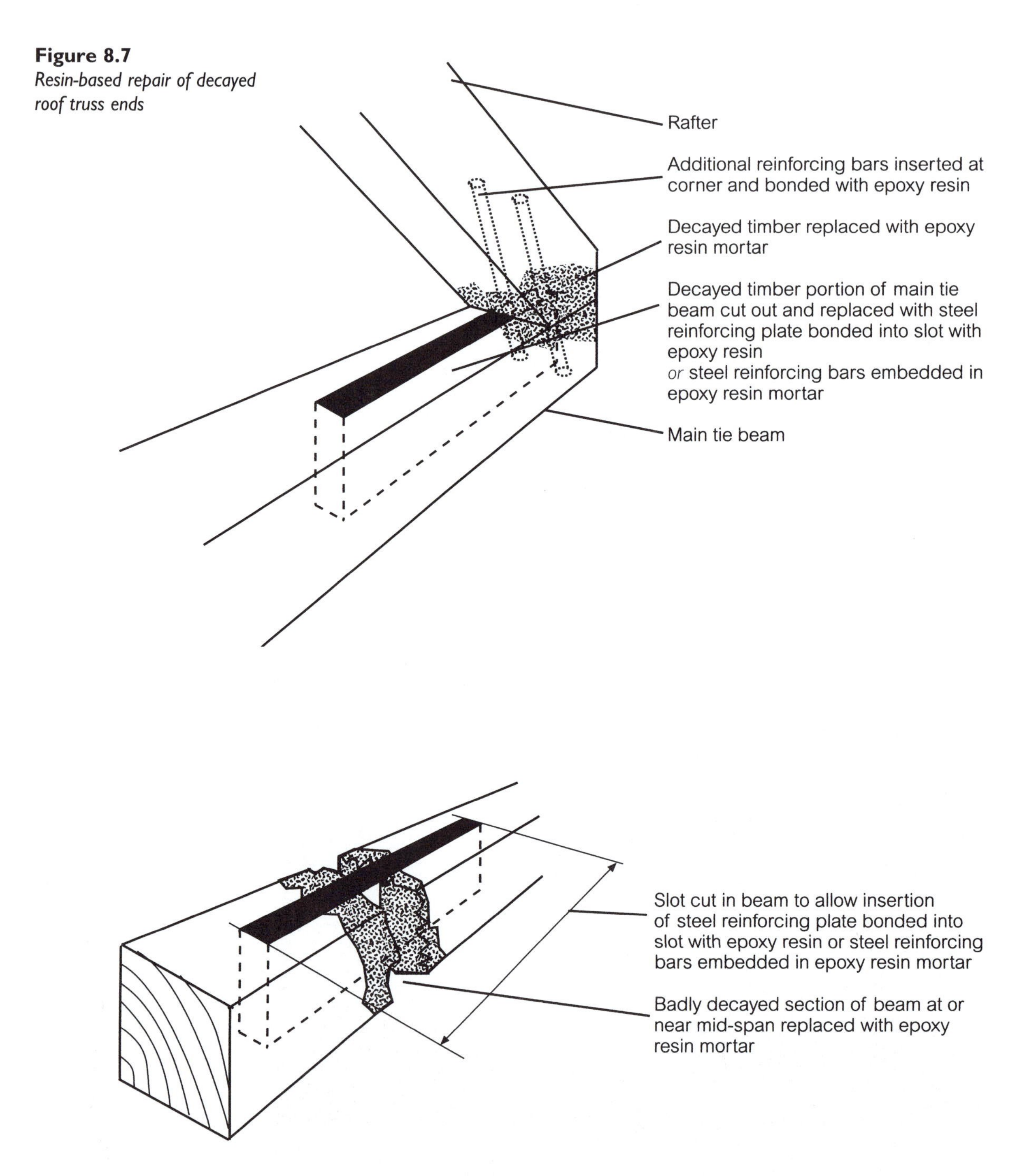

Figure 8.7
Resin-based repair of decayed roof truss ends

Figure 8.8
Resin-based repair of decayed beams

the main tie beam. It is essential that the reinforcing plate and bars are set and fully bonded deep into the sound portion of the timber if a totally effective repair is to be achieved.

Repair of decayed beams between supports

In some buildings that have suffered from neglect over a lengthy period, timber decay may be more widespread, and not merely restricted to those more vulnerable areas discussed above. Timber decay may occur in beams, rafters and the main ties of roof trusses between their supports, and epoxy-resin repair techniques are effective enough to be capable of repairing structural timbers that have virtually completely rotted through. A repair technique employed for this purpose is illustrated in Figure 8.8 and involves inserting steel bar or plate reinforcement into a specially cut slot, which extends deep into the sound timber of the beam on each side of the decayed area. The plate is fully bonded into the slot with epoxy resin, and epoxy-resin mortar is used to fill the void left after cutting out the area of decayed timber.

8.9 Traditional joints and structural timber replacement

Where existing timbers are so severely damaged, the whole section of timber can be removed and replaced. With care, a new piece of timber can be shaped with a tenon that neatly fits into a mortise or rebate formed in the existing timber. To ensure that the joint is good, epoxy resin can be used; the resin will fill any gaps and bond the material. Where possible the new joint can be formed using the same methods as used originally; for example, the renovation of the barn shown in Photograph 8.14 used mortise and tenon joints secured with timber dowels.

Photograph 8.14
Existing structural timber can be removed and replaced

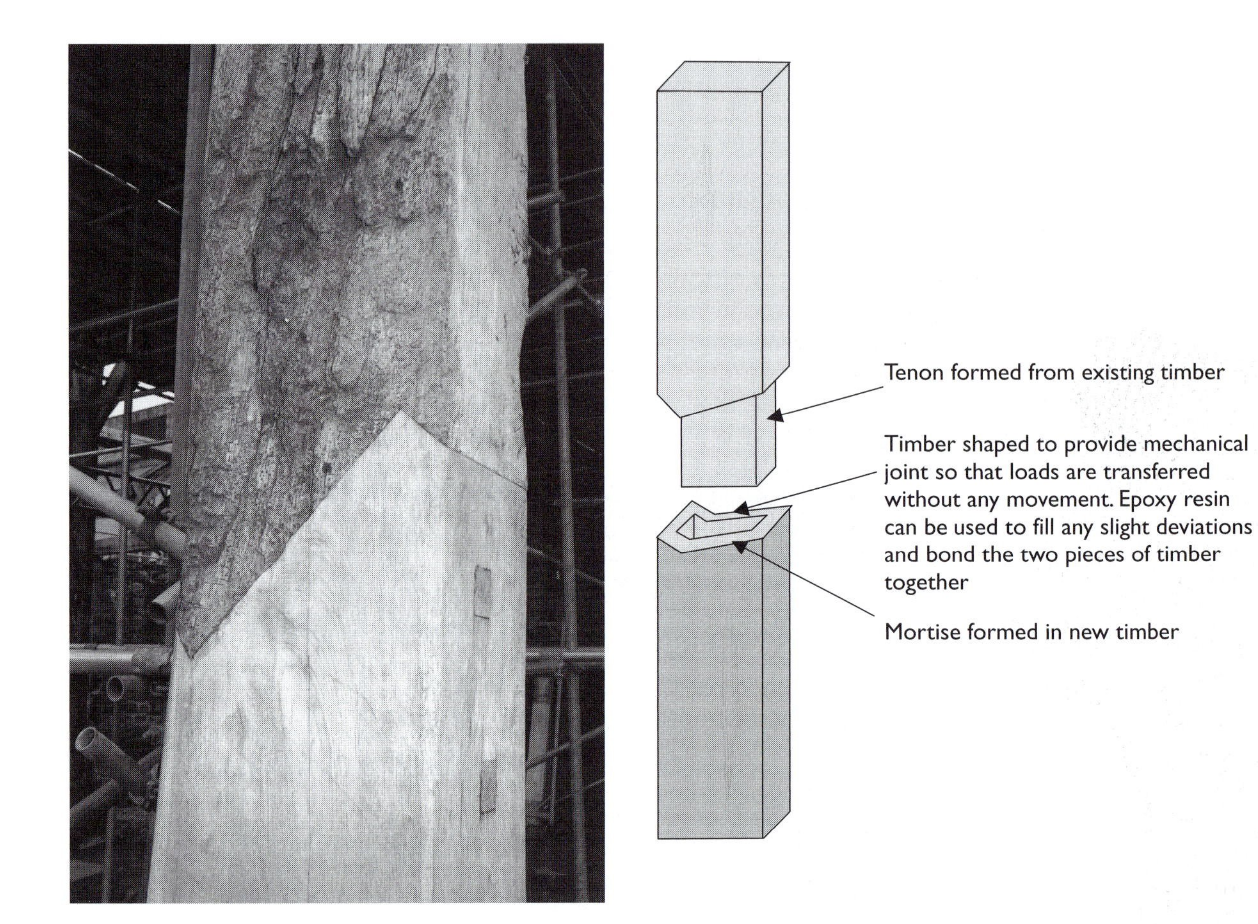
Tenon formed from existing timber
Timber shaped to provide mechanical joint so that loads are transferred without any movement. Epoxy resin can be used to fill any slight deviations and bond the two pieces of timber together
Mortise formed in new timber

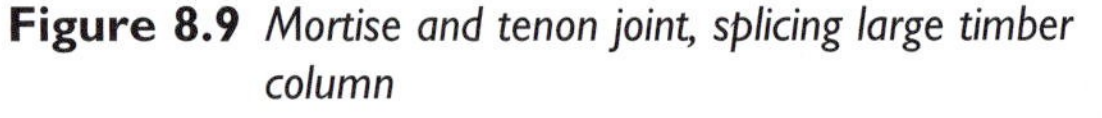

Photograph 8.15 *Pinned mortise and tenon joint*

Figure 8.9 *Mortise and tenon joint, splicing large timber column*

Photograph 8.15 shows a sizable oak column spliced with a new piece of oak. Where the timber had been attacked by mould and insect infestation, this has been removed and replaced with a new section of wood that is capable of acting as a fully supporting structural member. The existing timber has been neatly trimmed ready to receive the new section. A mortise is cut into the new section, and the old section is shaped to provide a tenon (see Figure 8.9). The tenon provides a central section of timber that fits into the rebate provided by the mortise.

Photograph 8.16 shows an oak column where a large section of timber has been removed. The new section of timber has been cut to the required shape and pinned in place. The existing footing for this column was a few large stones set on top of each other. To improve the stability of the column and foundation, the footings have been surrounded by structural concrete.

Photograph 8.16
Section of timber cut into existing column and pinned in place

Photograph 8.17
Spliced oak column

General repairs to decayed timber members

In addition to the repairs described above, epoxy resin-based repair techniques, using various types of reinforcement and epoxy-resin mortar, can also be used to effect permanent repairs to decayed timbers and joints in many other situations where fungal or insect attack has taken place. As stated previously, where this type of repair might be considered appropriate, a specialist contractor should be consulted in the first instance.

References

Building Research Establishment (1991) *Design of Timber Floors to Prevent Decay* (Digest 364), Watford: BRE.

Building Research Establishment (1993) *Wood Preservatives: Application Methods* (Digest 378), Watford: BRE.

Building Research Establishment (1996) *Reducing the Risk of Pest Infestation in Buildings* (Digest 415), Watford: BRE.

Building Research Establishment (1997) *Repairing Timber Windows: Parts 1 & 2* (Good Repair Guide 10), Watford: BRE.

Building Research Establishment (1997) *Wood Rot: Assessing and Treating Decay* (Good Repair Guide 12), Watford: BRE.

Building Research Establishment (1998) *Wood-boring Insect Attack Identifying and Treating Damage: Parts 1 & 2* (Good Repair Guide 8), Watford: BRE.

BSI (2003) *Recommendations for the Preservation of Timber*, BS8417: 2003, London: BSI.

HSE (2001) *A Guide to Good Practice and the Safe Use of Wood Preservatives*, London: HMSO.

Property Care Association (2008) *Remedial Timber Treatment: Code of practice*, Huntingdon: Property Care Association, www.property-care.org.

Richardson, B.A. (1991) *Defects and Deterioration in Buildings*, London: E. & F.N. Spon.

Ridout, B. (1998) 'The durability and decay of oak', in J.Taylor (ed.), *The Building Conservation Directory 1998*, Tisbury: Cathedral Communications Ltd: 97–99.

University of Bath, Department of Architecture and Building Engineering (1985) *Building Appraisal Maintenance and Preservation: Symposium* Proceedings, University of Bath.

Strengthening of existing timber floors

CHAPTER 9

9.1 General

In certain refurbishment and alteration schemes, the existing floors may need to be strengthened in order to cater for increased loadings imposed by the proposed new use. The most common examples occur where existing buildings are converted to office use. In such cases, the existing timber floors may not be capable of carrying the excessive localised loads imposed by modern filing and storage systems, office machinery and equipment.

A number of solutions are available where an existing timber floor needs to be strengthened, and the most common techniques are described below.

9.2 Replacing with new timber or steel sections

The existing floor beams are replaced with new timber or steel sections capable of supporting the increased loads. This solution usually involves major disruption to the existing structure and often necessitates removal and reinstatement of the existing floorboards, joists and ceiling. In view of the considerable expense and inconvenience involved, the method is not, therefore, generally recommended, and an alternative solution should be considered.

Photograph 9.1 shows a new supporting floor. The existing floor was removed, the floorboards were taken up but retained for re-use as the new floor surface, new floor joists were used, and additional steel beams were used as intermediate supports to reduce the deflection and increase support.

9.3 Strengthening with new steel channel sections or timber joists

The existing floor beams are strengthened with new steel channel sections or timber joists fixed to both sides to cater for the increased floor loadings. This technique is also very disruptive and expensive, requiring the removal of floorboards and ceilings, the cutting back of floor joists and reinstatement of their ends onto the new steel channels. Access holes may also need to be formed through the existing external walls to allow insertion of the steel channels or timber joists.

Photograph 9.1
Replacement floor structure with intermediate supporting steel beam

Photograph 9.2
New joists positioned alongside existing joists provide extra strength to the floor

Photograph 9.3
New joists bolted to existing joists. Galvinised steel wall straps give additional stability to the wall tying the wall and the floor structure together

Photograph 9.2 shows the new timber joists positioned alongside the existing joists. Bolts have been inserted at 400 mm centres, ensuring that both joists carry the floor. Although the height of the floor is neither lowered nor raised, the floor is still quite labour intensive and expensive. Photograph 9.3 shows the wall straps fixed to the floor joists, which add lateral stability to external walls.

9.4 Stiffening with steel or timber

The existing floor beams are stiffened with steel or timber fixed to their top or bottom surfaces. Figure 9.1 shows the application of this solution using epoxy-resin bonding and dowelling techniques to increase the depth of a floor beam. The additional depth of timber may be built up in laminations, bonded to the existing beam and each other using epoxy adhesive. Alternatively, a single piece of timber may be added and bonded to the existing beam using epoxy adhesive and reinforcing dowels set in epoxy-resin mortar.

Clearly, this solution will either raise the floor level or lower the ceiling level, and it will also involve partial disruption of the existing floor or ceiling. In view of these factors, therefore, this method of floor strengthening may not be appropriate in certain circumstances.

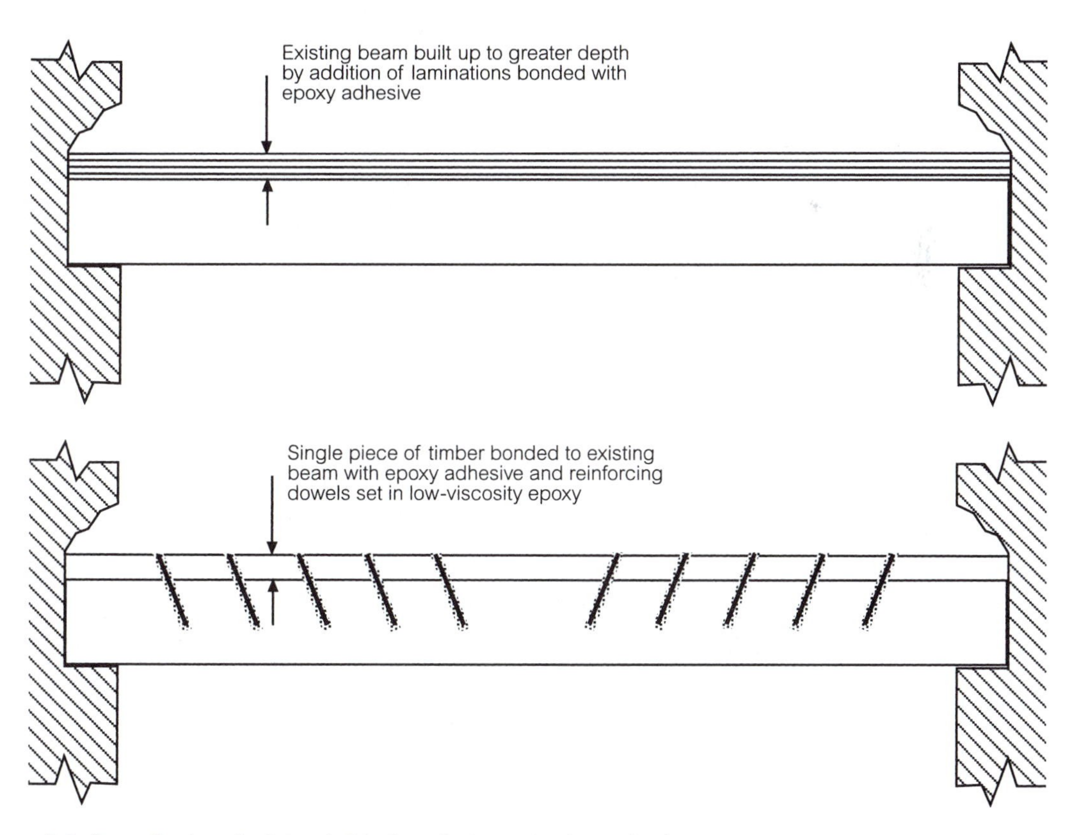

Figure 9.1 *Strengthening of existing timber floors by increasing beam depths*

9.5 Stiffening with steel plates

The existing floor beams are stiffened by means of steel plates fixed to both sides. Where supported members (for example, secondary beams and/or floor joists) are seated on top of the member being strengthened, this technique will produce an effective and economic method of strengthening. However, where supported members are connected to the sides of the member being strengthened, the detailing and fixing can be cumbersome, time-consuming and expensive.

9.6 Strengthening with steel stiffening reinforcement

The existing floor beams are strengthened by inserting steel stiffening reinforcement, embedded in low-viscosity epoxy, within their thickness.

This technique, invented and developed by RTT Restoration Ltd, is illustrated in Figure 9.2 and involves cutting a slot out of the centre of the beam and inserting a number of steel reinforcing bars embedded in, and bonded to, the existing timber with low-viscosity epoxy. The size of the slot, and the number and diameter of the reinforcing bars, are designed to suit the particular circumstances, an increase of 50 per cent in load-carrying capacity being possible in the majority of cases. The advantages of this technique over those previously described are:

- access is required only from above the beam;
- disturbance of the existing floor is minimised;
- disturbance of the existing ceiling is avoided, a particularly important advantage where the ceiling is ornate or has to be preserved as part of a listing requirement;

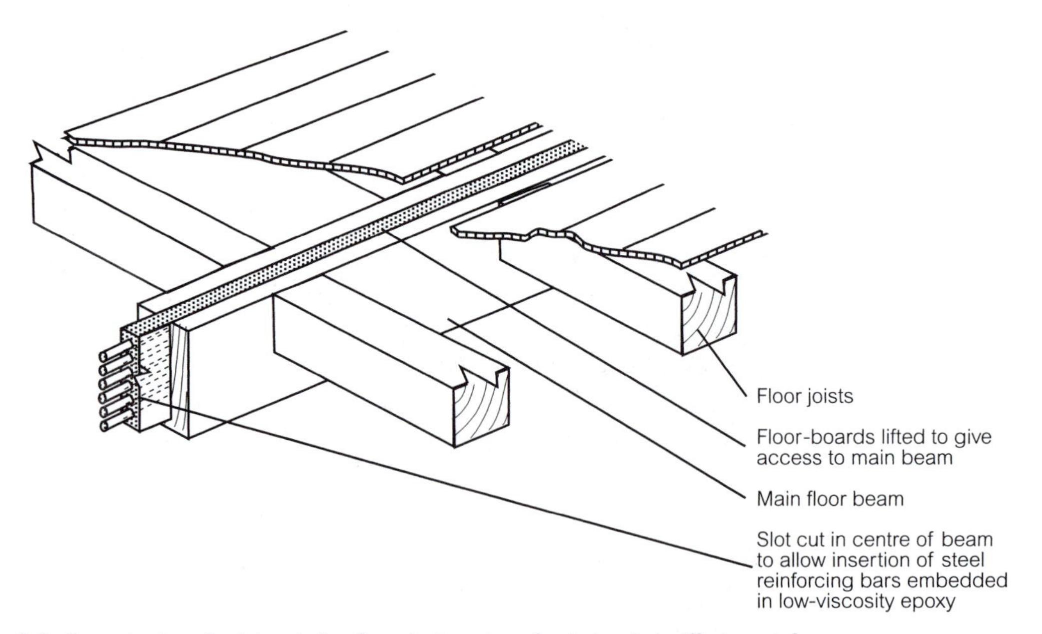

Figure 9.2 *Strengthening of existing timber floors by insertion of resin-bonded stiffening reinforcement*

Photograph 9.4
Floor-beam strengthening using steel plate embedded in low-viscosity epoxy

Photograph 9.5
Temporary support to beam during strengthening work

Photograph 9.6
Floor-beam strengthening using steel plate embedded in low-viscosity epoxy

- the existing components are fully retained;
- any existing distortion can be accommodated, since the reinforcing bars can be bent to match the beam's deflected shape if necessary;
- it avoids the need to form holes in existing walls to introduce replacement members;
- it obviates the need to manhandle heavy materials or components;
- it does not adversely affect the fire-resistance of the existing construction.

As an alternative to the use of steel reinforcing bars, as described above, steel plate may be inserted into the slot and embedded in low-viscosity epoxy, as shown in Photographs 9.4–9.6.

References

University of Bath, Department of Architecture and Building Engineering (1985) *Building Appraisal Maintenance and Preservation: Symposium Proceedings,* University of Bath.

Wihide, E. (2005) *The Essential Source Book for Planning, Selecting and Restoring Floors*, New York: Ryland Peters & Small.

Heavy-lifting systems

CHAPTER 10

10.1 General

Occasionally, building refurbishment work necessitates the use of highly specialised skills and equipment to carry out major structural alterations that involve the large-scale movement or lifting of entire buildings or elements.

10.2 Movement of complete buildings

One of the most publicised recent examples of the movement of a complete building was the moving of the Belle Tout lighthouse at Beachy Head, East Sussex, in early 1999. Continuous erosion of the chalk cliffs had left the 850-tonne granite lighthouse within 5 m of the cliff-edge, having originally been built some 30 m back from the edge. To save the lighthouse, which had been converted some years earlier to a dwelling, the entire structure was jacked up and 'slid' 17 m further inland, away from the cliff-edge and onto a newly constructed foundation, using specialist computerised sliding equipment This involved providing a cradle of new, reinforced concrete beams beneath the structure, jacking the whole building up by 600 mm using twenty-two hydraulic jacks and then using six hydraulic rams to push the structure 17 m, along four concrete tracks, before lowering it down onto its new foundation.

This project was carried out by Abbey Pynford, a company specialising in the movement of complete buildings and other heavy-lifting/movement applications (www.abbeypynford.co.uk).

10.3 Movement of building elements

Heavy-lifting technology of the type described above can also be applied to the movement of structural elements of buildings to facilitate their refurbishment and re-use. One such project, also carried out by Abbey Pynford, involved raising the height of the complete roof structure of the Granary Building, a Grade II* listed, canal-side warehouse in Leeds.

The Granary Building, shown in Photograph 10.1, was built *c.*1778 for the Leeds and Liverpool Canal Company as the canal's eastern terminus warehouse at its junction with the River Aire, near to the Leeds city centre. The building is four

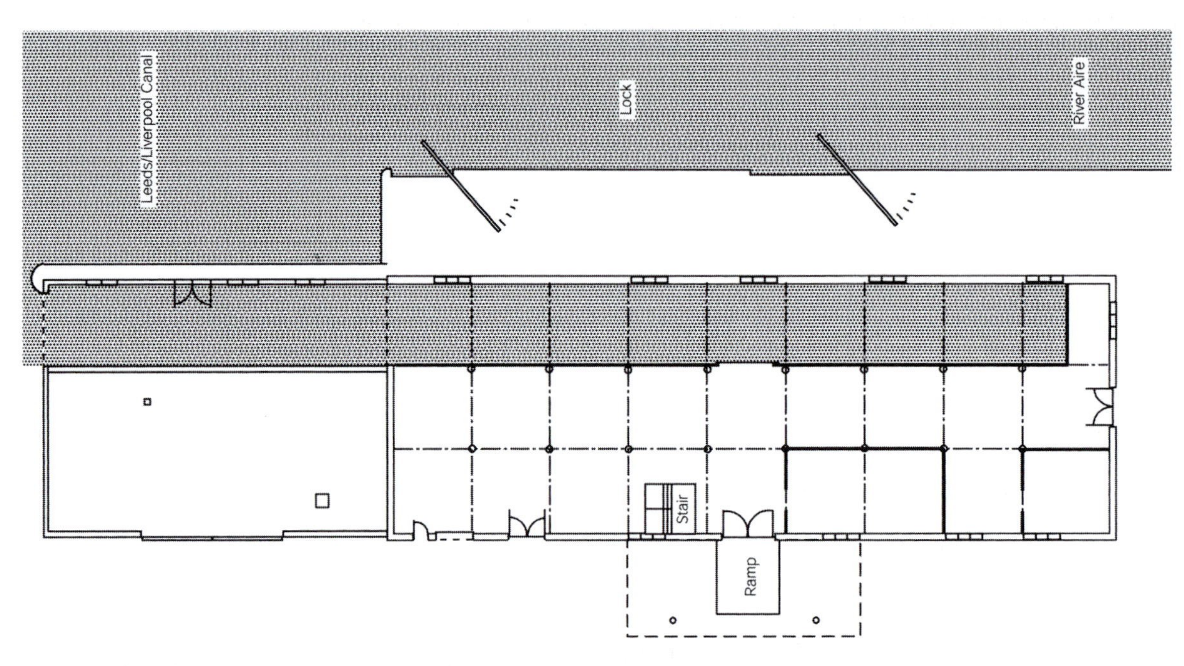

Figure 10.1 *Granary Wharf, Leeds: ground-floor plan as existing*

storeys high, built of coursed, squared stone, with a graduated stone slate roof. The canal channel extended into the building, as shown in Figures 10.1 and 10.3, to enable boats to be loaded and unloaded inside. The massive timber roof structure comprises cross-beams supporting queen posts, clasping a collar, x-braces and six rows of purlins. The interior was remodelled in the mid to late nineteenth century to give a safer, fire-proof construction, the original timber floors being replaced by concrete vaults supported on two rows of cast-iron columns. Other additions at this time included a gantry and slate roof canopy and a lower extension block at the western end of the building (see Figure 10.2).

By the early 1990s the building, which had been empty and neglected for a long period, had fallen into a state of disrepair, but it was given a new lease of life in

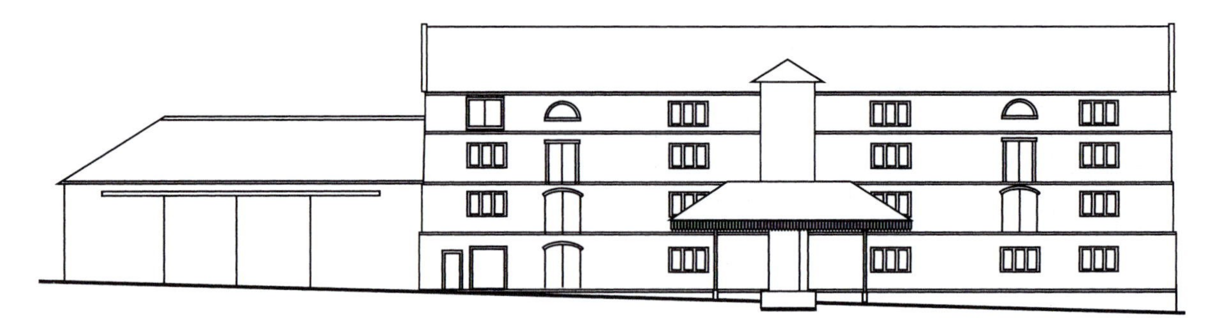

Figure 10.2 *Granary Wharf, Leeds: south elevation as existing*

1995 through its restoration and refurbishment to produce modern office accommodation. The existing structure, including the roof and concrete vaulted floors, was still structurally sound and therefore completely retained. The original canal channel, inside the building, was drained and covered with a new floor slab, and new raised floors were provided over the existing floors to accommodate services. This enabled the existing vaulted floor soffits to remain exposed as a feature in the new design.

A major design problem with the Granary Building was that the fourth storey did not provide sufficient clear headroom beneath the existing exposed queen-post roof structure to enable its use as office space, meaning that 25 per cent of the building's potential floor area was effectively unusable (see Figures 10.3 and 10.4). However, in view of the building's prime location in a rapidly growing and attractive

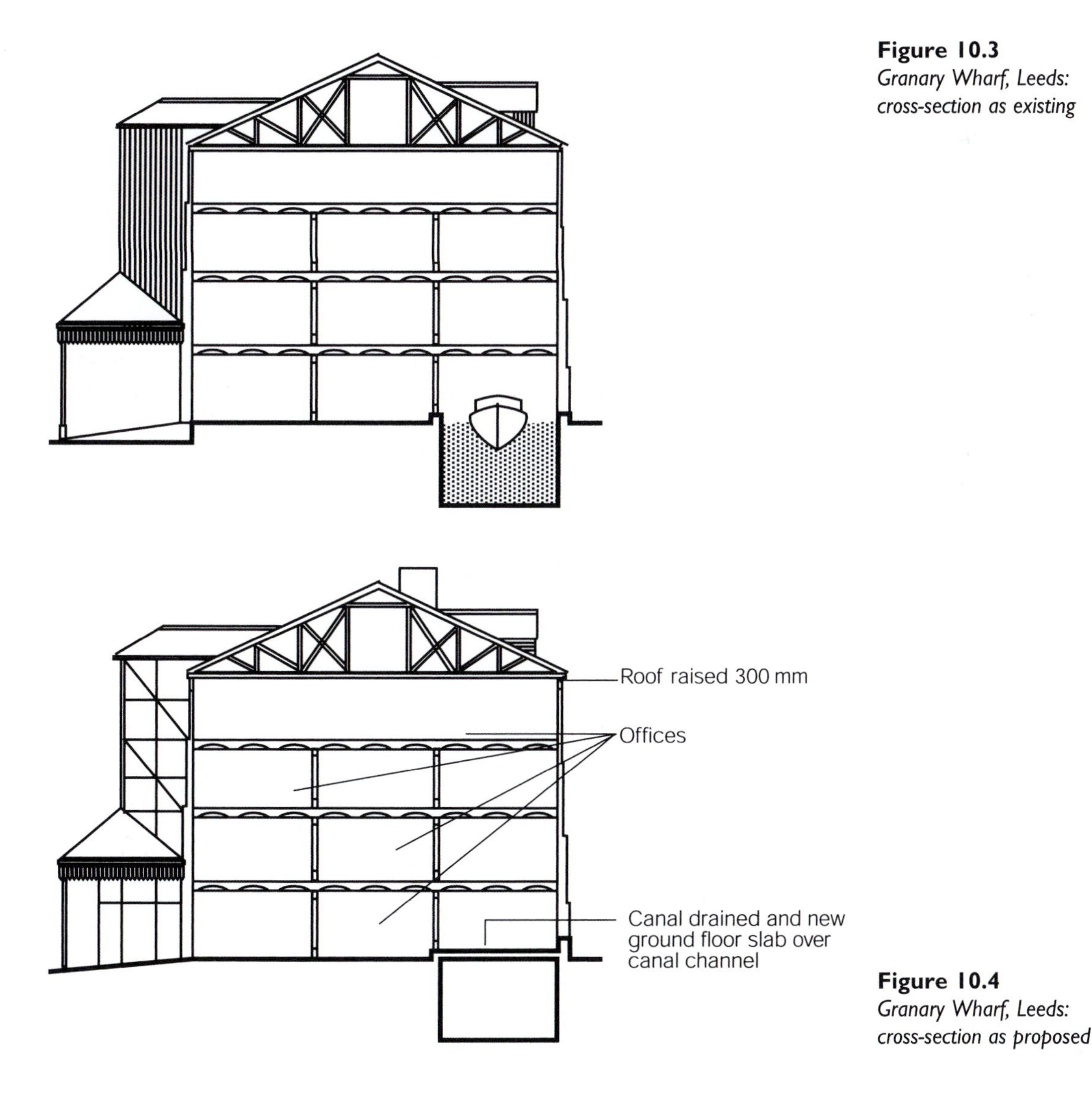

Figure 10.3
Granary Wharf, Leeds: cross-section as existing

Figure 10.4
Granary Wharf, Leeds: cross-section as proposed

commercial area, the developer decided that it would be feasible to incur the significant cost of physically raising the existing roof height, using specialist heavy-lifting technology, to enable the fourth storey to be converted into office accommodation (see Figure 10.4). The existing clear headroom of 2.1 m beneath the existing roof structure had to be increased by 300 mm to 2.4 m to enable the upper storey to be converted into office space (see Figure 10.4), and this was achieved using a specialised, hydraulically operated and computer-monitored jacking system provided by Abbey Pynford.

The sequence of operations for lifting the roof of the Granary Building was as follows:

- The stone-slate roof covering was removed to reduce the loading on the roof structure.
- Temporary weatherproof sheeting was applied over the roof.
- Temporary steelwork supports to tops of walls were installed internally and externally, immediately below the roof level, to prevent walls spreading outwards during the works. The internal components of these supports are shown in Photographs 10.2–10.4.
- A temporary steel jacking infrastructure was installed, with jacks located under the ends of each queen-post truss (see Photograph 10.2).
- Stonework was removed from the ends of the roof trusses to 'free' the trusses prior to jacking.
- Hydraulic jacks were connected to the central console, and all trusses were simultaneously jacked up by 300 mm (see Photograph 10.3).
- New supporting padstones were inserted under the ends of the roof trusses (see Photograph 10.4).
- New stonework was built up around the padstones and roof truss ends (see Photograph 10.5).
- New roof insulation was installed, and slate battens were laid.
- The stone-slate roof covering was replaced.

Photograph 10.1
The Granary Building, Leeds: north and east elevations

Photograph 10.2
The Granary Building, Leeds: temporary supports and jacks prior to lifting of roof

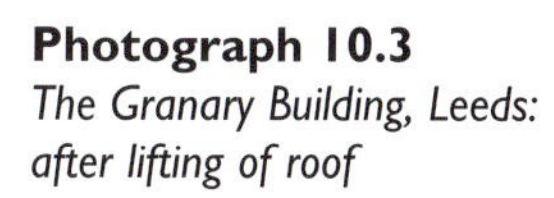

Photograph 10.3
The Granary Building, Leeds: after lifting of roof

Photograph 10.4
The Granary Building, Leeds: after lifting of roof, showing new padstones inserted to support roof trusses

Photograph 10.5
The Granary Building, Leeds: after lifting of roof, showing new padstone and new infill stonework

Underpinning systems

CHAPTER 11

11.1 General

The purpose of underpinning is to take the level of an existing foundation to a deeper, firmer stratum by adding a new foundation construction beneath it, and this may be necessary for a range of different reasons, the most common of which are:

- where excessive settlement of the existing foundation has occurred and has caused, or threatened, structural damage to the building;
- to permit the level of the adjacent ground to be lowered, for example for the construction of a new, adjacent building with a basement extending deeper than the existing foundations;
- to increase the load-bearing capacity of the existing foundation, for example because of increased floor loadings or the construction of an additional storey;
- to prevent settlement, where the load of adjacent developments has potential to impact on the strata below existing foundations, meaning that the distribution of loads from existing buildings also needs to be altered.

11.2 Precautions prior to and during underpinning

The operation of underpinning inevitably involves excavating beneath the existing foundation in order to construct the new foundation beneath it, and this temporarily puts the stability of the existing structure at risk. It is essential, therefore, that the following precautions are taken to minimise any risk caused by the underpinning excavations:

- Existing loads on the structure should be reduced by temporary removal of the building's contents where possible, especially where they are imposing excessive loading.
- Where the existing structure is weak, additional temporary works should be carried out to stabilise the existing building, such as raking shores to the walls being underpinned, internal flying shores, pinning the structure through the walls above the foundation, and dead shoring.
- During the underpinning works, a constant check should be kept for any movement of the existing structure, using calibrated tell-tales, heavy plumb-bobs or laser monitoring devices.

Photographs 11.1 and 11.2 show glass tell-tales used to determine whether a crack in an existing building is increasing or remaining stable during new construction works. Other measuring devices can be used to measure any movement. With glass tell-tales, even small movement will result in the glass breaking, indicating that movement has taken place.

Photograph 11.1
Glass tell-tale, fixed to the building with epoxy resin; if any movement occurs, the glass breaks

Photograph 11.2
Glass tell-tales used to measure additional movement of cracked structure

11.3 Underpinning techniques

Techniques used for underpinning range from traditional methods, using brickwork or mass concrete, to more sophisticated methods involving needling and piling. The following sections describe some of the underpinning systems currently in use.

Generally, traditional brickwork and mass concrete underpinning can be carried out by the builder or general contractor, while the more sophisticated methods, especially those involving piling, are carried out by specialist sub-contractors such as Roger Bullivant (www.roger-bullivant.co.uk), which offer complete survey, design and construction packages.

It should be noted that it is rarely necessary to underpin the whole of a building, since most settlement and subsidence problems are due to failure of only part of the existing foundation.

Brickwork underpinning

Brickwork underpinning is the most traditional of all of the techniques used and has largely been replaced by other methods. However, it is still useful for small underpinning works to brick and masonry structures. With brickwork underpinning, much of the new foundation is constructed from concrete blocks or bricks, built under the existing foundation (Figure 11.1).

The construction of brickwork underpinning, and its sequence of operations are similar in all respects to mass concrete underpinning described below; also see Figure 11.2. A vertical cross-section of brickwork underpinning is shown in Figure 11.1.

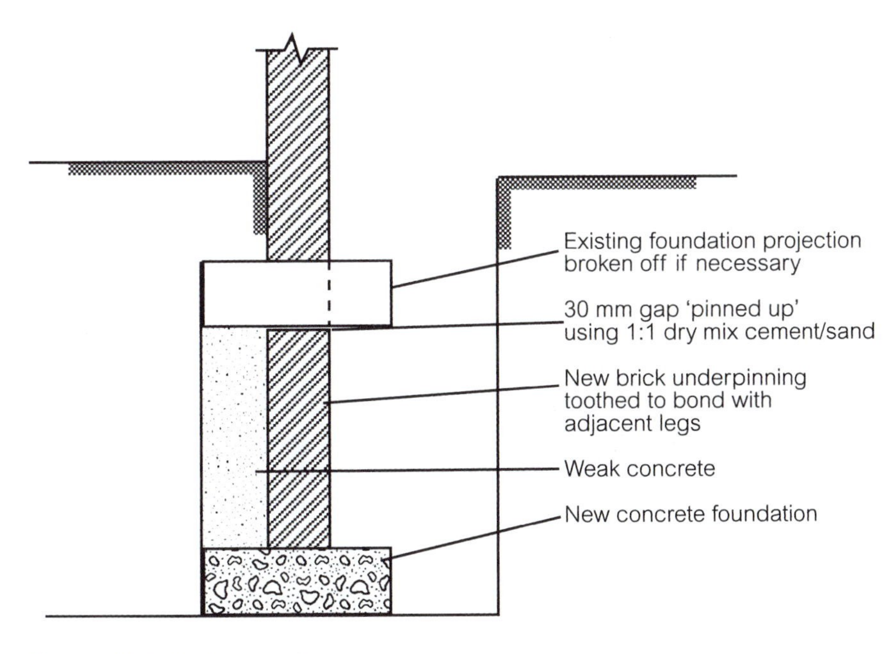

Figure 11.1 *Brickwork underpinning*

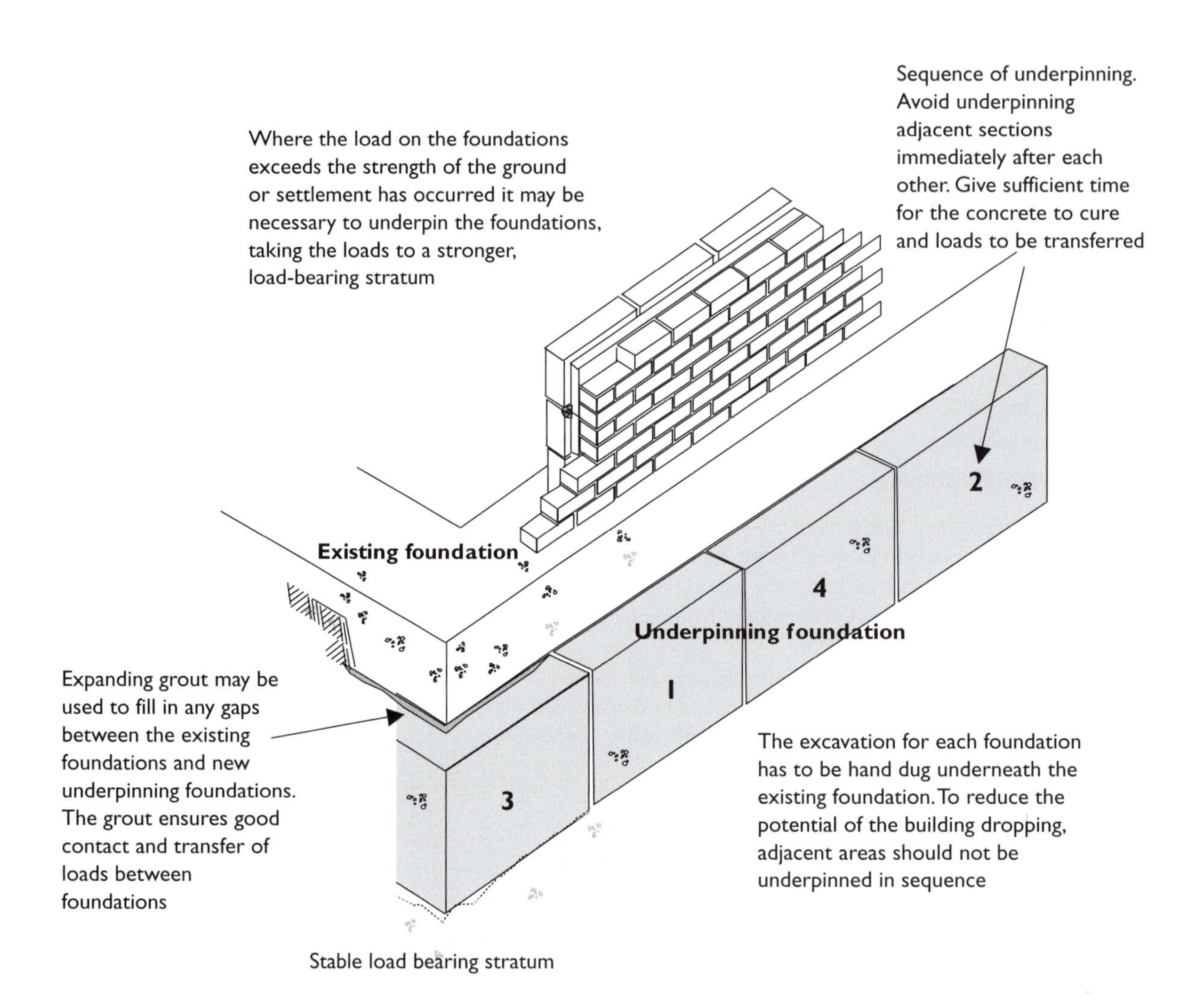

Sequence of underpinning. Avoid underpinning adjacent sections immediately after each other. Give sufficient time for the concrete to cure and loads to be transferred
Where the load on the foundations exceeds the strength of the ground or settlement has occurred it may be necessary to underpin the foundations, taking the loads to a stronger, load-bearing stratum
2
Existing foundation
4
Underpinning foundation
Expanding grout may be used to fill in any gaps between the existing foundations and new underpinning foundations. The grout ensures good contact and transfer of loads between foundations
1
3
The excavation for each foundation has to be hand dug underneath the existing foundation. To reduce the potential of the building dropping, adjacent areas should not be underpinned in sequence
Stable load bearing stratum

Figure 11.2 *Underpinning foundations using mass concrete: sequence of operation*

Mass concrete underpinning

Photographs 11.3 and 11.4 show mass concrete underpinning to an existing building. The stonewalls were originally built directly onto the ground. These walls, which also acted as foundations, had been in place for over 200 years and had seen little movement. The underpinning was necessary as the floor of the building, which was attached to the front of the building, was to be lowered. The new floor level would have undermined the existing foundation. Each section of concrete was dug out separately, and the concrete was inserted and left to cure before any attempt to remove adjacent sections. The undulating and irregular shape of the concrete clearly shows where each section of concrete has been cast and poured.

Photograph 11.3 *Mass concrete underpinning*

Photograph 11.4 *Mass concrete underpinning*

Photograph 11.5
Mass concrete underpinning of an existing wall

Photograph 11.5 illustrates the precarious nature of underpinning. The stonewall shown needed support while the mass concrete underpinning was excavated by hand and poured. Where once the stonewall rested directly on the ground, the loads are now transferred to the mass concrete underpinning, which has created a strip foundation.

Mass concrete underpinning is one of the most commonly used methods for both small and large buildings. The key factors in the design and construction of mass concrete underpinning are identified as follows (and illustrated in Figure 11.3):

- The underpinning must be carried out in 'legs' (sections) in order to leave the greater proportion of the existing foundation fully supported throughout the operations.
- The maximum length of each leg depends on the stability of the existing structure, its loadings and the subsoil conditions. The leg is generally between 0.9 and 1.5 m in length.
- The total length of the unsupported legs of the existing foundation, at any one time during the underpinning operation, should not generally exceed one-sixth of its total length for unstable structures carrying heavy loads; one-quarter of its total length for larger buildings with good structural stability; and one-third of its total length for small buildings with good structural stability.
- When one set of legs is completed and providing full support, the next set of legs is constructed. New legs should not be constructed immediately adjacent to legs that have just been completed.
- Each leg of mass concrete must be properly keyed to the next leg by a 'joggle' joint or by hacking/scabbling the previous leg prior to casting the next.
- The mass concrete underpinning is constructed to within 75 mm of the underside of the existing foundation and, on achieving its required strength, is

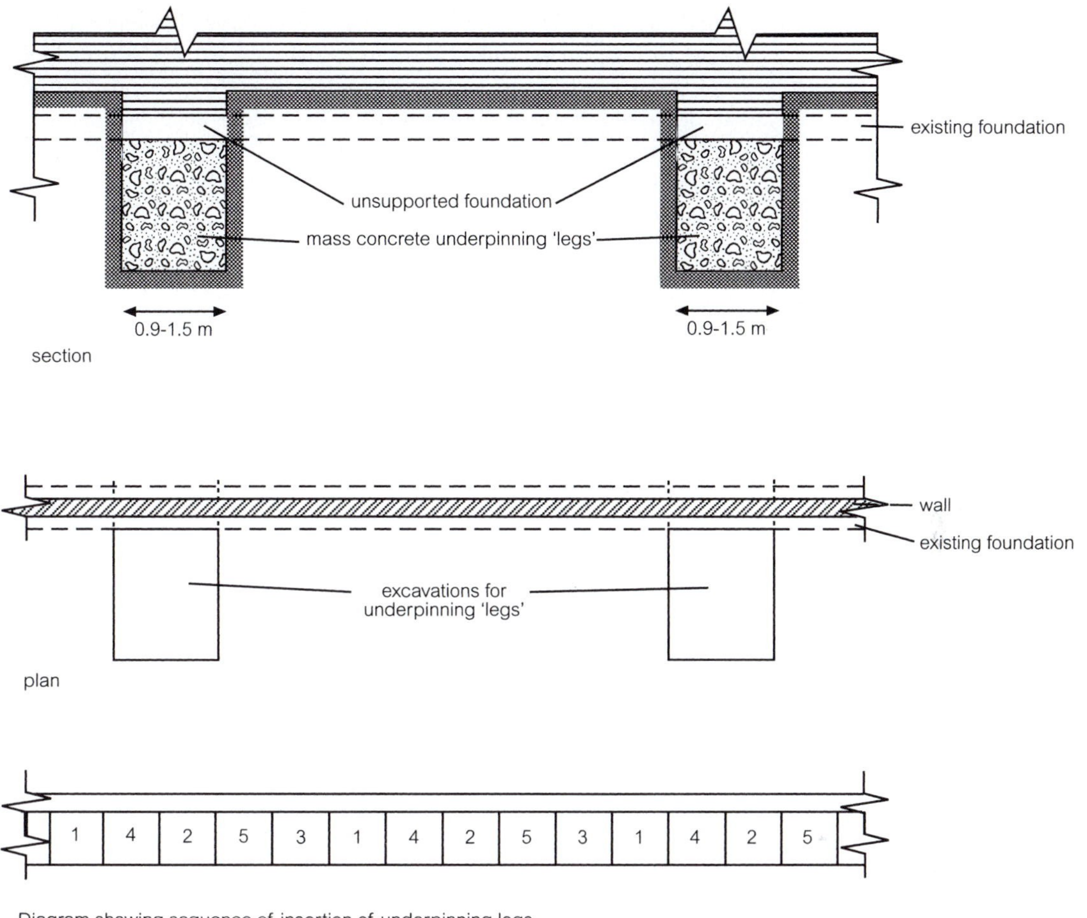

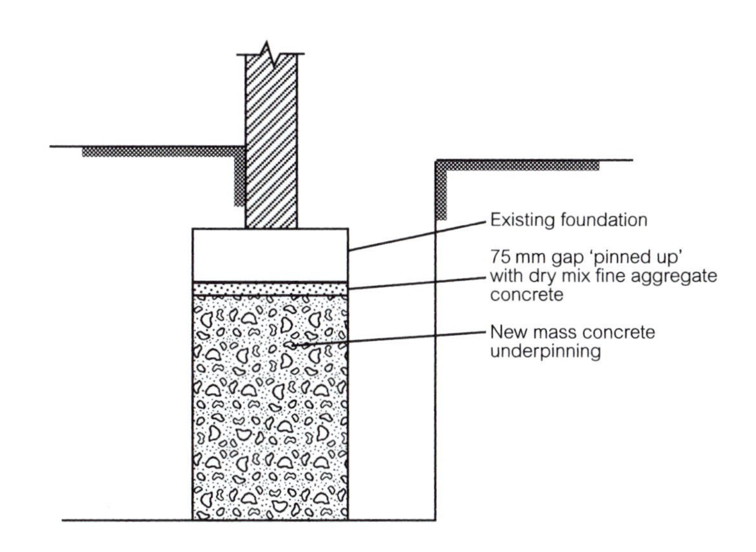

Figure 11.3
Mass concrete underpinning

'pinned up'. This involves either ramming a non-shrink, dry-mix concrete into the gap or using an expanding grout to fill the void. Where an expanding non-shrink grout mix is used, the gap will be less than 75 mm and consistent with that described in the manufacturer's instructions. Suppliers of non-shrink grouts include ROM (www.rom.co.uk) and ARCON (www.arcon.supplies.co.uk) amongst others. Non-shrink cement should be used in dry mixes as well as the grouts, preventing gaps developing during curing. Where a dry mix is used, it should comprise one part cement to one part fine aggregate, with a maximum particle size of 10 mm, into the 75 mm space. The purpose of this is to enable the mass concrete to undergo its initial, and most significant, drying shrinkage prior to achieving a positive structural connection between the existing foundation and its underpinning. Failure to carry out this operation would result in undesirable, minor settlement of the underpinned foundation.

Beam and pier underpinning

Beam and pier underpinning, illustrated in Figure 11.4, comprises a reinforced concrete beam, inserted either directly above or below the existing foundation, supported by mass concrete piers constructed at 2.5–3.0 m centres. The function of the reinforced concrete beam is to transfer the loads from the wall being underpinned to the piers, which, in turn, carry the loads to a deeper, firmer stratum.

The sequence of operations is as follows:

- Excavation for mass concrete piers: The relatively small plan area of each pier is such that only a limited length of the existing foundation is undermined, therefore maintaining the stability of the structure.
- Construction of mass concrete piers to underside of existing foundation.
- Removal of masonry above existing foundation, between piers, for construction of reinforced concrete beam.
- Construction of reinforced concrete beam between piers.

As with traditional underpinning techniques, it is essential that the total length of unsupported/ undermined existing structure at any one time is kept to a minimum. This is achieved by ensuring that the maximum total unsupported length is between 0.9 and 1.5 m (this may need to be less, depending on the condition of the building and ground).

The principal advantage of beam and pier underpinning, compared with a traditional underpinning system, is that excavation beneath the existing structure is reduced from the entire length of wall to the piers only, thereby reducing the risk to the building's stability during the works.

Pile and needle underpinning

Pile and needle underpinning, illustrated in Figure 11.5, comprises reinforced concrete needles inserted through the existing wall, above foundation level, and supported at each end by small-diameter piles that transmit the building's loads to a deeper, firmer stratum. The needles are inserted at approximately 1.5 m centres

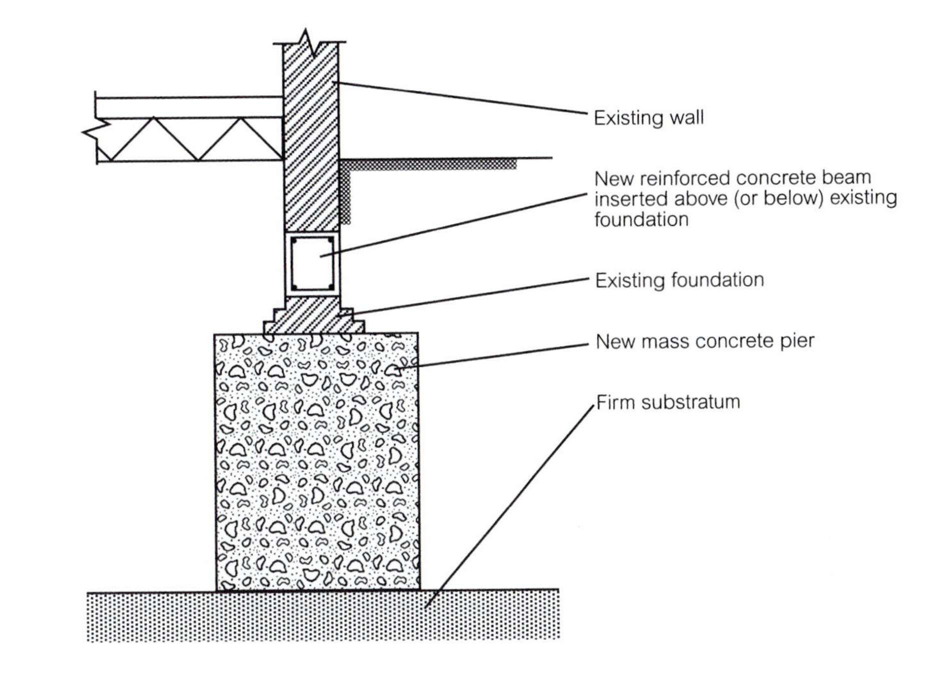

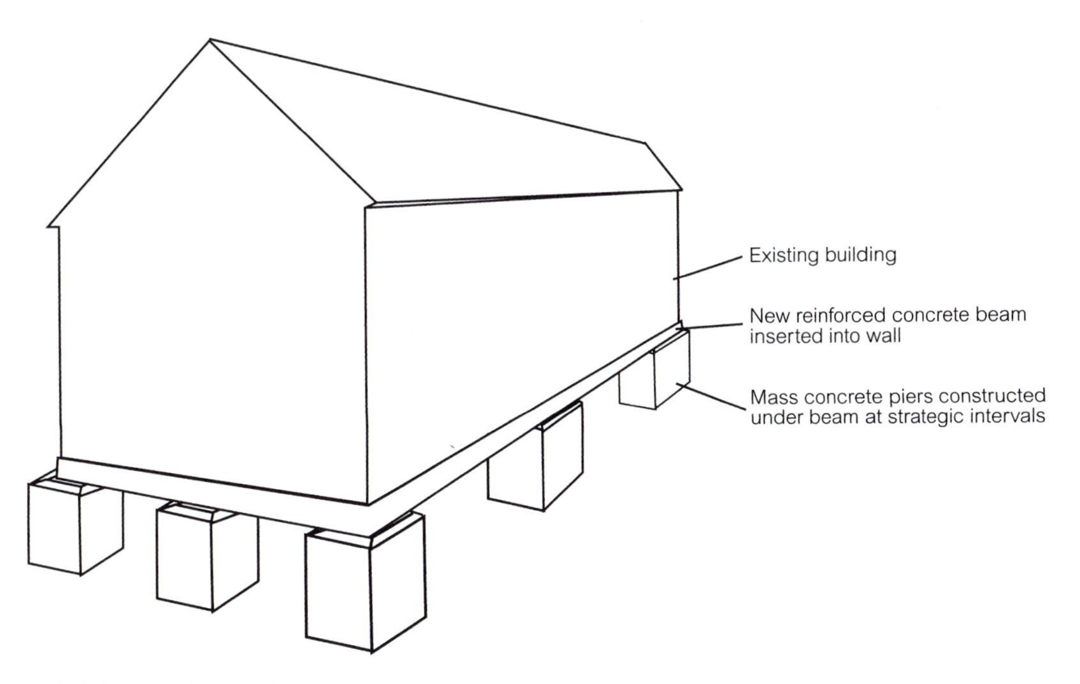

Figure 11.4 *Beam and pier underpinning*

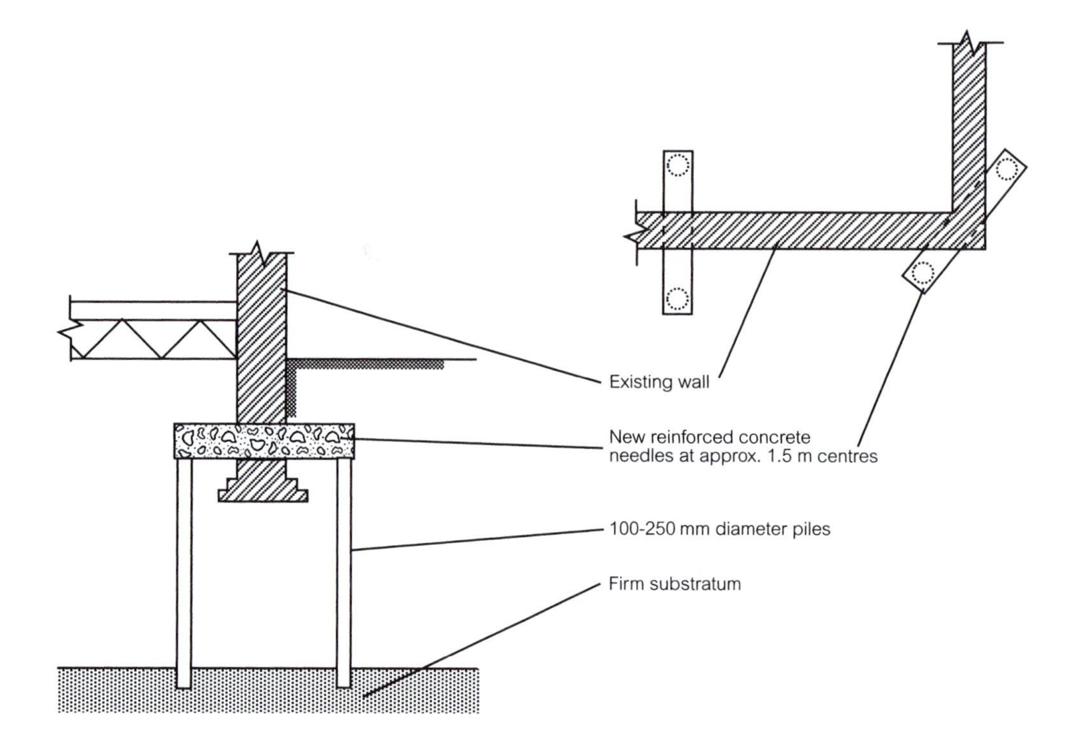

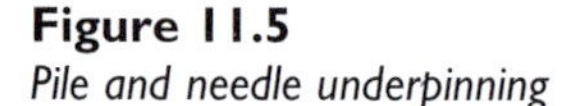

Figure 11.5
Pile and needle underpinning

along the length of the wall being underpinned, their function being to transmit loading from the wall to the piles.

The actual needle centres and pile diameters and depths will be determined by the stability of the building, the loads to be transmitted and the subsoil conditions. The 100–250 mm diameter piles are driven or bored using compact piling rigs, which require only 1.80 m headroom and 2 × 1.5 m working space, especially important when installing piles inside the building, where space may be limited.

The principal advantage of the pile and needle system is that there is no direct undermining of the existing foundation, and only small areas of masonry need to be removed to construct the needles. This method is also faster than traditional systems involving bulk excavation. It should be borne in mind that pile and needle underpinning involves inserting piles from inside, as well as outside, the building, which may cause disturbance to the occupants and day-to-day functioning of the building throughout the works. This, however, can be overcome by using pile and cantilever needle underpinning, where the needles, supported by two piles installed outside the building, function as cantilevers, ruling out the need for any internal work. This method is illustrated in Figure 11.6.

Cantilever ring beam underpinning

Cantilever ring beam underpinning, illustrated in Figure 11.7, comprises steel I-section cantilever needles that transmit the loads from the wall to a deeper bearing stratum by means of a reinforced concrete ring beam and mini-piles.

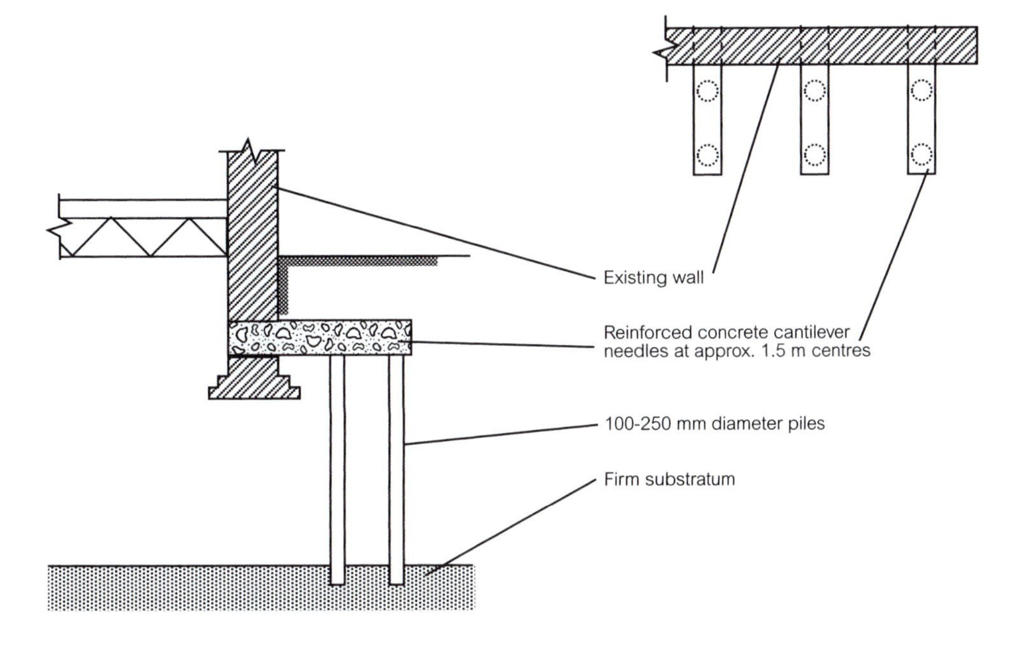

Figure 11.6
Pile and cantilever needle underpinning

Two staggered lines of mini-piles, 90–250 mm in diameter, are first driven, drilled or augered at approximately 1 m centres alongside the wall being underpinned. The inner line of piles gives direct support to steel I-section needles, and the outer line supports the reinforced concrete ring beam between the needles.

Following construction of the mini-piles, the steel needles are positioned, their inner ends located in pockets formed in the masonry, and the reinforced concrete ring beam is cast. The ring beam encases the needles and is connected to the tops of the piles to produce a fully integrated support.

The system is designed so that the loads from the underpinned wall place the inner line of piles in compression and the outer line of piles in tension. This method, by providing a direct connection between the needles and ring beam, gives continuous support to the wall along its entire length, rather than at intervals as with pile and needle underpinning. It is therefore better suited to less stable walls, in a more serious state of structural distress.

The principal advantages of cantilever ring beam underpinning are similar to those for pile and cantilever needle underpinning: no direct undermining of the existing foundation is necessary; only small areas of masonry need be removed for insertion of the needles; no bulk excavation is required; and all of the work can be executed from outside the building. The system is also faster than traditional 'dig-out' underpinning methods.

Double angle mini-pile underpinning

The double angle mini-pile underpinning system, illustrated in Figure 11.8, involves the installation of small-diameter piles in pairs formed at an angle *through* the existing foundation, at between 1.0 and 1.5 m intervals.

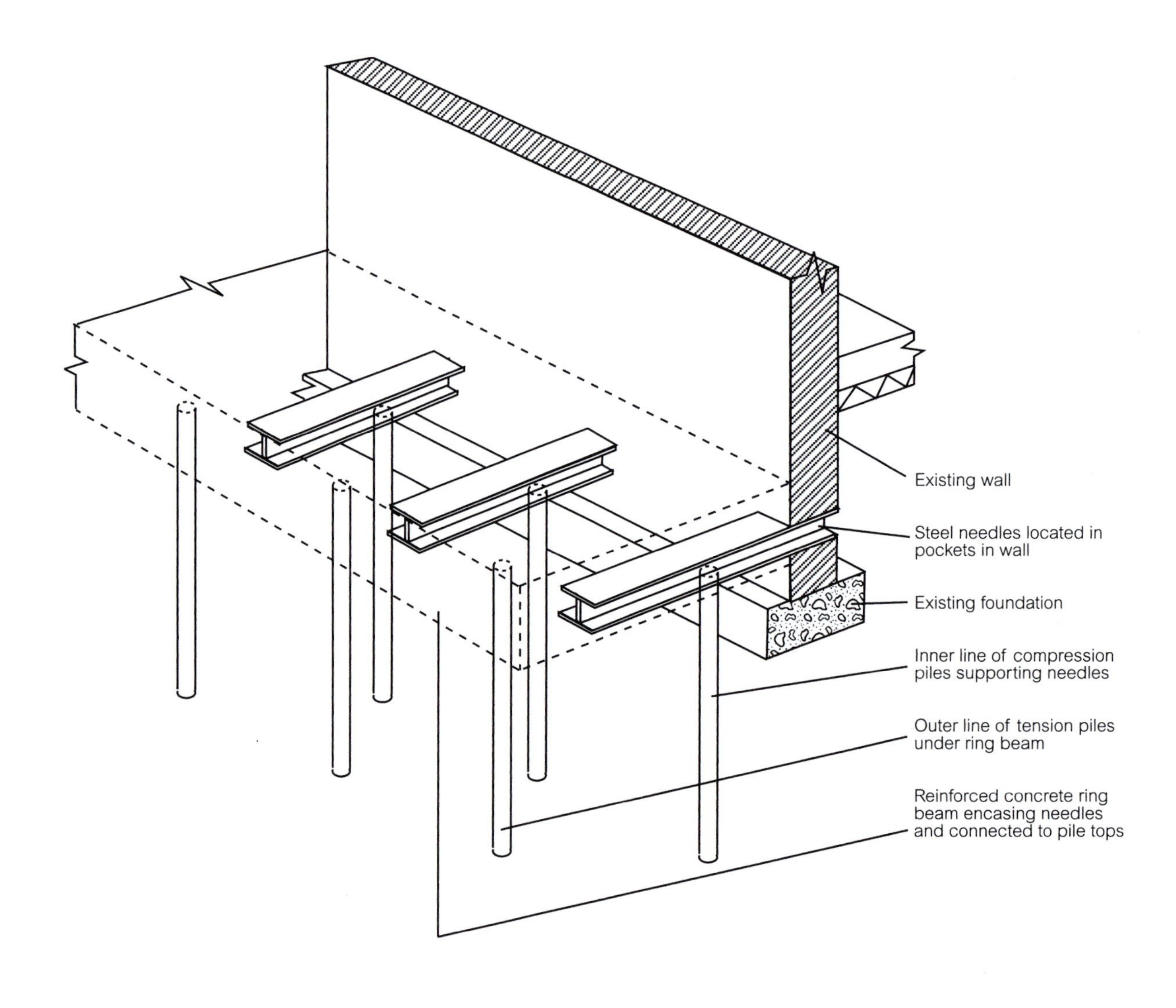

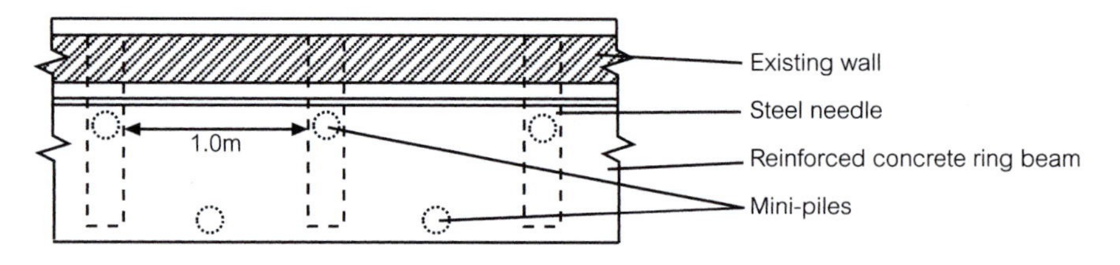

Figure 11.7 *Cantilever ring beam underpinning*

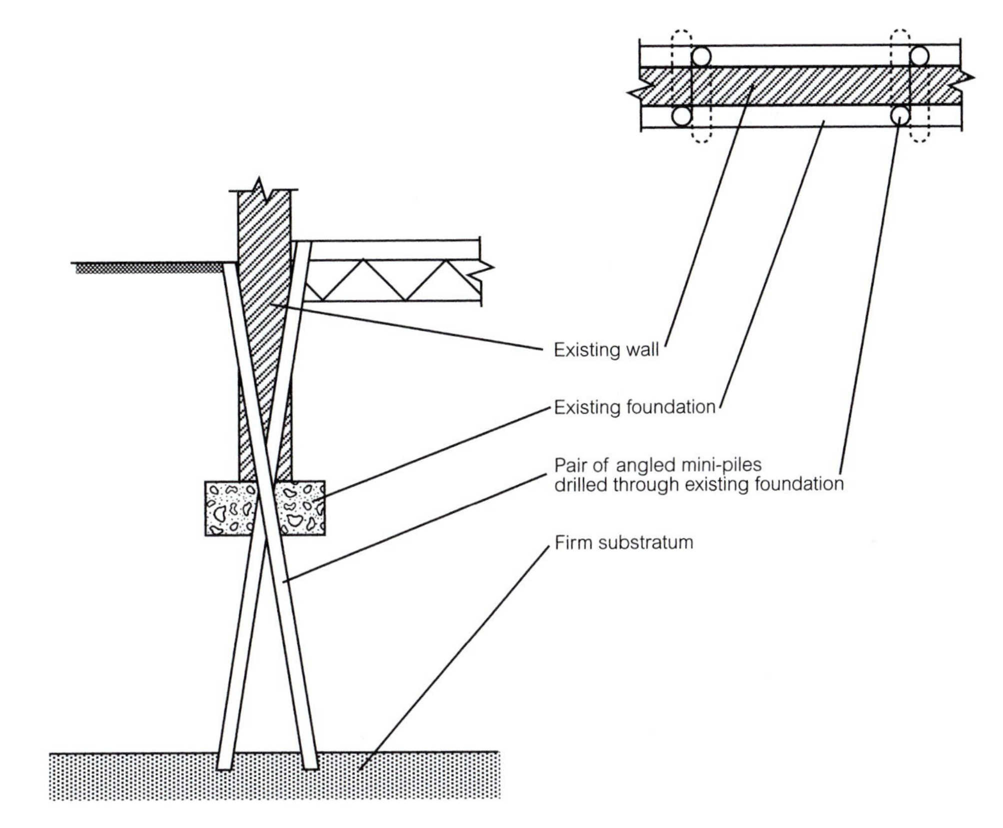

Figure 11.8
Double angle mini-pile underpinning

The existing foundation is pre-drilled using air-flushed rotary percussive equipment, with drilling heads capable of drilling through brickwork, masonry, concrete and steel as necessary. Permanently cased steel-driven or solid- or hollow-stem augered piles are then installed through the pre-drilled holes, with the casings terminated at the underside of the existing foundation. The piles are then concreted and reinforced up through the existing foundation. Double angle mini-pile underpinning involves less excavation and disruption than most of the alternative methods. Undermining of the existing foundation is negligible, especially when compared with traditional 'dig-out' techniques. In addition, the construction of the piles through the existing foundation achieves a more direct physical connection between the existing structure and the new underpinning. Other advantages of this system are its speed of installation and its high load capability, achieved using piles from 90 to 250 mm in diameter, penetrating to any reasonable depth.

In cases where access to the interior of the building for pile construction is a problem, single piles may be installed from the outside only, at closer intervals of between 0.9 and 1.2 m.

All of the underpinning systems described, together with other underpinning and foundation stabilisation systems, are carried out as full survey, design and construct packages by Roger Bullivant, one of the UK's leading specialist foundation and underpinning engineers (www.roger-bullivant.co.uk).

References

Building Research Establishment (1990) *Underpinning* (Digest 352), Watford: BRE.
Bullivant, R. A. (1996) *Underpinning: A Practical Guide,* Oxford: Blackwell Science.
Hunt, R., Dyer, R. H. and Driscoll, R. (1991) *Foundation Movement and Remedial Underpinning in Low Rise Buildings,* Watford: BRE.
University of Bath, Department of Architecture and Building Engineering (1985) *Building Appraisal Maintenance and Preservation: Symposium Proceedings,* University of Bath.

Strengthening existing walls

CHAPTER 12

12.1 Stabilising cavity walls

For a number of reasons, it may be necessary to strengthen existing walls when undertaking refurbishment works. Cavity ties are a particular problem in buildings constructed before 1981. Galvanised and mild steel ties used before this date did not have a good resistance to corrosion; subsequently, the quality of wall ties has improved. Stainless steel, galvanised and plastic ties are now much more resistant to corrosion. Where cavity ties are corroded or there are no ties inserted in the wall, the external leaf of the cavity will tend to bulge (Figure 12.1). The internal leaf of the cavity wall normally remains plumb and stable as it is tied into the structure at the floor and roof junctions. If galvanised wall straps have been installed at the floor and roof junctions, these will help considerably with the walls stability.

Some polyurethane foams, which have highly adhesive properties, can be used to help stabilise cavities. The foam insulation can be injected into the cavity where it quickly fills the cavity, bonding to the masonry and stabilising the wall; Isofoam CRF is one such product (www.mpinsulations.co.uk). The additional insulation also improves the thermal performance of the wall.

12.2 Rebuilding the external leaf

If bulging is considerable, the external leaf will need to be taken down and rebuilt (see Figure 12.2). Angle wall ties can be installed as the new leaf of the external cavity is built. The angle wall ties are usually plugged and screwed into the existing wall. To ensure a solid fixing, the ties are often fixed to the internal block rather than the existing bed joint.

12.3 Remedial wall ties

Where the structure is sound but the existing wall ties have failed, remedial wall ties need to be installed. For masonry walls that are structurally sound, mechanical fixing ties can be fitted into holes that are drilled into the wall.

If walls have not deformed or can be repositioned, the stability of the wall can be improved by installing new cavity wall ties. Expanding rawl bolts made with stainless steel, brass or neoprene expanders can be used to provide a good mechanical key (Figure 12.3).

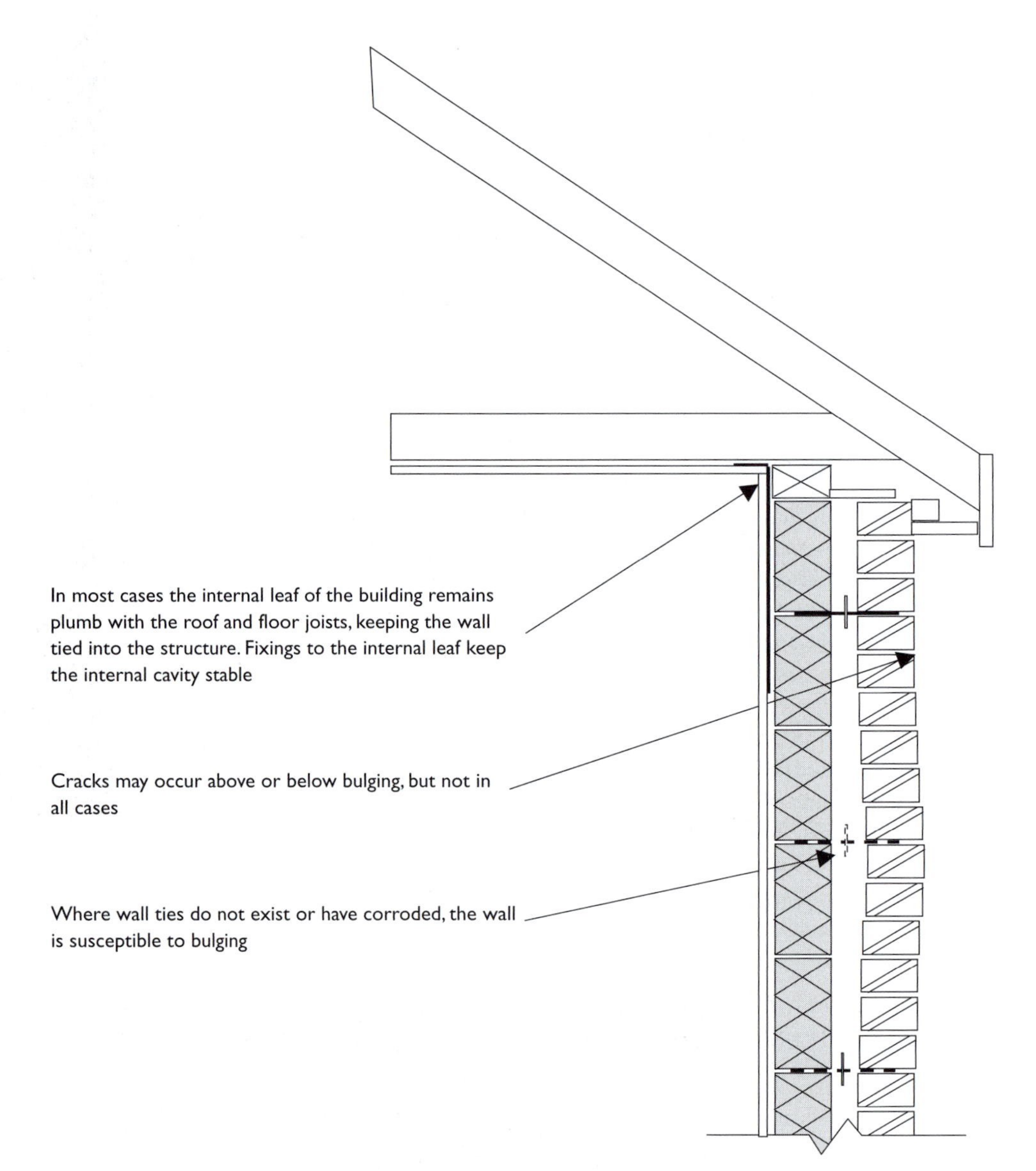

In most cases the internal leaf of the building remains plumb with the roof and floor joists, keeping the wall tied into the structure. Fixings to the internal leaf keep the internal cavity stable

Cracks may occur above or below bulging, but not in all cases

Where wall ties do not exist or have corroded, the wall is susceptible to bulging

Figure 12.1
Bulging cavity due to no cavity ties or cavity tie failure

Brass expanders are more malleable than stainless steel and help to improve the key between the tie and the concrete blocks. Where the point loads caused by hard steel and brass would cause the concrete or clay to crack, ties with neoprene expanding sleeves can be used (Figure 12.4). The softer neoprene expands into the concrete, gripping the structure, rather than just compressing it. Pull out test tools can be used to test the strength of each tie.

Figure 12.5 shows a cross-section of a wall where expanding wall ties have been installed to strengthen the wall, and external insulation has been applied. Photograph 12.1 shows a rendered wall with replacement cavity wall ties installed; the wall is now ready to be clad with insulation and render.

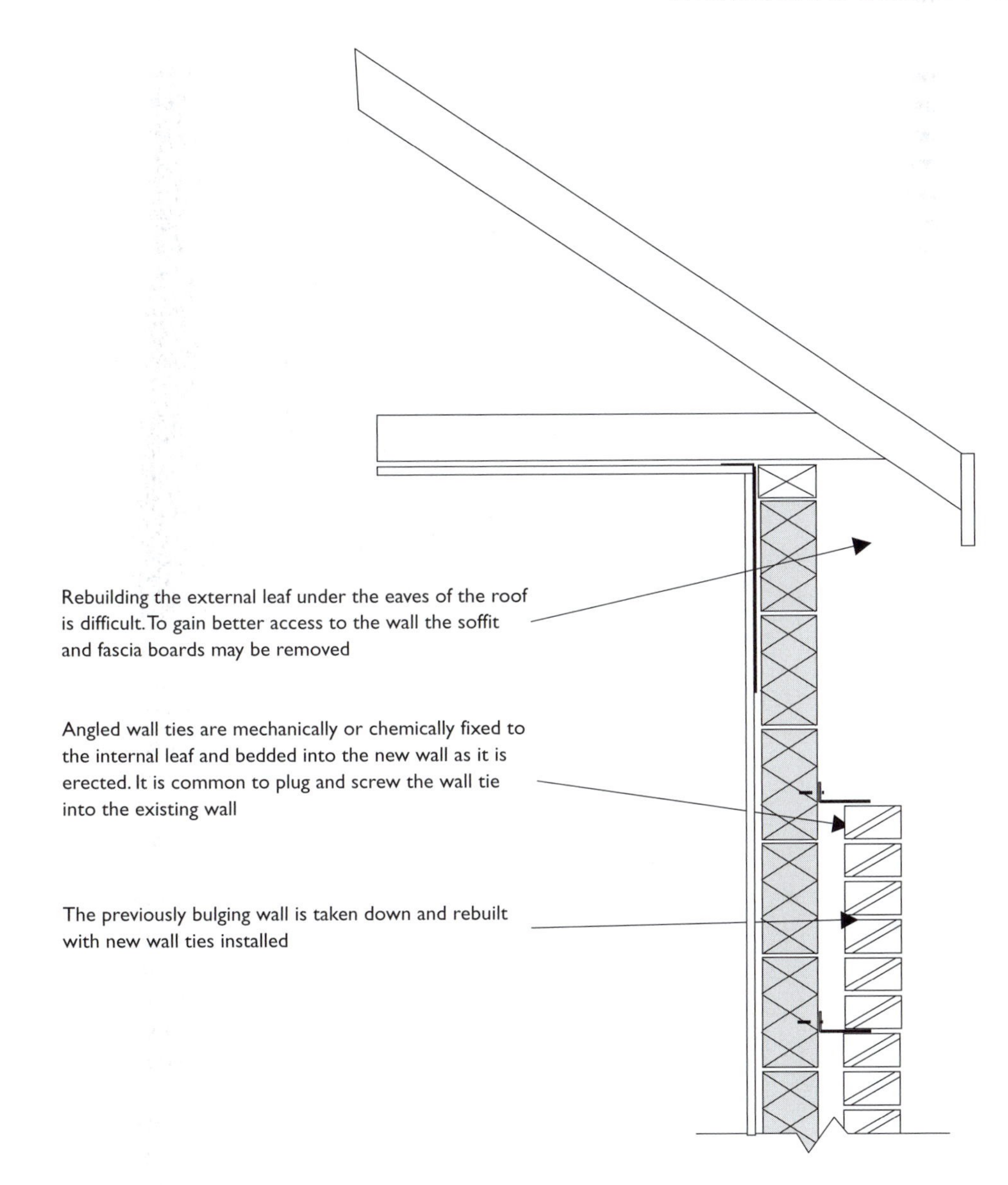

Rebuilding the external leaf under the eaves of the roof is difficult. To gain better access to the wall the soffit and fascia boards may be removed
Angled wall ties are mechanically or chemically fixed to the internal leaf and bedded into the new wall as it is erected. It is common to plug and screw the wall tie into the existing wall
The previously bulging wall is taken down and rebuilt with new wall ties installed

Figure 12.2
Rebuilding the external leaf of a cavity wall

Photograph 12.1 shows examples of the areas that have been repaired with epoxy resin. In the photograph, remedial cavity ties have been inserted to strengthen the cavity wall. To reduce the risk of accidents due to the protruding cavity tie, yellow pain has been sprayed around the bolts, highlighting their presence and reducing the risk of injury.

For safety reasons, it is important that the insulation is fixed as soon as possible, thus reducing the risk of the building occupants and workers injuring themselves on the protruding bolts; alternatively, the bolts can be ground flush once properly installed. Where sharp or dangerous protrusions are exposed for a period of time, protection devices must be fitted to prevent people injuring themselves.

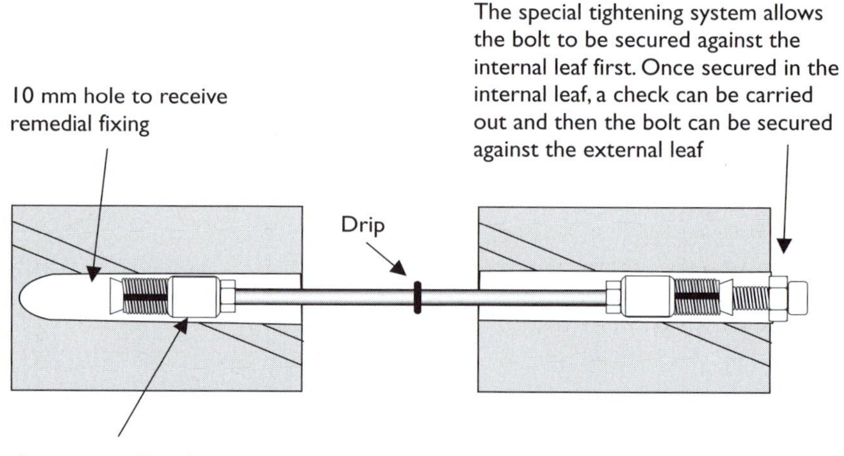
The special tightening system allows the bolt to be secured against the internal leaf first. Once secured in the internal leaf, a check can be carried out and then the bolt can be secured against the external leaf
10 mm hole to receive remedial fixing
Drip
Brass expanding sleeve
Brass expanding sleeve is more malleable than steel and provides a better key with the brickwork or blockwork

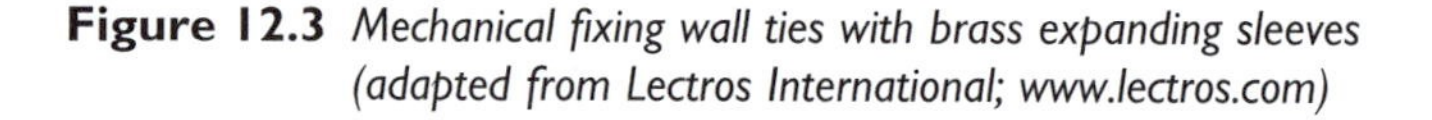
Figure 12.3 *Mechanical fixing wall ties with brass expanding sleeves (adapted from Lectros International; www.lectros.com)*

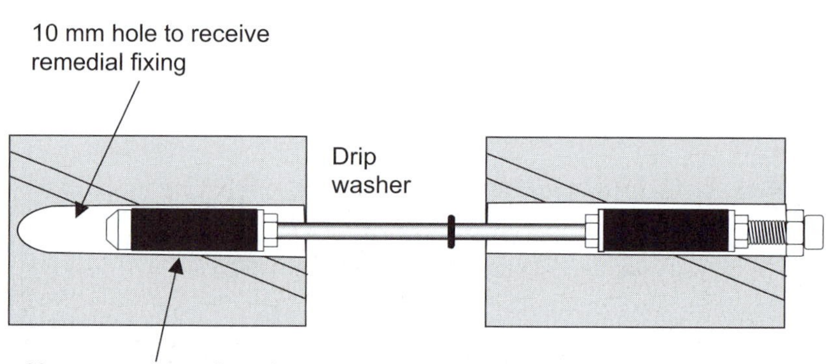
10 mm hole to receive remedial fixing
Drip washer
Neoprene expanding sleeve
The neoprene expands and moulds into the ceramic material without crushing it. As the neoprene forces itself into the masonry it grips the surface

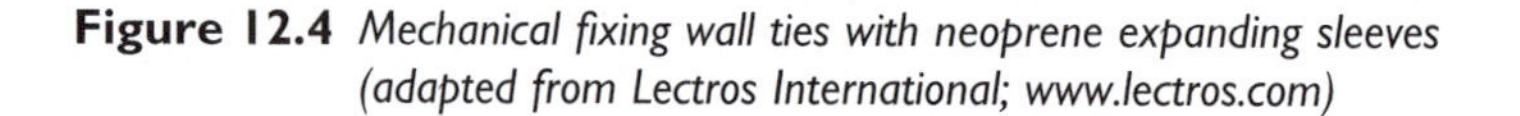
Figure 12.4 *Mechanical fixing wall ties with neoprene expanding sleeves (adapted from Lectros International; www.lectros.com)*

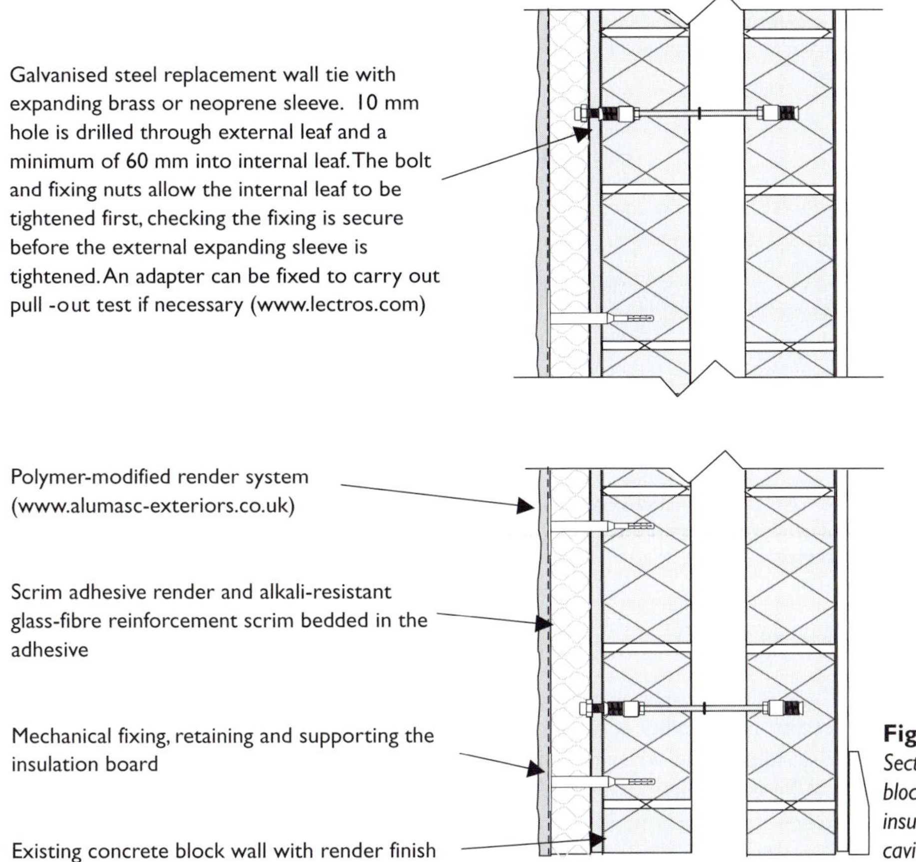
Galvanised steel replacement wall tie with expanding brass or neoprene sleeve. 10 mm hole is drilled through external leaf and a minimum of 60 mm into internal leaf. The bolt and fixing nuts allow the internal leaf to be tightened first, checking the fixing is secure before the external expanding sleeve is tightened. An adapter can be fixed to carry out pull -out test if necessary (www.lectros.com)
Polymer-modified render system (www.alumasc-exteriors.co.uk)
Scrim adhesive render and alkali-resistant glass-fibre reinforcement scrim bedded in the adhesive
Mechanical fixing, retaining and supporting the insulation board
Existing concrete block wall with render finish

Figure 12.5
Section through concrete block cavity wall with external insulation and replacement cavity ties

Photograph 12.1
Remedial cavity ties inserted: cracks in the render and structure have been filled and sealed with epoxy resin, ready to receive insulation and render

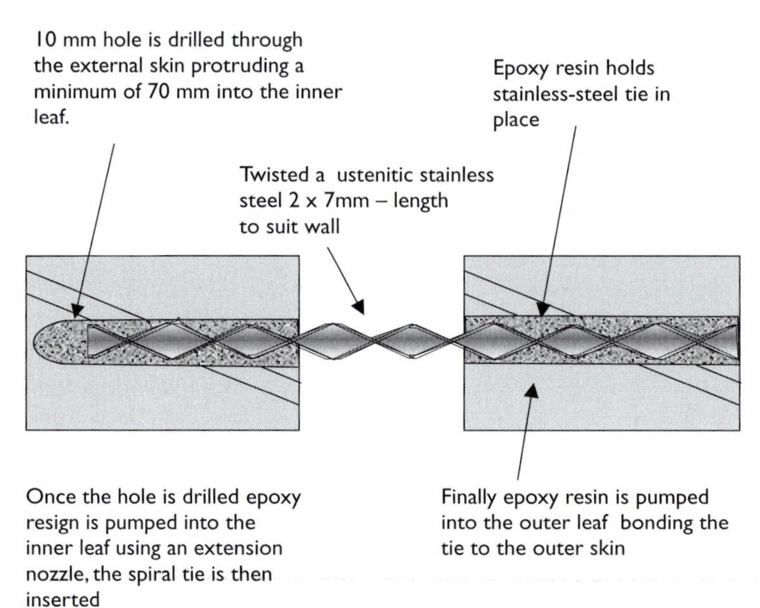
10 mm hole is drilled through the external skin protruding a minimum of 70 mm into the inner leaf.
Epoxy resin holds stainless-steel tie in place
Twisted a ustenitic stainless steel 2 x 7mm – length to suit wall
Once the hole is drilled epoxy resign is pumped into the inner leaf using an extension nozzle, the spiral tie is then inserted
Finally epoxy resin is pumped into the outer leaf bonding the tie to the outer skin

Figure 12.6
Helix spiro remedial wall ties, bonded in epoxy resin (adapted from www.helixfixings.co.uk)

12.4 Strengthening walls

Where masonry walls are in a particularly poor condition, or proposed work could threaten the stability of a wall, remedial action may need to be taken to improve the condition and strength of a wall.

Where it is difficult to achieve a good mechanical key, epoxy resin is used to provide a chemical bond for the wall ties (Figure 12.6). Photograph 12.2 shows chemically bonded spiral ties used to strengthen a two-brick-thick wall where an

Photograph 12.2
Two-brick-thick solid wall reinforced with spiral cavity ties to prevent delamination of the brick wall

opening has been made in an existing façade. To add strength to the remaining wall, the spiral tie stainless steel reinforcement has been introduced. The ties will prevent the skins of the wall delaminating. The photograph clearly shows that it would have been difficult to use mechanical fixing bolts effectively in this situation.

12.5 Removing buildings and supporting the remaining properties

Where parts of the structure are removed that may be providing support to an existing wall, it may be necessary to introduce additional support to ensure all remaining walls are stable. It is important that bracing, brackets, pins and support are in place before work is carried out on the supporting structure. In the schematic below (Figure 12.7), which is based on a real situation, a terraced property was removed, and the two adjacent properties needed additional support. In one situation, a property was being redeveloped, and as part of the party wall agreement between the redeveloped property and the property being demolished it was agreed that galvanised steel wall straps would be used to provide additional lateral support to the party wall. With the addition of wall straps, this meant that when the adjacent property was demolished no additional support was required. The property to the other side of the development was still occupied. As part of the occupiers' party wall agreement, they wanted no internal disruption. The meant that flying shores and raking support needed to be in place throughout the demolition and construction phase.

Figure 12.7
Different methods of providing wall support where buildings are removed

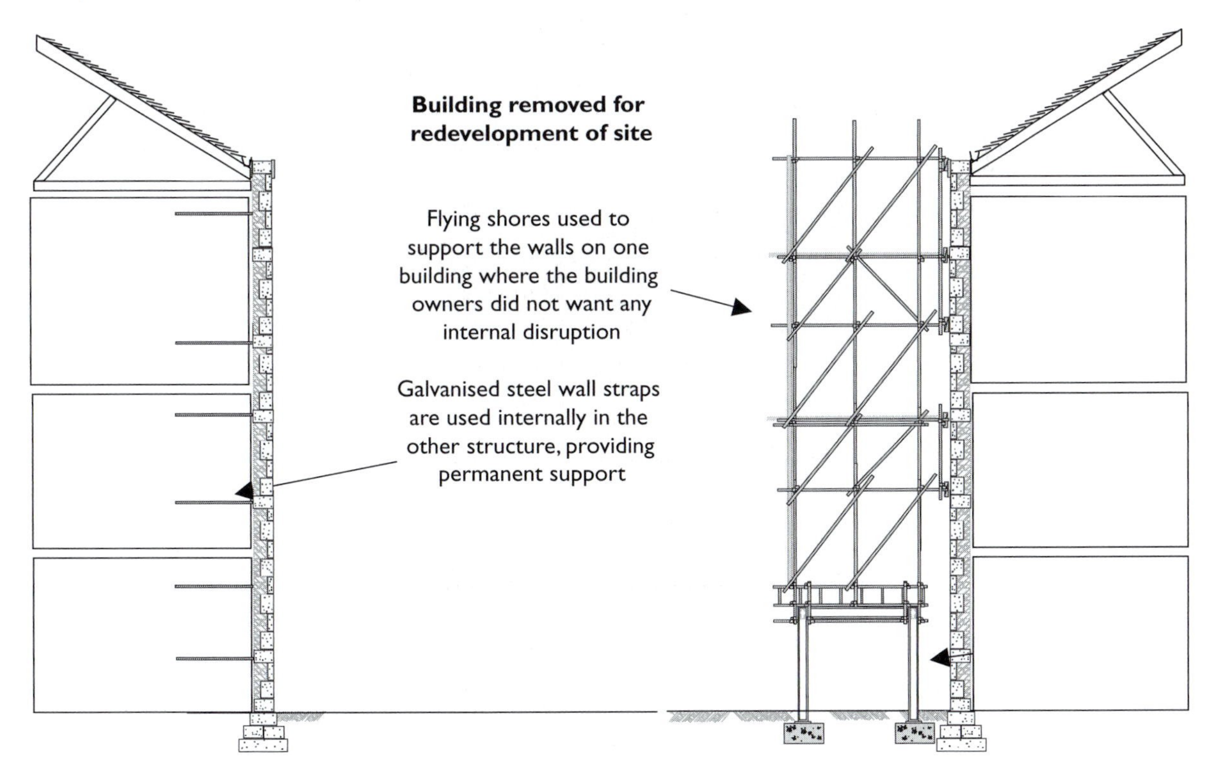
Building removed for redevelopment of site
Flying shores used to support the walls on one building where the building owners did not want any internal disruption
Galvanised steel wall straps are used internally in the other structure, providing permanent support

Photograph 12.3
In addition to the wall straps, galvanised steel straps were also secured to the floor joists and party wall

Photograph 12.4
Galvanised steel angle brackets used to provide support to a party wall

Photographs 12.3 and 12.4 show galvanised steel straps used to provide additional support to a party wall. The supporting straps and the padstones were positioned prior to the adjacent structure and parts of the party wall being removed. The party wall was only half a brick thick, so it was essential that the work was undertaken with care, and support was in place prior to alterations taking place.

The occupier of the structure on the other side of the property did not want any disruption to their building use, thus no internal works could be undertaken. As part of the party wall agreement, flying shores were used to support the party wall (see Photograph 12.5). Photograph 12.6 shows flying shores used to support

Photograph 12.5
Flying shores with scaffolding boards and plywood sheets provide support to a party wall

Photograph 12.6
Flying shores with scaffolding boards and plywood sheets provide support to a party wall; abutting wall constructed to provide lateral support

a party wall while the new steel frame of the refurbished building is being constructed. An abutting pier wall has been constructed from engineering bricks to provided additional lateral support to the building and support for the flying shores.

CHAPTER 13 Façade retention

13.1 General

Refurbishment of a building normally involves keeping most of the existing structure and fabric and restoring, repairing and upgrading it to provide accommodation that meets current standards in terms of comfort, amenity and, in most cases, building legislation. This enables owners and developers to re-use what otherwise might be considered as obsolete, redundant or outdated buildings.

A more extreme form of building re-use, which keeps considerably less of the existing structure and fabric than 'low-key' refurbishment, involves retaining only the external façade and constructing an entirely new structure behind it. 'Façade retention', as it has become known, may involve retaining only one elevation if the building is part of a 'row' of buildings forming a street frontage (Photographs 13.1 and 13.2); two elevations if it is a corner building; three elevations if it forms the end of a block (Photographs 13.3 and 13.4); or, more rarely, all four elevations if it is an isolated building. All of the external wall is retained in Photograph 13.9; however the structure is circular on plan rather than square. Whichever form of façade retention scheme is used, the entire interior and roof are normally demolished, leaving only the external wall(s) standing to form the preserved external elevation(s) to a completely new structure erected behind.

Façade retention has the benefit of retaining the historic or important façade feature, but does not score at all well when issues of sustainable construction are considered. Often much of the structure that could be used is removed. Where at all possible, the materials that are not retained within the new structure should be systematically removed and segregated, so that the greatest possible chance of re-use or recycling is achieved.

The retention of existing façades in this way, in what might be regarded as the most drastic form of building refurbishment/re-use, short of total demolition, has become increasingly common during the last 25 years and requires special solutions to the technological problems that it presents. The principal problems met with all façade retention schemes include providing temporary support to the façade throughout the works; permanently tying back the façade to the new structure erected behind it; allowing for differential settlement between the new structure and the retained façade; and ensuring that the new structure's foundations do not impair the stability of the retained façade.

Photograph 13.1
Front façade of a row of terraces retained

Photograph 13.2
Rear view of façade to row of terraces; shows new steel frame inserted behind façade

13.2 Temporary support systems

In all buildings comprising load-bearing external façade walls, there is an interdependency between those walls and the elements they carry. While the external façade walls provide structural support to many internal elements (floors, roof structure and some cross-walls), these internal elements, in turn, provide lateral support to the façade. Thus, when this lateral support to the façade is totally removed by demolishing the building's interior, it becomes necessary to provide some means of temporary support to the façade until the new structure is constructed and the

Photograph 13.3
Façade retention to corner plot

Photograph 13.4
Rear view of two-sided façade retention scheme

retained façade is tied back to it. A major consideration in the design of a temporary support system is that it must provide the façade with stability and resistance against wind-loads from both sides, to which it will be subjected for an extensive period during the works while the building is opened up to the elements.

Generally, temporary support systems to retained façades fall into three categories: wholly external, located entirely outside the façade; wholly internal, located behind the façade within the zone of the existing and new structures; or part internal/part external, some of the supporting elements being located behind, and some outside, the façade.

External support systems have the important advantages of not interfering with demolition or subsequent construction work, but they often obstruct adjacent footpaths and roadways, and, in many schemes, have not been permitted for that reason. Internal support systems, on the other hand, leave adjacent thoroughfares unobstructed but place severe constraints upon the progress and efficiency of both the demolition of the existing structure and the new construction work. Part internal/part external systems combine the advantages and disadvantages of both.

It is essential, whatever category of temporary support system is used, that it is installed, and capable of giving total support to the façade, before any demolition of the building's internal structure takes place, and that it remains in position until the façade is permanently tied back to the newly erected structure behind. It is clear, therefore, that if an internal support system is used, its design will be complicated by having to ensure that it does not conflict either with elements of the existing building or of the new building. In addition, some members of the internal support system will inevitably interfere with the demolition operations and erection of the new structure. In certain cases, the nature of the scheme may require that the temporary support system is partly internal and partly external. The main difficulty with a part internal/part external support system is that it combines the disadvantages of both: adjacent footpaths, and possibly roadways, may be obstructed by the external elements, and demolition and construction operations are interfered with by the internal elements.

The temporary support provided to the façade can be achieved by either:

- a tubular scaffold system (Photographs 13.5 and 13.6);
- a proprietary support system (Photographs 13.7–13.10);
- a temporary system utilising universal rolled steel, designed and fabricated specifically for the purpose (Photographs 13.11 and 13.12);
- utilisation of a new steel frame as façade support during demolition and construction (Photographs 13.13).

Tubular façade retentions schemes

Tubular façade retention schemes have the benefit of being adjustable and flexible to suit the structure being retained. In the majority of cases they are erected by specialist scaffolding companies, require regular checking and should always be checked and re-secured after adverse weather conditions.

Clamps and wedges are used to secure and brace the wall (Photograph 13.6); owing to the flexibility of the tubular system, these can be easily located in positions

Photograph 13.5
Tubular façade retention scheme, East Brook Hall, Bradford

Photograph 13.6
Internal view of façade retention scheme at East Brook Hall, Bradford (cropped)

Photograph 13.7
Temporary external façade support – proprietary system

Photograph 13.8
Adjustable façade retention system

Photograph 13.9
Proprietary adjustable façade retention system

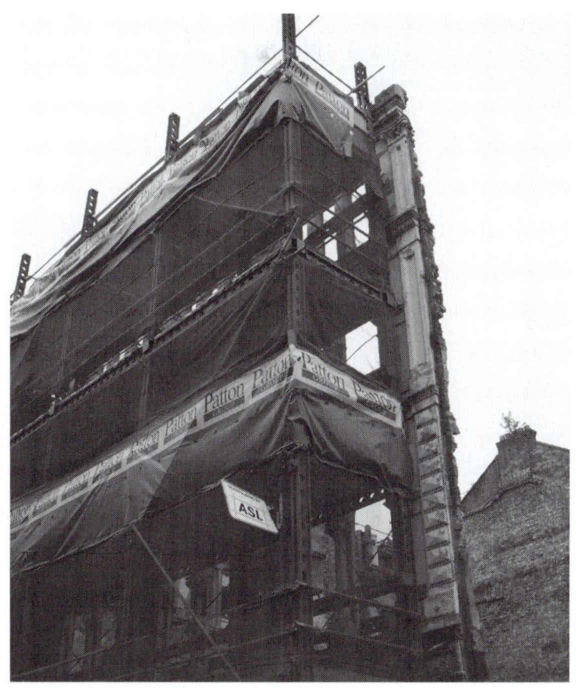

Photograph 13.10
Proprietary façade retention system

Photograph 13.11
External façade retention scheme using universal rolled steel sections

Photograph 13.12
External façade retention scheme using universal rolled steel sections; rear of façade with rolled steel pins protruding a short way into the internal part of the structure

that best suit the façade. Decorative architectural features can easily be supported by tubular systems (see Photographs 13.30 and 13.31). However, most of the proprietary schemes have been designed to be flexible and adjustable and to hold the walls and features securely. In most cases, the choice of scheme comes down to economics and which specialist sub-contractor can offer the best deal. The degree of bracing and specialist support systems required can have a significant bearing on the choice of system. Where there is need for heavy bracing, propping, jacking and pinning, proprietary systems are becoming the preferred option. The specialist companies that have developed the systems will often be used to design and construct the support systems, ensuring experts are used at all stages.

Proprietary systems

Proprietary systems are often expensive but often provide systems that are flexible and easily tightened and adjusted to hold the façade. With adjustable props and shores, any movement can be easily accommodated. Some manufacturers of proprietary systems utilise hydraulic jacks and supports (www.rmdkwikform.

Photograph 13.13 *Use of new steel frame to provide support; wholly internal façade support system*

net). If the façade needs to be loaded, jacked up or held in place, pressure can be applied.

Other proprietary façade support systems and their manufacturers include the Megasor, super slim soldiers (www.interserveplc.com), Kwickstage (www.rmdkwikform.net) and the Mass 50 and Mass 25 systems (www.mabeysupport.co.uk).

Rolled steel support systems

Photographs 13.11, 13.12 and 13.14 show an external façade retention scheme using universal rolled steel that has been fabricated and designed specifically for the use as temporary support. Such schemes are quick to assemble and may prove cost-effective for projects that require support for some time. The rolled steel systems are usually purchased specifically for the project, rather than the proprietary or tubular scaffold retention systems, which are normally hired. The longer the duration of the project and the requirement for the temporary façade retention system, the greater the hire cost. With rolled steel sections, the initial outlay will be high, but this could pay off if the project has a long duration. Some of the cost

Photograph 13.14
Façade secured by rolled steel façade support system, through-ties and clamps over the top of the façade

can be recouped when the rolled steel is recycled or re-used. Some façades and properties are braced and supported for years until the money becomes available to develop the project. In some cases, developers wait until the economic conditions of the country make the scheme worth developing. In a few cases, planners and developers become locked in planning and design discussions that prolong the development periods excessively.

To ensure that the steel is fabricated correctly, it is essential that the survey from which the design of the support system is prepared is accurate. Systems that are fabricated for a specific building have a limited degree of adjustment and flexibility. Some adjustment using slots and multiple connection points can be designed into the system prior to fabrication.

Securing the façade to the temporary support structure

Figure 13.1 and Photograph 13.15 show a typical tubular scaffolding external support system used on a scheme at East Parade, Leeds. The independent tied scaffold, erected off a steel gantry to allow the pavement to remain open, acted as a vertical cantilever to which the façade was tied by means of horizontal and vertical scaffold tubing, timber wall plates and folding wedges (see detail in Figure 13.1).

Photograph 13.16 shows a tubular steel scaffold façade retention system. The system is three bays deep and extends out onto the road, where the kentledge is positioned to resist any lateral force applied to the façade. The detailed drawings of the façade retention system are shown in Figures 13.2–13.5. The kentledge shown in Figure 13.4 and Photograph 13.17 provides an anchor, a downward force, that prevents the façade falling inwards, and the three-bay scaffold supports and braces the wall, preventing it falling towards the road. The through-ties that clamp and hold the façade make use of ladder beams, shown in Photograph 13.18 and Figure 13.5. The ladder beams add extra rigidity and stiffness to the support, ensuring the façade is held securely. The scaffold on the external face of the façade is butted

Photograph 13.15
Tubular scaffolding external façade support system, East Parade, Leeds

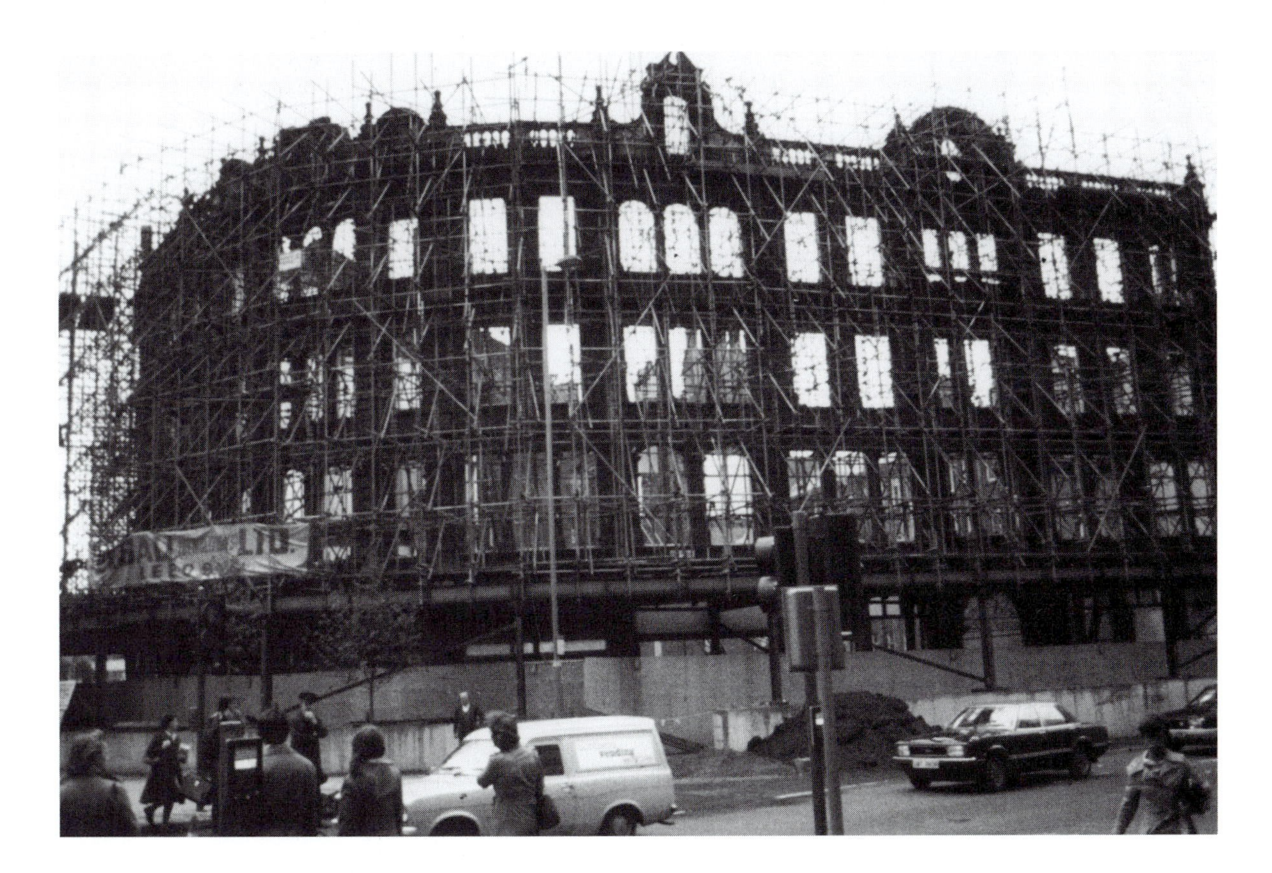

Figure 13.1
Typical example of wholly external temporary support system using tubular steel scaffolding components

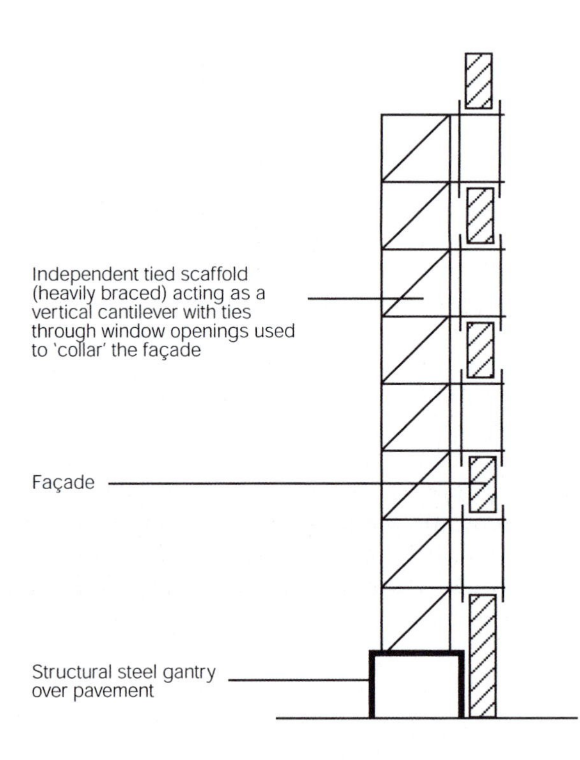

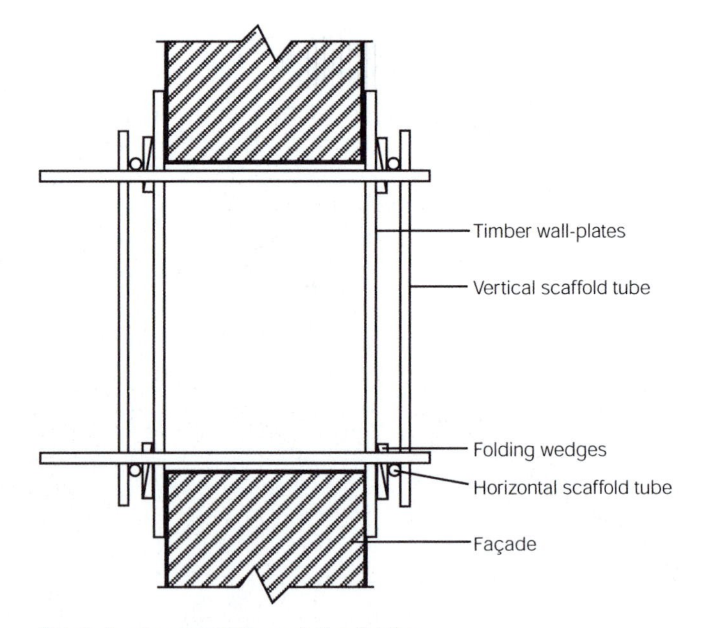

Detail showing use of through-ties, folding wedges and wall plates to 'collar' the façade

Photograph 13.16
(facing page, top)
York Place, Leeds: tubular steel façade retention, kentledge positioned on the road behind hording (courtesy of Paul Smith ROK Developments)

Photograph 13.17
(facing page, foot)
Concrete block kentledge for steel façade retention project (courtesy of Paul Smith ROK Developments)

QUARMBY
CONSTRUCTION COMPANY LTD
ROK
ROK
ROK

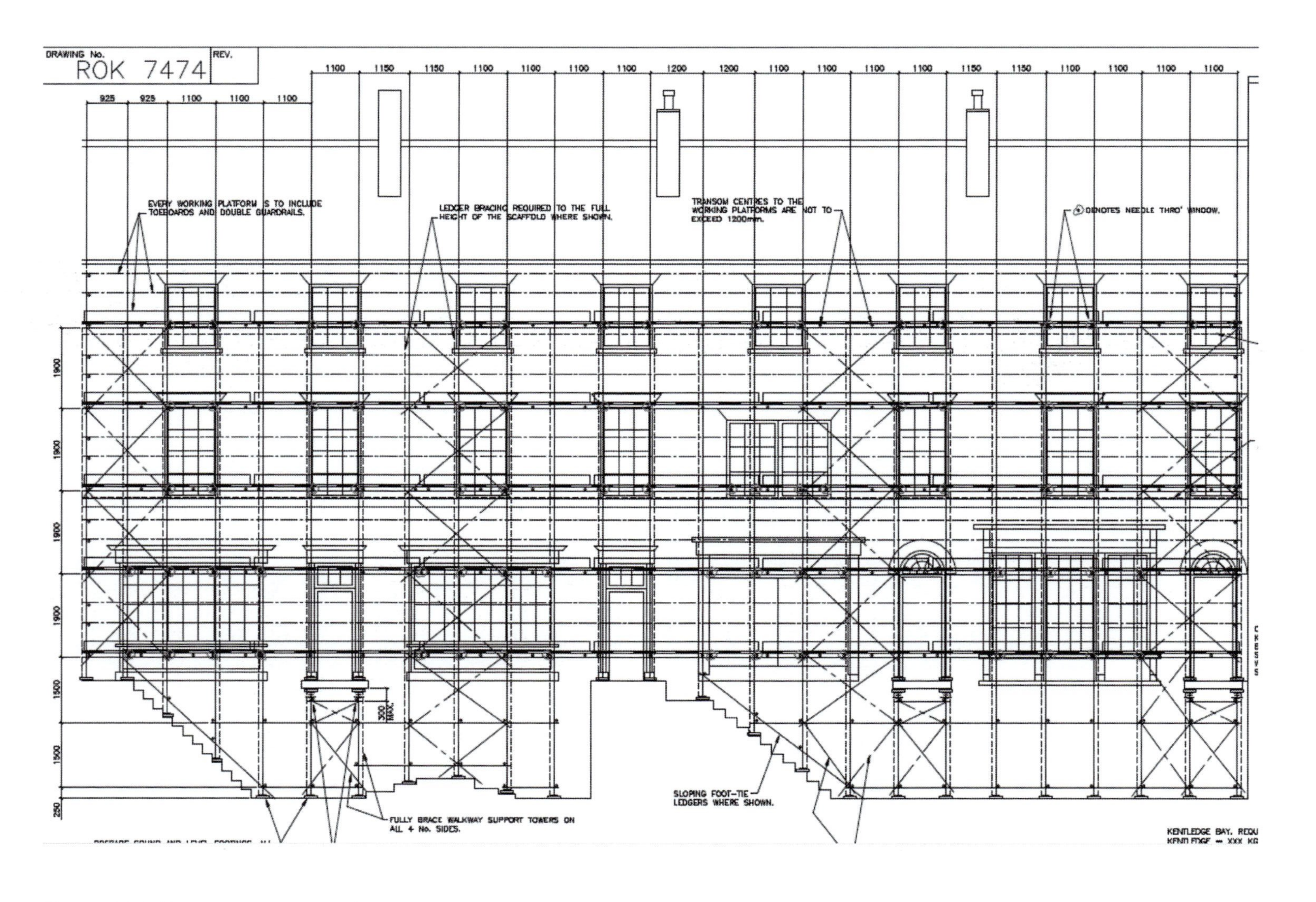

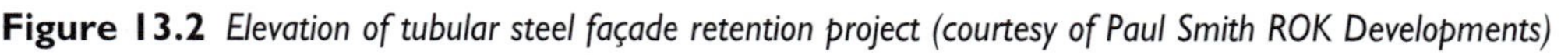

Figure 13.2 *Elevation of tubular steel façade retention project (courtesy of Paul Smith ROK Developments)*

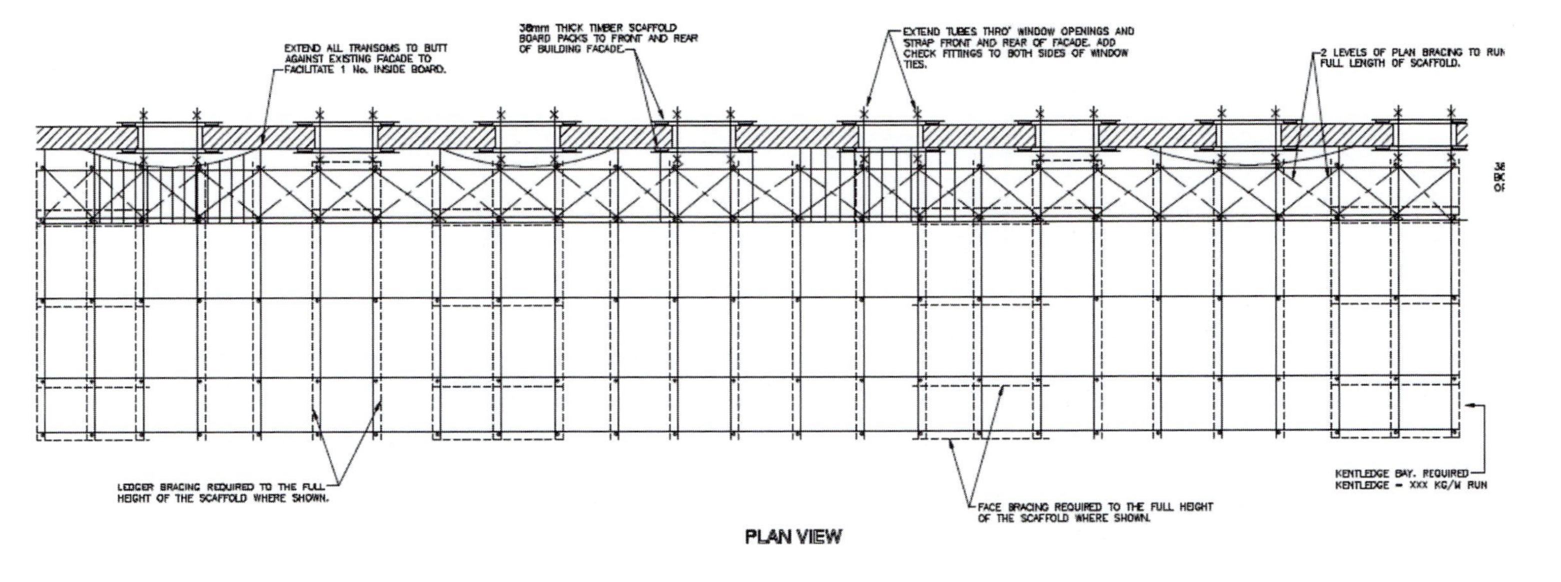
EXTEND ALL TRANSOMS TO BUTT AGAINST EXISTING FACADE TO FACILITATE 1 No. INSIDE BOARD.
38mm THICK TIMBER SCAFFOLD BOARD PACKS TO FRONT AND REAR OF BUILDING FACADE.
EXTEND TUBES THRO' WINDOW OPENINGS AND STRAP FRONT AND REAR OF FACADE. ADD CHECK FITTINGS TO BOTH SIDES OF WINDOW TIES.
2 LEVELS OF PLAN BRACING TO RUN FULL LENGTH OF SCAFFOLD.
LEDGER BRACING REQUIRED TO THE FULL HEIGHT OF THE SCAFFOLD WHERE SHOWN.
FACE BRACING REQUIRED TO THE FULL HEIGHT OF THE SCAFFOLD WHERE SHOWN.
KENTLEDGE BAY. REQUIRED KENTLEDGE = XXX KG/M RUN
PLAN VIEW

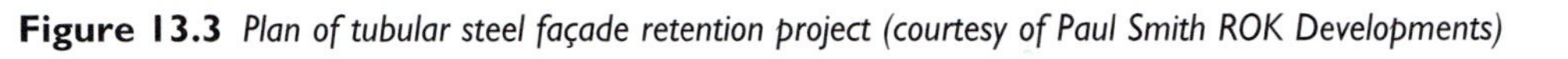
Figure 13.3 *Plan of tubular steel façade retention project (courtesy of Paul Smith ROK Developments)*

Figure 13.4
Cross-section of tubular steel façade retention project (courtesy of Paul Smith ROK Developments)

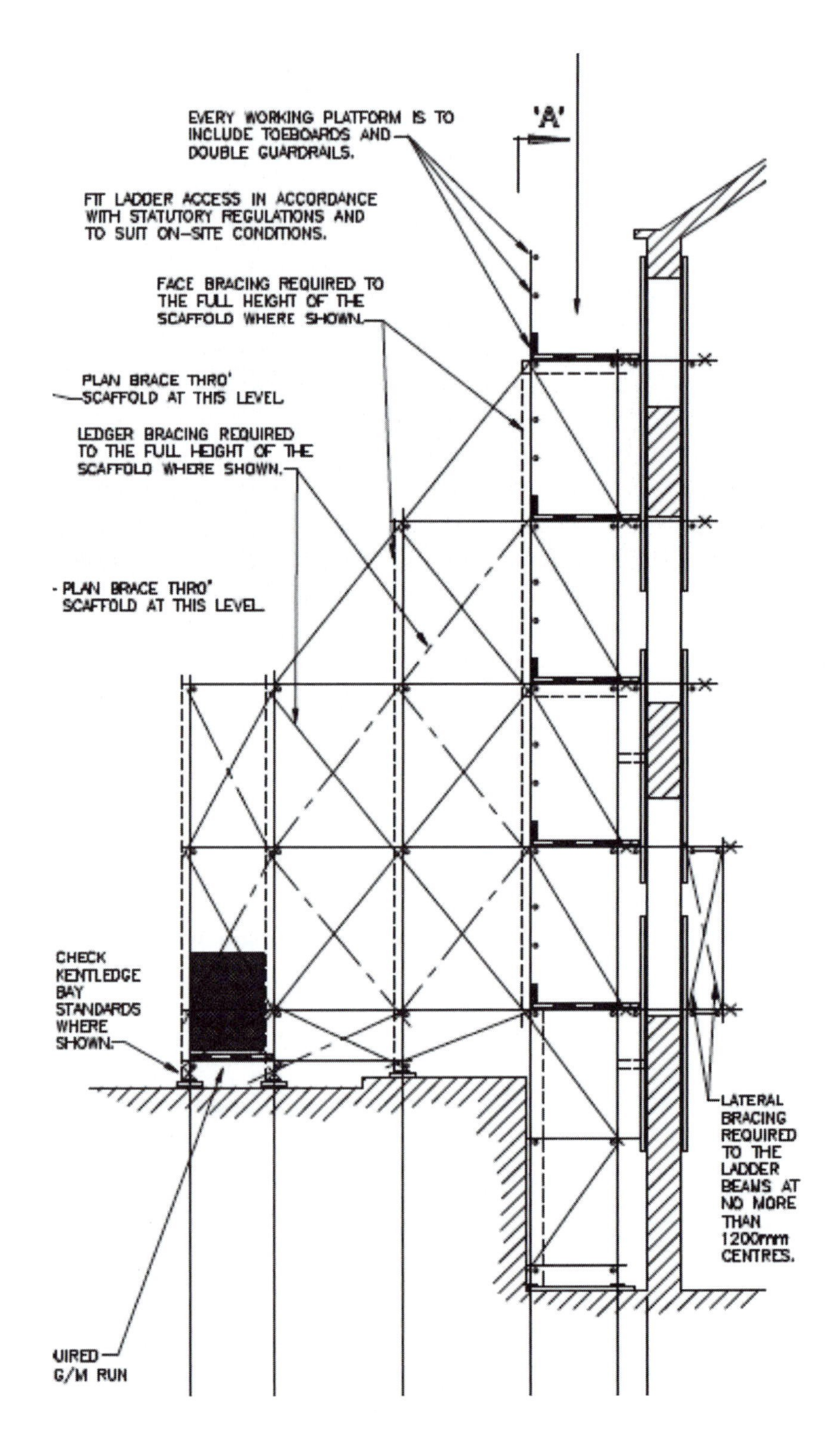

'A'
EVERY WORKING PLATFORM IS TO INCLUDE TOEBOARDS AND DOUBLE GUARDRAILS.
FIT LADDER ACCESS IN ACCORDANCE WITH STATUTORY REGULATIONS AND TO SUIT ON-SITE CONDITIONS.
FACE BRACING REQUIRED TO THE FULL HEIGHT OF THE SCAFFOLD WHERE SHOWN.
PLAN BRACE THRO' SCAFFOLD AT THIS LEVEL
LEDGER BRACING REQUIRED TO THE FULL HEIGHT OF THE SCAFFOLD WHERE SHOWN.
PLAN BRACE THRO' SCAFFOLD AT THIS LEVEL
CHECK KENTLEDGE BAY STANDARDS WHERE SHOWN.
LATERAL BRACING REQUIRED TO THE LADDER BEAMS AT NO MORE THAN 1200mm CENTRES.
UIRED
G/M RUN

Photograph 13.18
Through-tie and strap using deep steel ladder beams (courtesy of Paul Smith ROK Developments)

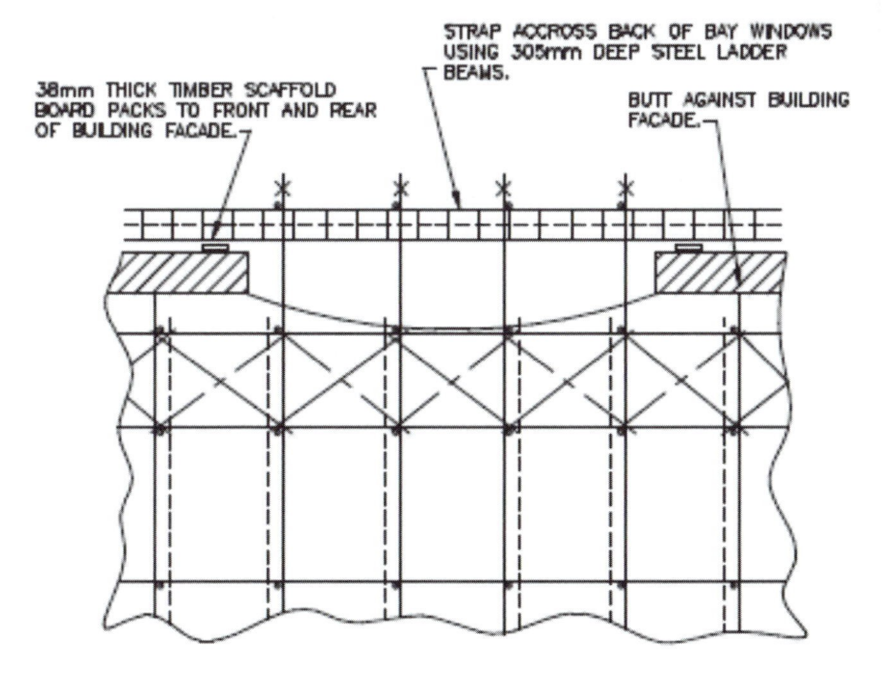
STRAP ACCROSS BACK OF BAY WINDOWS USING 305mm DEEP STEEL LADDER BEAMS.
38mm THICK TIMBER SCAFFOLD BOARD PACKS TO FRONT AND REAR OF BUILDING FACADE.
BUTT AGAINST BUILDING FACADE.

Figure 13.5
Through-tie and strap using deep steel ladder beams (courtesy of Paul Smith ROK Developments)

Photograph 13.19
Threaded eye, mechanically fixed into façade, restrains horizontal movement but allows vertical movement; however, it disrupts the face of the façade

up directly to the face of the brickwork (see Photograph 13.19). Photograph 13.19 shows a threaded eye that has been fixed into the face of the façade: the eye is slotted over a scaffold tube to prevent any load being transferred from the scaffold to the façade, but also to ensure that any lateral movement is resisted.

Figure 13.6 and Photographs 13.20 and 13.21 show the use of a new structure's new steel frame as a wholly internal 'temporary' support system on a scheme at Colmore Row, Birmingham. This is a fairly common method of providing internal support to a façade during a project and has the advantage of utilising part of the new structure, rather than a wholly temporary support system, to hold up the façade. The first bay of the new steel frame is erected by 'threading' it through openings made in the existing structure before demolition takes place. The façade is then tied back to the frame (see detail in Figure 13.6 and Photograph 13.21), following which demolition can take place. The remainder of the new steel frame is then erected.

Figure 13.7 shows an RMD proprietary external façade support system that uses standard 'slimshor' components. The principal advantage of the RMD system, in comparison with the alternatives, is that it uses far fewer components and is much quicker to erect and dismantle. The bracing of the wall is achieved by horizontal walings. The walings are tied through wall openings using threaded bar and collars. RMD Ltd also provides a full survey, design and erection service. Photographs 13.22 and 13.23 show a similar proprietary system in use on a façade retention scheme in Belfast.

Figure 13.8 and Photograph 13.24 show a wholly internal temporary support system used at St Paul's House, Park Square, Leeds. The local authority would not permit any encroachment onto the narrow footpaths and streets on three sides of the retained façade, and the contractor, therefore, had no choice but to design an internal temporary support system. This comprised four structural steel 'military trestle' towers in the centre of the building, which supported tubular steel box-

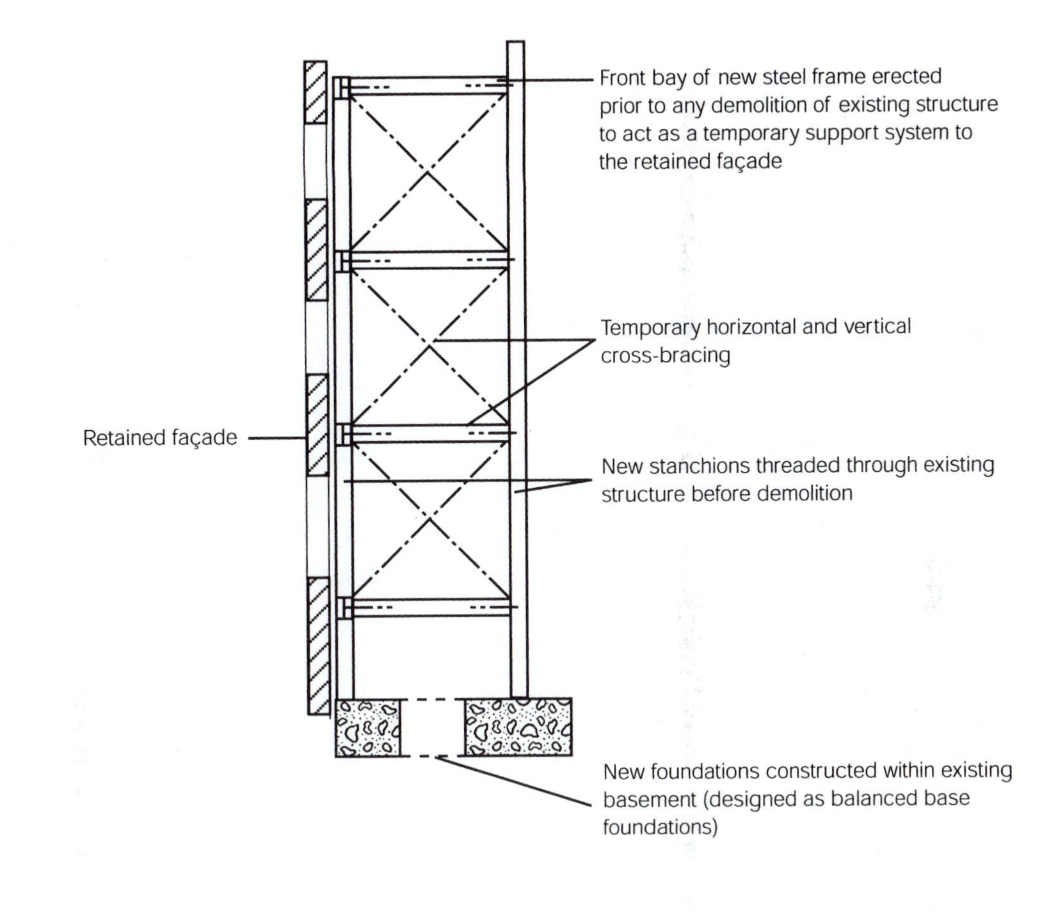

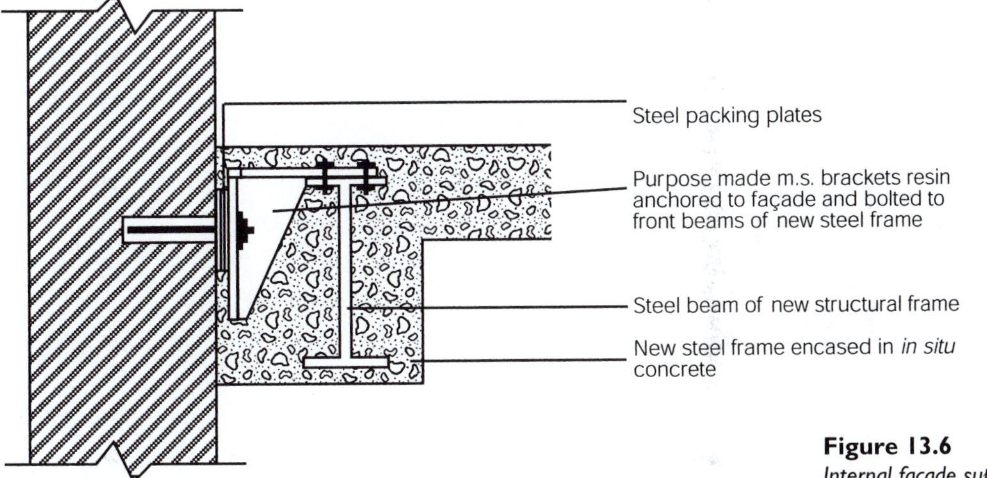

Detail showing connection between steel frame and façade

Figure 13.6
Internal façade support system employing part of new steel frame

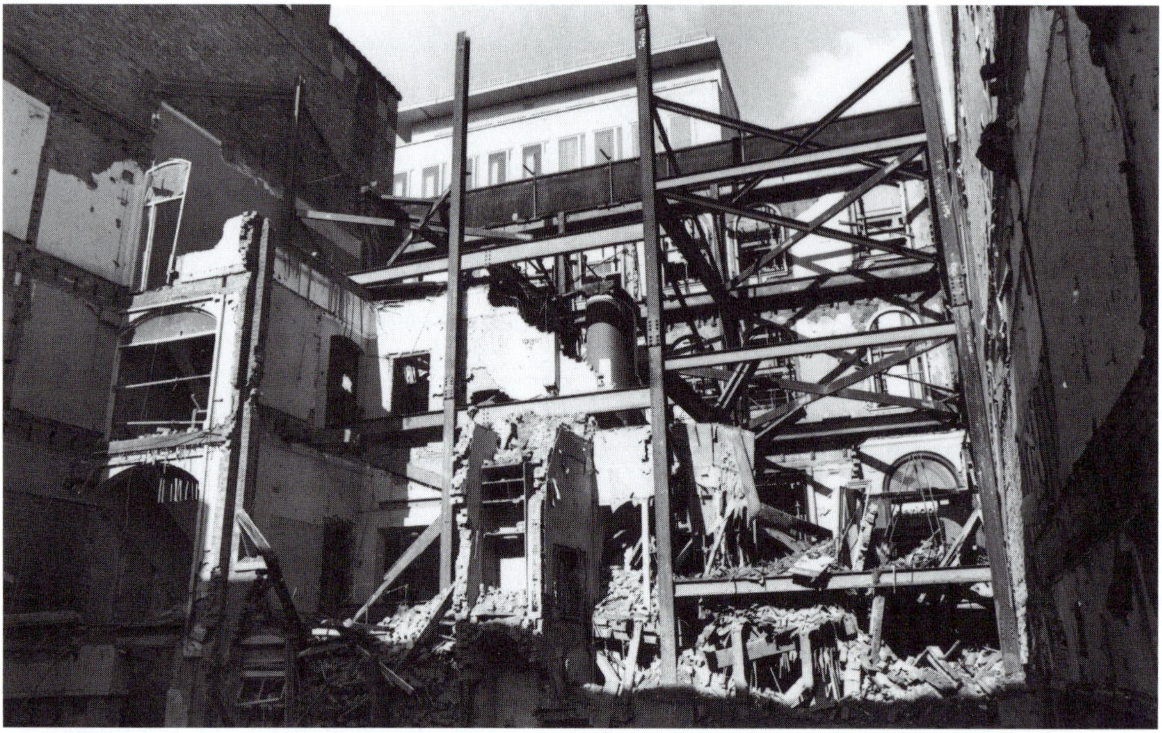

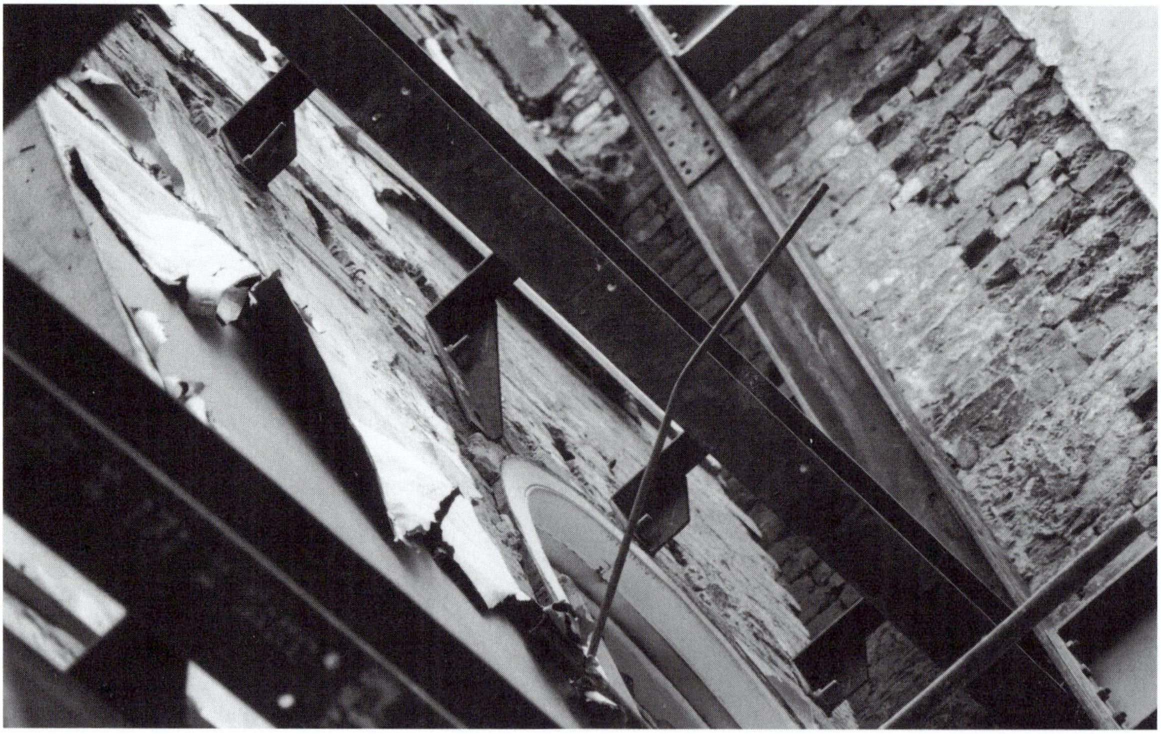

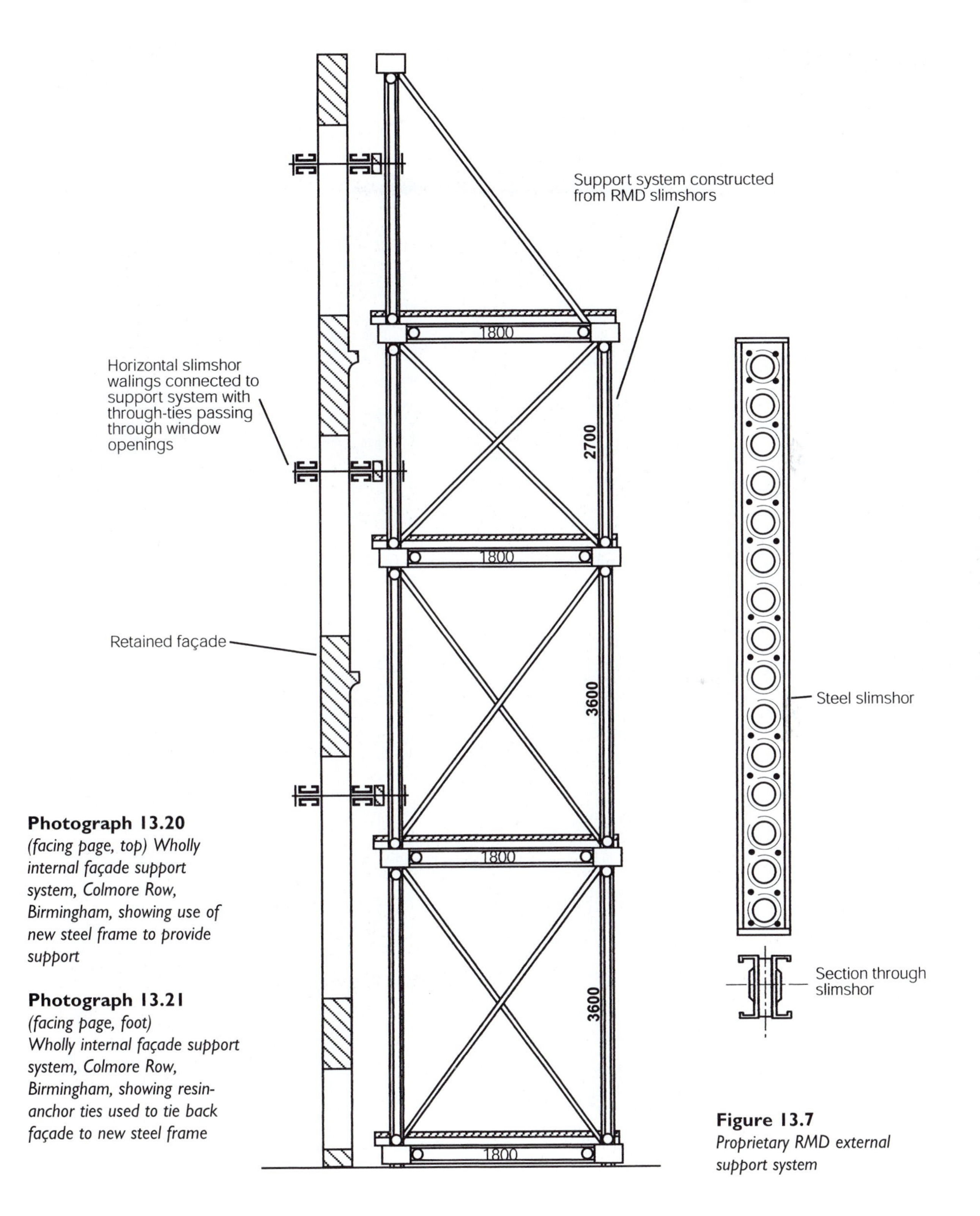

Photograph 13.20
(facing page, top) Wholly internal façade support system, Colmore Row, Birmingham, showing use of new steel frame to provide support

Photograph 13.21
(facing page, foot) Wholly internal façade support system, Colmore Row, Birmingham, showing resin-anchor ties used to tie back façade to new steel frame

Figure 13.7
Proprietary RMD external support system

Photograph 13.22 *Proprietary external façade support system, Belfast*

Photograph 13.23 *Proprietary external façade support system, showing through-ties and straps across the internal face of the building, Belfast*

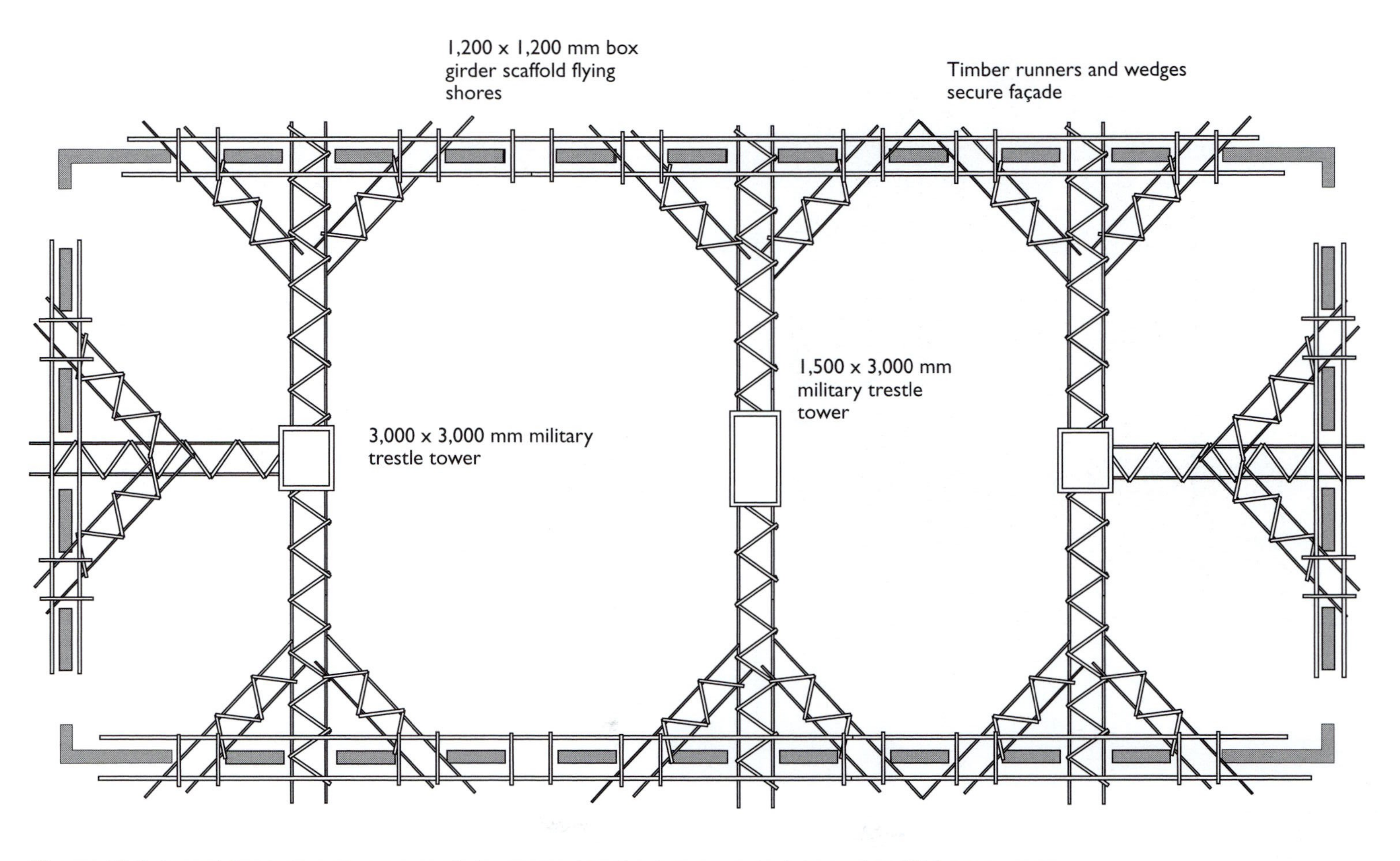

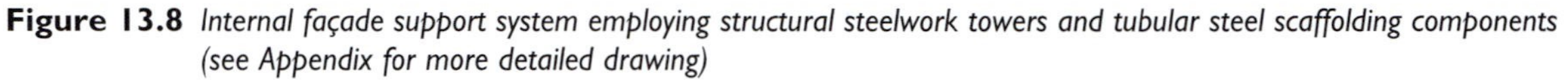

Figure 13.8 *Internal façade support system employing structural steelwork towers and tubular steel scaffolding components (see Appendix for more detailed drawing)*

Photograph 13.24
Wholly internal temporary support system, St Paul's House, Park Square, Leeds (note resin-anchored façade-tie angle in bottom left corner)

section flying shores at two levels. A combination of twin timber walings, folding wedges and steel through-bolts at the outer ends of the flying shores provided a supporting 'collar' to the façade throughout the project (see Figure 13.8). As previously stated, such a system places severe constraints on both the demolition and new construction works, since the temporary supports must be erected prior to commencement of the demolition and remain in place until the new structure is complete and the façade is tied back to it.

13.3 Permanent façade ties

A problem common to all façade retention schemes involves permanently tying back the retained façade to the new structure erected behind it. The lateral support formerly provided by the original internal structure must be replaced by some form of mechanical tie system between the façade and the new structure. These façade ties must fulfil a number of important functional requirements:

- They must effectively hold back the retained façade and prevent any outward movement away from the new structure.
- They must not transmit any vertical loads from the new structure to the façade (since the façade should not normally act as a load-bearing element in the new design).

- They must be capable of accommodating any predicted differential settlement between the new structure and the retained façade without causing damage to the ties themselves, the façade or the new structure.

The most widely used solution to this problem is to employ some form of resin anchor system, which involves anchoring steel tie-bars into the façade masonry with a rapid-setting resinous mortar and then connecting them directly or indirectly to the new structure. The anchoring of the tie-bars into the façade masonry may be executed using either a resin cartridge or pre-mixed resin. With the resin cartridge method, a pre-formed plastic or glass phial, containing the unmixed ingredients capable of forming the rapid-setting resinous mortar, is inserted into a pre-drilled hole in the masonry. The tie-bar is then spun into the hole using a drilling tool, breaking the cartridge and mixing its ingredients to form the mortar, which anchors the bar firmly into the masonry. The alternative method is to pump pre-mixed resinous mortar into a pre-drilled hole and either spin or push the tie-bar into it. The resin-anchoring procedure is illustrated in Figure 13.9.

The connections between the projecting tie-bars and the new structure can be effected in a number of ways (Figure 13.10). Methods used include casting the projecting tie-bars directly into the edges of the new floor slabs, or using indirect connections where steel angles, bolted to the edges of the new structure, are fastened to the projecting resin-anchored tie-bars using locknuts. Figure 13.10 (bottom left-hand corner) shows a typical indirect steel angle façade-tie system used on a scheme at St Paul's House, Leeds.

As an alternative to using resin-anchored tie-bars, various forms of through-tie may be used. These comprise steel bars passing completely through the façade and secured to steel plates on the external face. The inner ends of the tie-bars project from the internal face of the façade and are secured to the new structure directly

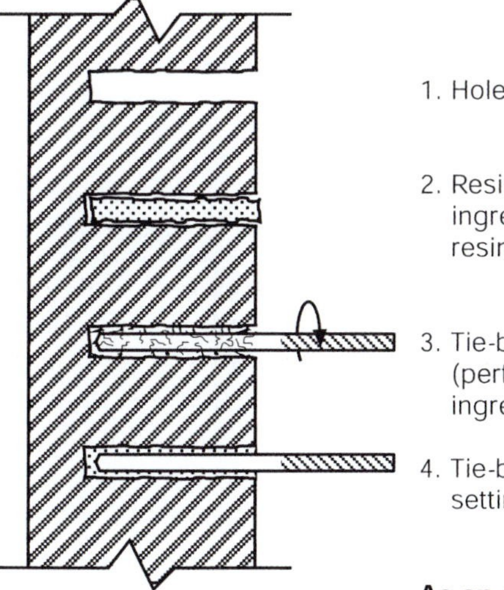
1. Hole drilled into masonry
2. Resin cartridge (containing unmixed ingredients required to form rapid-setting resinous mortar) inserted into hole
3. Tie-bar spun into hole with drilling tool (perforates cartridge and mixes ingredients)
4. Tie-bar anchored into masonry by rapid setting resinous mortar
As an alternative to resin cartridges pre-mixed resinous mortar may be pumped into hole and tie-bar pushed in

Figure 13.9
Resin-anchoring technique for installing façade ties

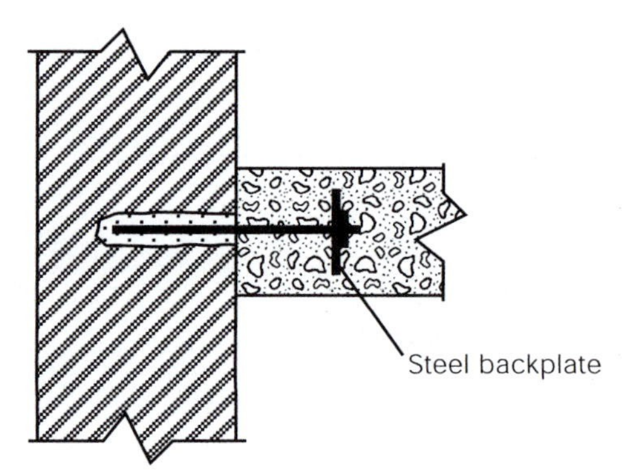

Projecting resin anchored tie-bars cast directly into new concrete floor slabs with steel backplates to give lateral restraint

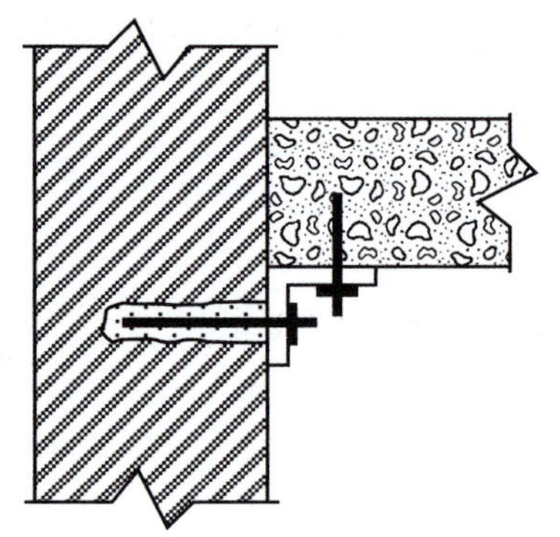

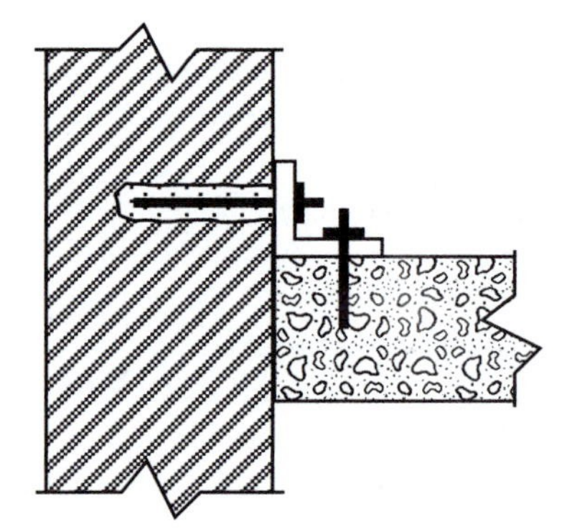

Underside of slab

Upper surface of slab

Steel angles bolted to new floor slab and resin-anchored to façade masonry

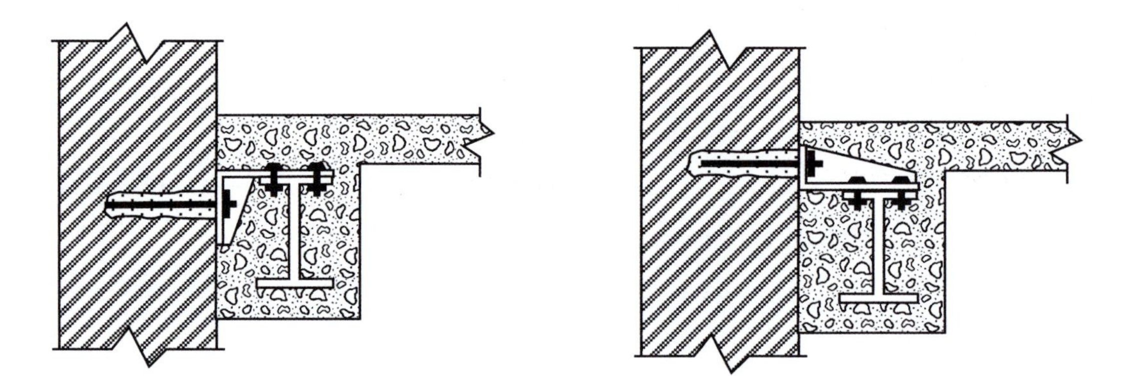

Purpose made steel angles bolted to beams of new steel frame and resin-anchored to façade masonry

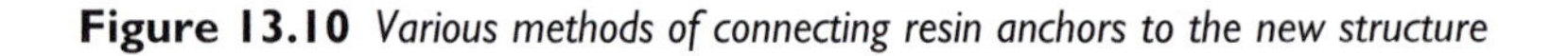
Figure 13.10 *Various methods of connecting resin anchors to the new structure*

Photograph 13.25
In situ concrete floor tied to façade. Façade secured by bolts fixed by resin into the wall and restraining channels cast into the floor, see Figure 13.11.

or indirectly in the same way as resin-anchored tie-bars. One problem with this through-tie method involves the concealment of the outer ends of the tie-bars and anchor plates, which is relatively easy with stuccoed or rendered façades, but more difficult with masonry or brickwork.

As already stated, the permanent fixing must allow vertical movement so that the new internal structure can settle or move independently of the façade, while at the same time still restraining any horizontal movement of the façade. Halfen produce a range of fixings that are designed to accommodate such movement (www.halfen.co.uk). Photographs 13.26–13.28 show similar fixings being used on the East Brook Hall, Bradford, façade retention project. Figure 13.11 shows how the bolts lock into the channel, providing the horizontal and lateral restraint while still allowing vertical movement.

A hole is drilled into the façade, and resin is inserted. The bolts are then pushed into the resin-filled hole and rotated to ensure a full covering of resin is received by the bolt. The head of the bolt or stud is left proud so that compression board can be positioned between the structure and façade, and the channel can be inserted over the bolt. When the channel is inserted over the bolt, insulation is pushed into the channel to ensure that concrete does not fill the slot when the concrete structure is poured around the fixing.

A video is available of the East Brook Hall project, narrated by Architect Ed Jagger (/www.youtube.com/watch?v=bPOiSSaA9KI).

Photograph 13.27
(below) Expansion board and restraint fixings in place prior to concrete floor being cast

Photograph 13.26 *(above) Halfen channel movement tie: the channel slots over bolt or stud heads. The polystyrene is inserted into the channel to stop concrete and debris entering the channel and preventing movement*

Photograph 13.28
The concrete floor will encase the legs of the fixings and tie the façade back to the structure

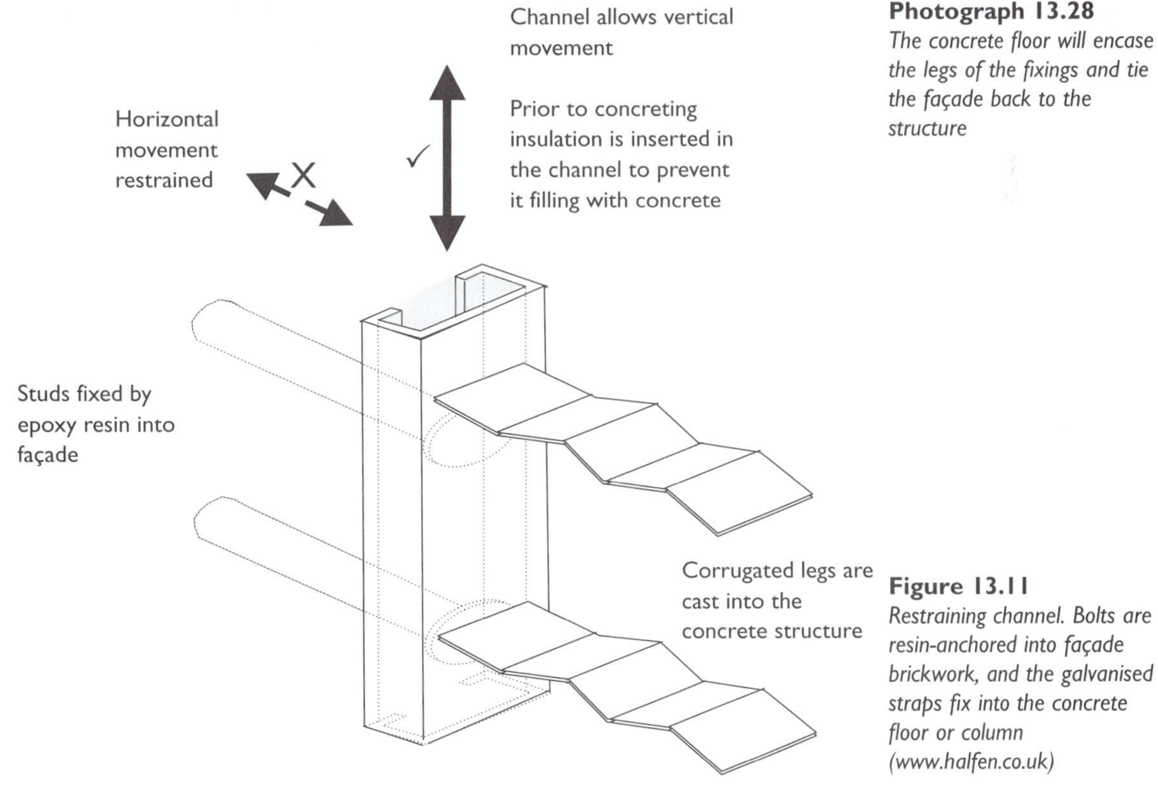
Channel allows vertical movement
Prior to concreting insulation is inserted in the channel to prevent it filling with concrete
Horizontal movement restrained
X
✓
Studs fixed by epoxy resin into façade
Corrugated legs are cast into the concrete structure

Figure 13.11
Restraining channel. Bolts are resin-anchored into façade brickwork, and the galvanised straps fix into the concrete floor or column (www.halfen.co.uk)

13.4 Differential settlement

It is essential in façade retention schemes that the detailing at the junction of the new structure and the existing façade allows settlement of the former to occur. A rigid connection between the new structure and the façade could result in potentially serious structural damage in the event of settlement. If such settlement has been predicted, its extent should be calculated using data from the subsoil investigations. The façade ties and interface detail between the façade and the new structure should then be designed to allow this settlement to take place without causing damage to either the structures or the façade ties themselves. One of the most effective ways of achieving this is to use an indirect steel angle tie with a vertical slotted hole through which the resin-anchored tie-bar passes, as shown in Figure 13.12. Alternatively the restraining slot shown in Figure 13.11 achieves the same objective. The slotted hole in the vertical leg of the angle enables the new structure to settle without causing damage.

The interface treatment between the new structure and the retained façade must also be considered where settlement of the former has been predicted. The interface between the new structure and the retained façade is usually at the outer faces of the new structure's columns and/or the edges of its floor slabs. The most common method of allowing for differential movement here is to provide some form of slip surface between the new and existing elements that will prevent bonding of the two surfaces. The slip surface may comprise single or multiple layers of dense polythene or similar material, or a thin layer of fibreboard. Another accepted interface treatment is to leave a narrow gap where the new structure meets the façade to ensure that settlement may take place without damage. Typical interface treatments are shown in Figure 13.13; also see Photographs 13.26 and 13.27 for an illustration of fibreboard positioned against the façade prior to the new concrete structure being poured.

Temporary façade restraint: bespoke design

The more ornate and original the façade, the more thought that is needed when restraining the existing wall (see Photographs 13.29–13.32). The design of façade

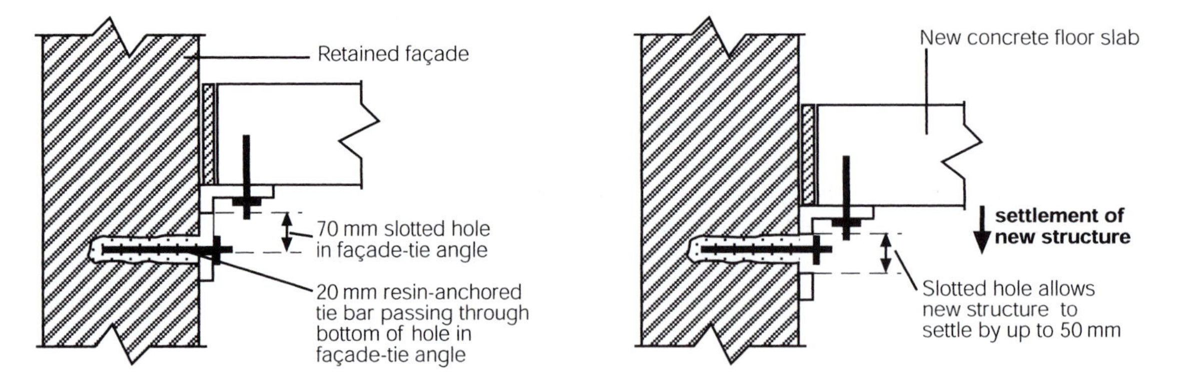

Figure 13.12 *Restraining channel. Bolts are resin-anchored into façade brickwork, and the galvanised straps fix into the concrete floor or column (www.halfen.co.uk)*

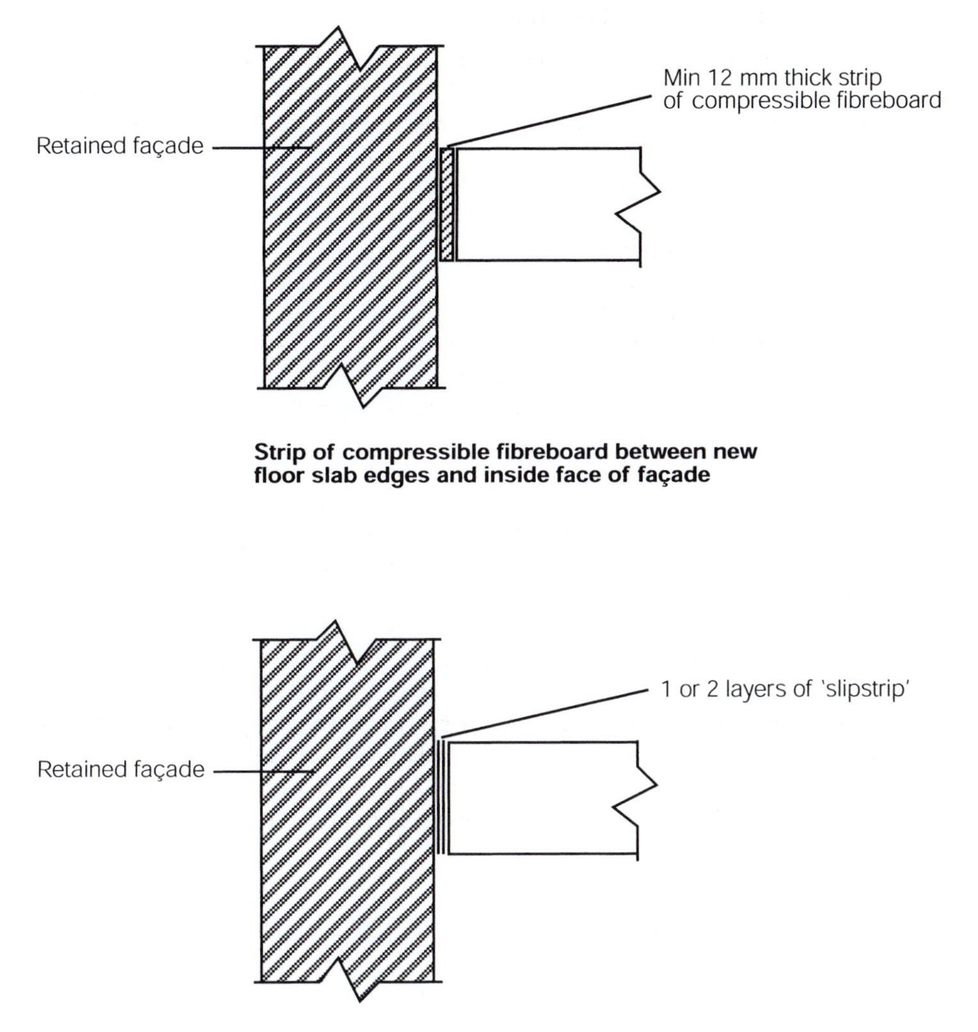

One or two layers of 'slipstrip' (e.g. dense polythene)
between new floor slab edges and inside face of façade

Figure 13.13
Interface details permitting settlement of new structure

retention systems is not typical, and each building needs to be considered in isolation. Features, details and nuances that give the building character must be taken into account. The best method of securing the structure is for the temporary support system to pass through windows and around features and to grip the top of the walls. Trapping the structure in this way reduces the possibility of damage. Drilling in through-bolts obviously disrupts the feature that the project is attempting to maintain.

The façade retention only restrains the façade from horizontal movement. In Photograph 13.33, there has been some differential movement at the base of the façade, which has resulted in cracks in the façade, movement affecting the stone sills and distortion of timber lintels, as well as some delamination of the solid brick

Photograph 13.29 *Tubular scaffold system: through-ties and clamps*

Photograph 13.30
East Brook Hall, Bradford: tubular scaffold system: clamps and through-ties secure ornamental features

Photograph 13.31
East Brook Hall, Bradford: each ornamental feature may need to be individually supported and braced

Photograph 13.32 *East Brook Hall, Bradford: scaffolding boards are packed and wedged to ensure a firm fixing*

wall. Where vertical movement is expected, heavy propping and pinning are necessary to restrain any vertical movement and distribute the load to strata that can support the wall. The part of the façade shown in Photograph 13.33 had to be taken down (Photograph 13.34) and then rebuilt (Photograph 13.35). The façade was carefully deconstructed brick by brick; each part of the façade was photographed, numbered and rebuilt. The cause of the differential settlement at the base of the façade was not properly determined. The scaffold façade retention system remained stable throughout the movement and did not fail. Possibly the vibrations of the new construction work or removal of ground too close to the footings, undermining the support, were the main cause. It is unusual for the façade to see such extensive movement. Once the floors, roof and other internal elements are removed, the loads on the façade are substantially reduced. Where it is expected that the footings of the building are inadequate, it is normal to underpin the foundations at an early stage in the construction programme, usually as soon as the façade support systems and any necessary propping are in place.

The façade support system needs to trap and restrain all of the features on the façade. Once the building structure is removed, no lateral support exists, apart from the retention system. The top of the façade must also be restrained. General disturbance through demolition operations and exposure to high winds can make the top of the façade particularly vulnerable. Photographs 13.36–13.39 show different methods of trapping and retaining the façade. There are often ornamental

Photograph 13.33
Movement at the base of the façade has caused settlement, cracks and delamination of the brickwork

Photograph 13.34 *Façade removed brick by brick*

Photograph 13.35 *Completed façade after rebuild*

Photograph 13.36
Specially made rolled steel façade retention scheme; clamps secure the top of the façade

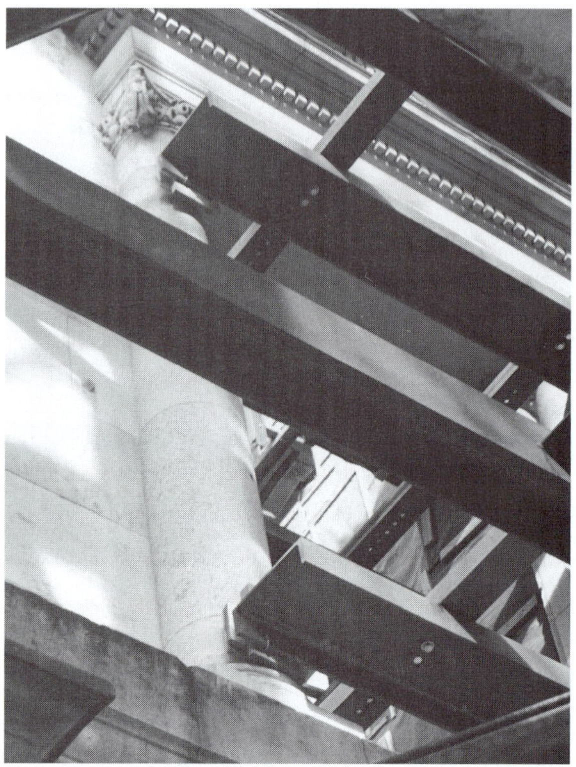

Photograph 13.37
Rolled steel façade retention adjustment holes in steel, and packing ensure a tight fit is maintained

features that sit on the top of the wall, such as coping or cornice stones. The wall may project out from the main face by means of corbel stones or bricks. The features on top of parapet walls need to be checked to ensure that they are structurally sound and should be held firmly in place by the façade retention system. To trap and support cornices or parts of parapet walls that corbel out, bespoke arrangements often need to be constructed, similar to those shown in Photographs 13.36 and 13.39. Legs of the rolled steel framework reach over the façade, securing the parapet; they also wrap around the ends of the wall, clamping the coping-stones. The legs that extend over the wall are also tied into the steel that penetrates through the window opening, making the top of the façade secure (Photograph 13.39).

With bespoke rolled steel façade retention systems, adjustment must be built into the supporting structure so that the façade can be properly clamped and supported. Support structures and scaffolding must be regularly checked. Wind and vibration can disturb the wall or support, and there may be a need to readjust and tighten the clamps and supports. If any movement does occur, the adjustment built into the support can be used. Photograph 13.37 shows multiple holes in the steel

Photograph 13.38
External rolled steel façade retention system

Photograph 13.39
Rolled steel façade retention system: protective building paper prevents moisture penetration and damage from frost; through-ties and clamps at the top of the façade can clearly be seen

that has been threaded through the wall opening. The adjustment created by multiple fixing points ensures that the structure can be trapped, secured and supported.

It is necessary to pack the space between the steel support and the façade. The timber packing serves to ensure firm contact between the temporary structure and façade. As the timber packing is softer than the brick and steel, it will deform slightly, helping to accommodate slight variations in the surface of the wall; this will help the packing grip the wall. The softer timber also prevents damage to the face of the façade. The hard surface of the steel would scratch or chip the façade if it came into direct contact with the surface.

13.5 Foundation design

It is essential that the design and construction of the new structure do not adversely affect the stability of the retained façade, and this is of particular importance at foundation level. The foundations to many historic buildings are often

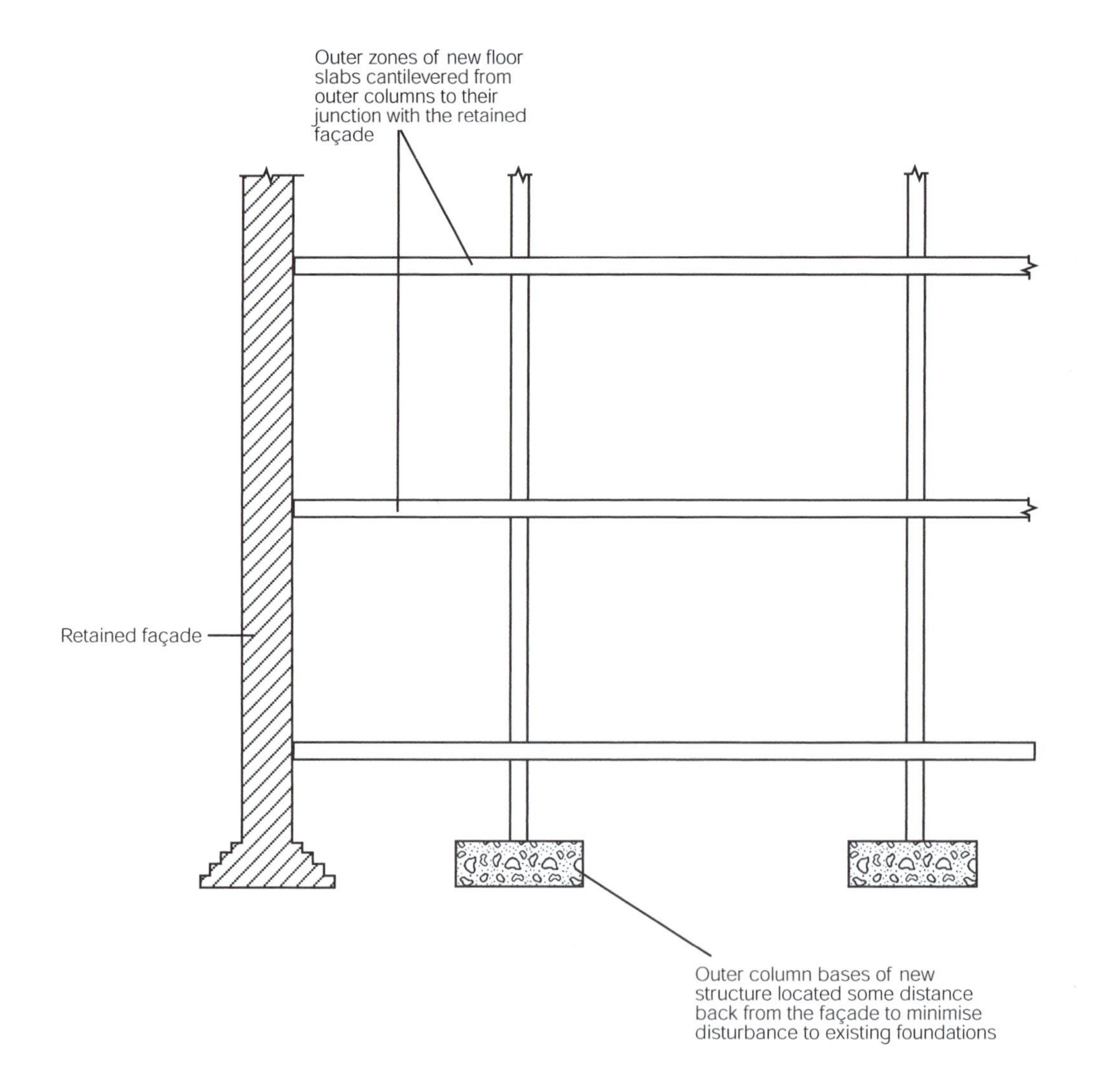
Outer zones of new floor slabs cantilevered from outer columns to their junction with the retained façade
Retained façade
Outer column bases of new structure located some distance back from the façade to minimise disturbance to existing foundations

Figure 13.14
Foundation design to prevent damage to the stability of the retained façade

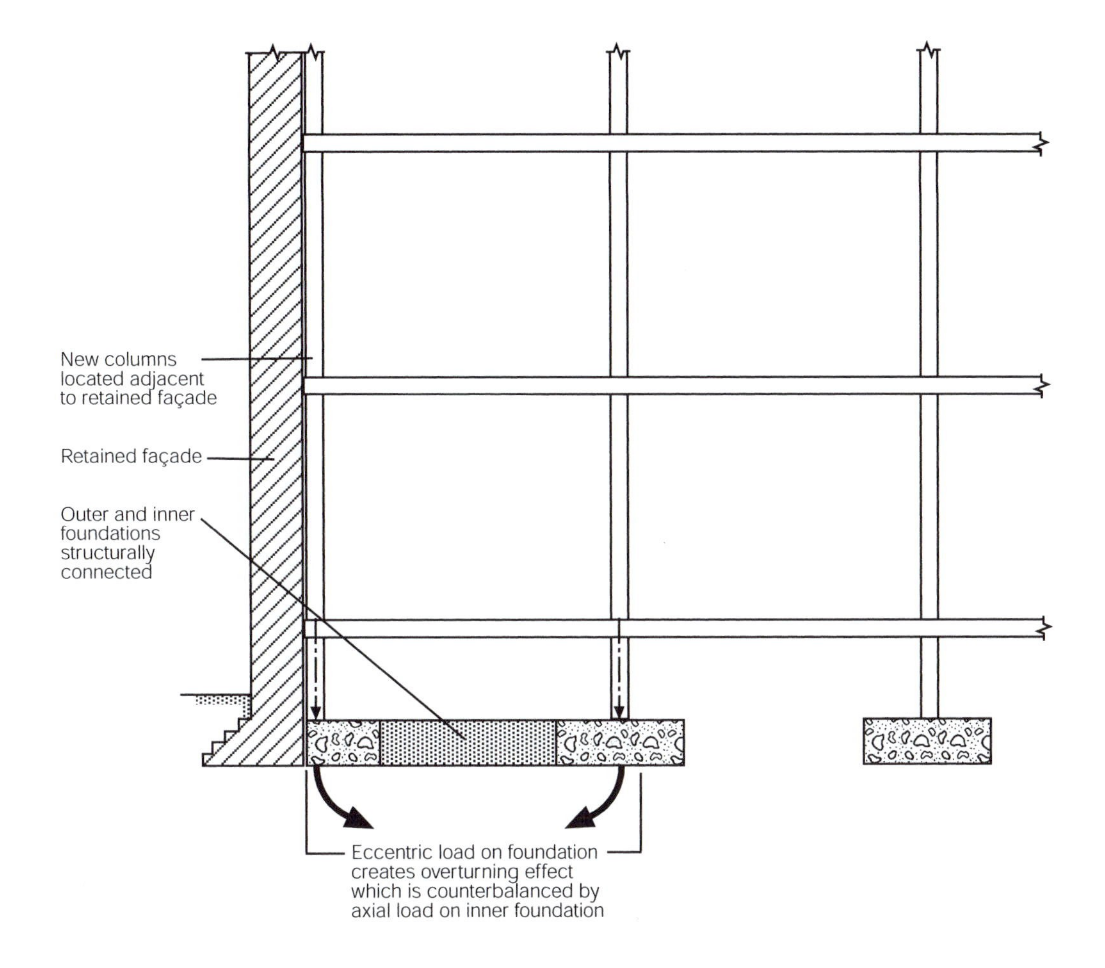
New columns located adjacent to retained façade
Retained façade
Outer and inner foundations structurally connected
Eccentric load on foundation creates overturning effect which is counterbalanced by axial load on inner foundation

Figure 13.15
Balanced-base foundations used to prevent damage to the stability of the retained façade

found to be weak and unstable and are therefore vulnerable to any disturbance caused by new construction works. The two most common solutions used to overcome this problem are described here.

The first, and probably the simplest, method is to locate the outer column bases of the new structure some distance back from the retained façade and to cantilever the new floors from these columns to their junction with the façade, as shown in Figure 13.14. This ensures that no new foundations are constructed adjacent to those of the façade, therefore ruling out any constructional disturbance and subsequent harmful effects caused by the new structure's loads.

The second method, illustrated in Figure 13.15, is employed when the design of the new structure requires that some columns must be located immediately adjacent to the retained façade. In order to minimise disturbance at the base of the façade, the new column bases are constructed immediately adjacent to it, but do not undermine it. The new columns, which are also immediately adjacent to the façade, inevitably subject these bases to eccentric loading and an overturning effect, which must be counteracted in some way if the foundations are not to fail. The eccentric loads are 'balanced' by structurally connecting these bases to the axially loaded bases of an inner line of columns. This use of 'balanced-base' foundations counteracts the overturning effect that the columns adjacent to the façade have on their bases.

In certain façade retention schemes, some undermining and underpinning of the existing foundations are inevitable. In these cases it is of paramount importance that the new foundations are designed, and their construction is phased, so as to minimise any adverse effects they may have on the stability of the retained façade.

References

Goodchild, S. L. and Kaminski, M. P. (1989) 'The retention of major façades', *The Structural Engineer*, 18 April: 67: 8.

Highfield, D. (1982) *The Construction of New Buildings Behind Historic Façades: The Technical and Philosophical Implications*, M.Phil thesis, University of York.

Highfield, D. (1991) *The Construction of New Buildings Behind Historic Façades*, London: E. & F.N. Spon.

Lazarus, D., Bussell, M. and Ross, P. (2003) *Retention of Masonry Façades – Best Practice Guide C579*, London: CIRIA.

Richards, J. (1994) *Façadism*, London: Routledge.

University of Bath, Department of Architecture and Building Engineering (1985) *Building Appraisal Maintenance and Preservation: Symposium Proceedings*, University of Bath.

Index of products and systems

CHAPTER 14

Fire protection

Supalux, Masterboard, Vermiculux, Vicuclad
Promat UK Limited, The Sterling Centre, Eastern Road, Bracknell, Berkshire
RG12 2TD
Tel: 01344 381300
Fax: 01344 381301
www.promat.co.uk

Mandolite CP2
Mandoval Douglas Drive, Catteshall Lane, Godalming, Surrey GU7 1JX

Aaronite, Aaronite House, Unit 1 Lakes Court, Newborough Road, Needwood,
Burton-on-Trent, Staffs, DE13 9PD
Web address not available
Tel: +44 (0) 1283 575901
Fax: +44 (0) 1283 575911
email: enquiries@aaronite.com
www.aaronite.com

Nullifire
Nullifire Ltd, Torrington Avenue, Coventry CV4 9TJ
Tel: 01203 855000
Fax: 01203 469547
email: protect@nullifire.com
www.nullifire.com

Foamed Perlite Insulation
Silvaperl, Albion Works, Ropery Road, Gainsborough DN21 2QB
Tel: 01427 610 160
Fax: 01427 811 838
www.william-sinclair.co.uk

Sprayed Limpet Mineral Wool – GP Grade
Thermica Ltd, Themework house, Vulcan Street, Clough Road, Kingston-upon-Hull HU6 7PS
Tel: 01482 348771
Fax: 01482 444626
www.thermica.co.uk

Finishes

Thistle Renovating Plaster and Finish, Thistle Dri-Coat, Thistle Finishes, Gyproc Wallboards, Gyproc Dri-Wall MF System, Dri-Wall Adhesive and Sealant, Gyproc Thermal Laminates, Gyproc Dri-Wall TL System, Gyproc Dri-Wall RF System, Gyproc Gypliner Wall Lining System
British Gypsum Ltd, East Leake, Loughborough, Leicestershire LE12 6HX
Tel: 08705 456 123 (+44 8705 456 123)
Fax: 08705 456 356 (+44 8705 456 356)
email: bgtechnical.enquiries@bpb.com
www.british-gypsum.com

Evo-Stik Floor Level and Fill
Bostik Ltd, Ulverscroft Road, Leicester LE4 6BW
Tel: 01785 272727
www.bostik.com

Thermal insulation

Thermalath
BRC Special Products, Carver Road, Astonfields Industrial Estate Stafford ST16 3BP
Tel: 01785 222288
Fax: 01785 240029
email: enquiries@brcsp.co.uk
www.brc-special-products.co.uk/

Isowool Timber Frame Batts, Gyproc Thermal Board, Gyproc Thermal Board Plus, Gyproc Thermal Board Super, Isowool General Purpose Roll, Gyproc Wallboard Duplex
British Gypsum Ltd, East Leake, Loughborough, Leicestershire LE12 6HX
Tel: 08705 456 123 (+44 8705 456 123)
Fax: 08705 456 356 (+44 8705 456 356)
email: bgtechnical.enquiries@bpb.com
www.british-gypsum.com

Roofmate SL, Roofmate LG
Dow Chemical Company Ltd, Dow Building Solutions, Diamond House, Lotus Park, Kingsbury Crescent, Staines TW18 3AG
Tel: 020 3139 4000
Fax: 020 3139 4013
www.dow.com *or* www.building.dow.com/europe/uk

Styroliner LK, Styrofloor
Panel Systems Ltd, Welland Close, Parkwood Industrial Estate, Rutland Road,
Sheffield S3 9QY
Tel: 0114 249 5635
Fax: 0114 278 6840
www.panelsystemsgroup.co.uk

Rockwool RockShield Rigid Slabs, Rockwool EnergySaver Cavity Wall Insulation, Rockwool Rollbatts, Rockwool EnergySaver Blown Loft Insulation, Rockwool Hard Rock
Rockwool Limited, 26–28 Hammersmith Grove, London W6 7HA
Tel: 0845 2412586
Fax: 0845 2412587
www.rockwool.co.uk

Expolath Polystyrene, Terratherm PSB, Terratherm PSM
Weber Ltd, Dickens House, Maulden Road, Flitwick, Bedford MK45 5BY
Tel: 0870 333 0070
Fax: 01525 71 8988
www.netweber.co.uk

Gyproc SoundBloc Wallboard, Isowool Batts, Gyproc SI Floor, Isowool General Purpose Mineral Wool Roll, Gyproc Independent Wall Lining System
British Gypsum Ltd, East Leake, Loughborough, Leicestershire LE12 6HX
Tel: 08705 456 123 (+44 8705 456 123)
Fax: 08705 456 356 (+44 8705 456 356)
email: bgtechnical.enquiries@bpb.com
www.british-gypsum.com

Acoustic insulation

Akustofloor
Panel Systems Ltd, Welland Close, Parkwood Industrial Estate, Rutland Road
Sheffield S3 9QY
Tel: 0114 249 5635
Fax: 0114 278 6840
www.panelsystemsgroup.co.uk

Gyproc Thermal Board, Gyproc Thermal Board Plus, Gyproc Thermal Board Super, Gyproc Duplex Wallboard
British Gypsum Ltd, East Leake, Loughborough, Leicestershire LE12 6HX
Tel: 08705 456 123 (+44 8705 456 123)
email: bgtechnical.enquiries@bpb.com
www.british-gypsum.com

Basements, tanking and waterproofing

Tough-Cote Superflex RW2

Carrspaints, 2E Eagle Road, North Moons Moat, Redditch, Worcestershire
B98 9HF
Tel: 01527 599 460
email: info@carrspaints.com
www.carrspaints.com

Evode Cementone Water Seal

Bostik Ltd, Ulverscroft Road, Leicester LE4 6BW
Tel: 01785 272727
www.bostik.com

Newlath 2000, Newton System 500

John Newton & Co. Ltd, 12 Verney Road, London, SE16 3DH
Tel: 020 7237 1217
www.newton-membranes.co.uk
www.newtonbasementwaterproofing.co.uk

Liquid Plastics K501, Monolastex Smooth, Monolastex Textured

Liquid Plastics Ltd, Iotech House, Miller Street, Preston, Lancashire PR1 1EA
Tel: 01772 255022
email: info@liquidplastics.co.uk
www.liquidplastics.co.uk

RIW Liquid Asphaltic Composition

RIW Ltd, Arc House, Terrace Road South, Binfield, Bracknell, Berkshire
RG42 4PZ
Tel: 01344 397777
www.riw.co.uk

RIW Flexiseal, Thoroseal

HSC UK Ltd, 19 Broad Ground Road, Lakeside, Redditch, Worcestershire
B98 8YP

Thoro Consumer Products

BASF Construction Chemicals, LLC, 23700 Chagrin Blvd, Cleveland, Ohio 44122
Tel: 01527 505100
Fax: 01527 510299
www.thoroproducts.com

Spry Seal System

Spry Products, 64 Nottingham Road, Long Eaton, Nottingham NG10 2AU
Tel: 01706 831223
www.spryproducts.co.uk

Timber treatment and ventilation

Glidevale Twist and Lock Soffit Ventilators, Glidevale Spring Wing Soffit Ventilators, Glidevale Universal Rafter Ventilators, Glidevale Tile and Slate Ventilators
Glidevale Limited, 2 Brooklands Road, Sale, Cheshire M33 3SS
Tel: 0161 962 71 13
www.glidevale.com

Ultra Tough Wood Filler System
Cuprinol, Wexham Road, Slough, Berkshire SL2 5DS
Tel: 0870 444 1111
www.cuprinol.co.uk

Wykamol Timber Injection System, Wykamol Boron Gel 40, Wykamol PJG Boron Rods
Wykamol Group, Unit 3, Boran Court, Network 65 Business Park, Hapton, Burnley, Lancashire BB11 5TH
Tel: +44 (0)845 4006666
www.wykamol.com

Timber Beam Strengthening System
RTT Restoration Ltd, Brooklands Approach, North Street, Romford, Essex RM 11 DX
Tel: 01708 725 127 *or* 01708 764576
Fax: 01708 746899

Permanent structural and temporary works

Heavy-lifting Systems
Abbey Pynford Plc, Second Floor, Hille House, 132 St Albans Road, Watford WD24 4AQ
Tel: 0870 085 8400
Fax: 0870 085 8401
www.abbeypynford.co.uk

Underpinning Systems
Roger Bullivant Ltd, Walton Road, Drakelow, Burton-on-Trent, Staffordshire DE15 9UA
Tel: 01283 511115
www.roger-bullivant.co.uk

Façade Support Systems
Interserve House, Ruscombe Park, Twyford, Reading, Berkshire RG10 9JU
Tel: 0118 932 0123
www.interserveplc.com
www.rmdkwikform.net

APPENDIX

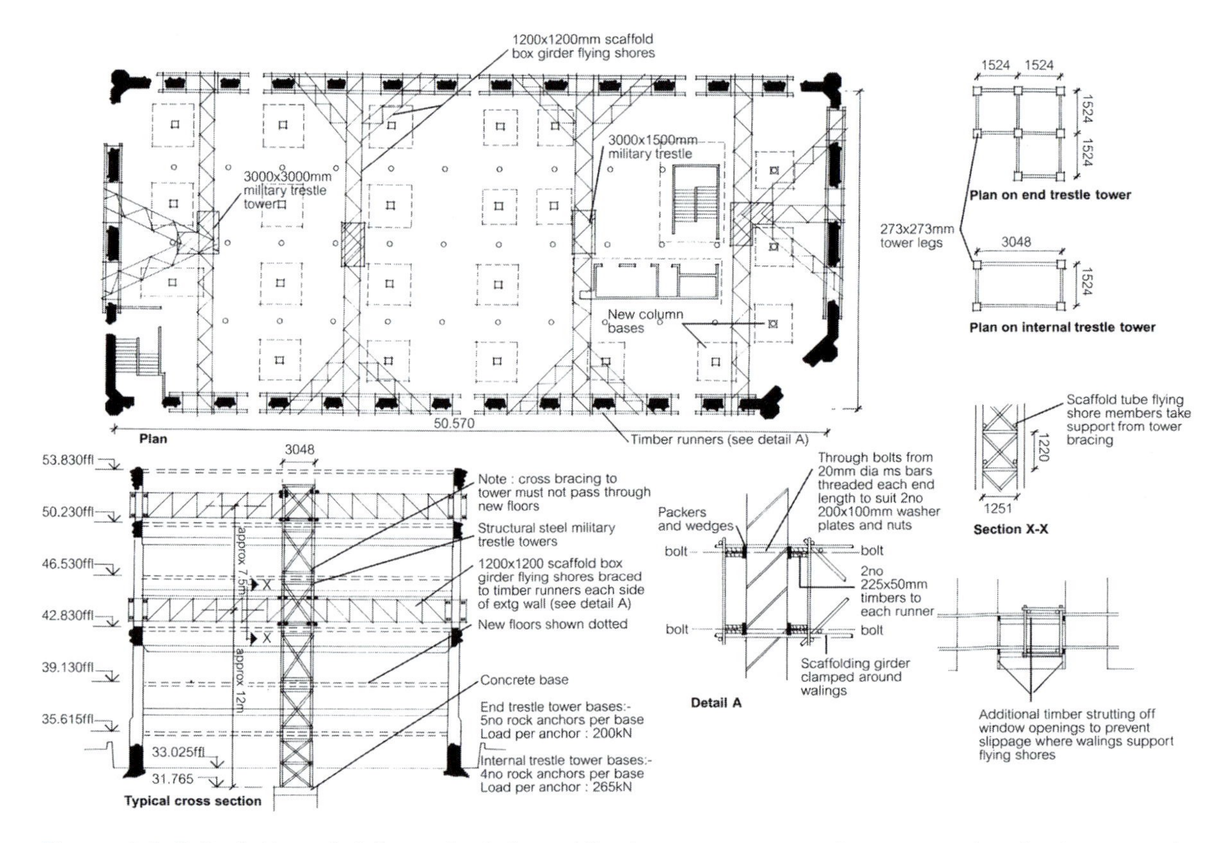

Figure A.1 *St Paul's House, Park Square, Leeds. Internal façade support system employing structural steelwork towers and tubular steel scaffolding components*

INDEX